ELEMENTS *of*
QUANTUM COMPUTATION
and QUANTUM
COMMUNICATION

ELEMENTS *of*
QUANTUM COMPUTATION
and QUANTUM
COMMUNICATION

ANIRBAN PATHAK

CRC Press
Taylor & Francis Group
Boca Raton London New York

CRC Press is an imprint of the
Taylor & Francis Group, an **informa** business
A CHAPMAN & HALL BOOK

CRC Press
Taylor & Francis Group
6000 Broken Sound Parkway NW, Suite 300
Boca Raton, FL 33487-2742

First issued in paperback 2019

ISBN-13: 978-1-4665-1791-2 (hbk)
ISBN-13: 978-0-367-37987-2 (pbk)

Library of Congress Cataloging-in-Publication Data

Pathak, Anirban.
  Elements of quantum computation and quantum communication / author, Anirban Pathak.
    pages cm
    Includes bibliographical references and index.
    ISBN 978-1-4665-1791-2 (hardback)
    1. Quantum computers. 2. Quantum communication. 3. Quantum theory--Mathematics. I. Title.

QA76.889.P37 2013
004.1--dc23   2013010831

Visit the Taylor & Francis Web site at
http://www.taylorandfrancis.com

and the CRC Press Web site at
http://www.crcpress.com

This book is dedicated to my mother, Mrs. Chandana Pathak, for the elementary and fundamental lessons of life that she taught me.

Contents

Preface

This introductory textbook is written primarily for undergraduate students of physics, mathematics, computer science and other related disciplines. It is also expected to be valuable to teachers as well as to researchers working in other domains, who are interested in obtaining an understanding of quantum computation and quantum communication. I used to offer a course on quantum information theory from 2002-2006. Later I offered a few short courses in different summer schools and workshops. This book is prepared mainly from those lectures. There are many excellent textbooks on quantum information theory. However, most of those books are either too technical for beginners or they are not complete. This was one of my reasons for writing this book. But more importantly, every teacher has his/her own way to present the subject and teachers are usually biased on that. I belong to that class of biased teachers and this book is an initiative to present the subject in my way. Another fact that played a very important role in the present initiative is that there are engineering students who hardly know anything about quantum mechanics and there are physics students who do not know what a Turing machine is. But students from both groups are equally interested in quantum computing and often they join the same course. Keeping both kinds of students in mind, this book aims to give a brief idea of quantum computation and quantum communication in a self-contained manner. It does not demand any prior knowledge of quantum mechanics or computer science. It is written in a lucid manner, and a large number of problems with detailed solutions are provided in each chapter (especially in Chapter 3). In addition, a set of thought-provoking cartoons is included to make the subject more attractive. This is an introductory textbook so it will not be so thick that readers get afraid of the volume of the book and leave it before they begin. The field of quantum information and quantum computation is rapidly growing. I have tried to give a flavor of the new developments and open questions in the field, but I could not accommodate all the flavors of this interdisciplinary subject. Specifically, I could not do justice to experimental techniques. I'll consider the book successful if the readers find it easily understandable, interesting and encouraging enough to read more advanced texts and journal papers.

I understand that many students and researchers do not have access to all the journals. Keeping them in the mind, I have tried to mention all such sources where one can get access to interesting articles and courses for free. Especially in the bibliography I have provided several references from arxiv.org, and in the "further reading" sections of each chapter I have mentioned sources where seminal papers related to the field covered can be read for free. I hope readers with restricted library access will find these sources useful.

I have tried my best to avoid typos and errors. Still there may be a few present in the book. I request readers to communicate any errors, typos and suggestions by kindly sending an email to anirban.pathak@gmail.com.

This book has been written over a considerable amount of time and most of the book originates from my lecture notes. I have tried to properly cite all sources that are used here but there is a possibility that some are unintentionally omitted. I am extremely sorry for any such occurrence.

In the process of writing this book, I received help, support and encouragement from many individuals and institutes at various stages and in various forms. I would like to thank all of them. To begin with I must mention that the first few words of this book were written in 2003. It took much longer than expected. In this long period, I have lost many people who were very close to me and would have liked to see this book. I have lost my grandmother, my father, my aunts, a few of my teachers and friends who would have been very happy to see this book in existence but who passed away before it was completed. I thank them for their interest and encouragement in all my academic activities including this book-writing project. I am thankful to Dr. Anindita Banerjee and Ms. Chitra Shukla who have helped me considerably by creating the figures, correcting the typos and giving their feedback. The front cover of this book shows the image of the processed output of spontaneous parametric down conversion (SPDC) process. In the center of the image we can see the pump beam. The experiment was carried out by my colleagues at the Joint Laboratory of Optics, Olomouc. I am specially thankful to my colleagues Dr. Martin Hamar and Mr. Radek Machulka for producing this image. My old friend Mr. Anshuman Das has drawn the cartoons included in this book. His kind help has made the text more attractive. Prof. Avijit Pathak and Dr. Subhashish Banerjee have carefully read part of the manuscript and have provided their valuable feedback. I am thankful to them. Prof. Ajoy Ghatak's interest in the book was a constant encouragement. It was he who advised me to include as many examples as possible. As I see the final manuscript, it appears that his suggestion has really made it a textbook that can be used for classroom teaching. I am thankful to him for his valuable advice. The manuscript took its final form during my one-year stay at Palacky University, Czech Republic. This visit provided me ample time and the perfect ambiance to complete the manuscript. Prof. J. Peřina, Prof. V. Peřinová, Dr. O. Haderka, Dr. J. Peřina Jr. and Prof. M. Hrabovský, whose collaboration and help made this visit possible, deserve special words of thanks. Without their kind support it would have been impossible to complete the book. My special friend, Dr. R. Srikanth, who was always awake late at night to share my concerns about the book, has helped me in many ways. My long late-night conversations with him have many direct and indirect contributions to this book. No word of thanks is enough for his help. I would also like to thank all my collaborators and students together with whom I have learned the subject discussed in this book. As I mentioned, to finish this book, I took a one-year leave from JIIT, India and visited Palacky University. During this period my wife Dr. Papia Chowdhury and my son Master Pustak Pathak were very cooperative. Their selfless encouragement and support made it possible. During most of the time while I was working on this manuscript, my research activities on quantum computing and quantum communication were supported by the Department of Science and Technology, India through project numbers SR/S2/LOP-0012/2010 and SR/FTP/PS-13/2004. My activities were also supported by the Operational Program Education for Competitiveness - European Social Fund project CZ.1.07/2.3.00/20.0017 and Operational Program Research and Development for Innovations - European Regional Development Fund project CZ.1.05/2.1.00/03.0058 of the Ministry of Education, Youth and Sports of the Czech Republic. Support obtained from these projects was the backbone of my book-writing project. I thank these agencies for their support. I am thankful to the administration of JIIT, Noida for granting me the sabbatical to complete the book. I am also thankful to IMSc, Chennai for offering me the associateship. I especially mentioned this because Ms. Aastha Sharma of CRC Press approached me with their proposal to write a textbook during my stay in IMSc. I had a half-written manuscript that had been gathering dust for a long time. This coincidence revived the project.

I am indebted to many more people for their indirect support to this book. I especially acknowledge the support and help of Mrs. Chandana Pathak, Mr. S. R. Chaudhuri, Ms. Dipti Ray, Mr. Kunal Jha, Mr. Sanjit Pathak, Mrs. Anindita Pathak, Dr. Gautam Sarkar, Dr. Y. Medury, Prof. K. C. Mathur, Prof. D. K. Rai, Prof. S. K. Kak, Prof. Swapan Mandal, Prof. P. K. Panigrahi, Prof. M. R. B. Wahiddin, Prof. Barry Sanders, Prof. Marco Genovese, Prof. J. Banerjee, Prof. Adam Miranowicz, Dr. Chiranjib Sur, Dr. Amit Verma, Dr. Biswajit Sen, Dr. B. P. Chamola, Dr. Somshubhro Bandyopadhyay, Mr. Aayush Bhandari, and Mr. Rishabh Jain.

Lastly many thanks to Ms. Aastha Sharma, Ms. Amy Rodriguez and their colleagues at CRC Press for their initiative to publish this book.

Olomouc, Czech Republic   Anirban Pathak
December 15, 2012

Author

Anirban Pathak is a theoretical physicist. He is a professor at Jaypee Institute of Information Technology (JIIT), Noida, India and a visiting scientist at Palacky University, Czech Republic. He received his Ph.D. from Visva Bharati, Santiniketan, India. Subsequently, he was a post-doctoral fellow at Freie University, Berlin. He joined JIIT in 2002. At present he is actively involved in teaching and research related to several aspects of quantum optics and quantum information. His group's recent research activities are focused on foundational aspects of quantum mechanics, secure quantum communication, quantum circuits and nonclassical states. His group has active research collaborations with several research groups in India, Czech Republic, Poland, Malaysia, Germany and Argentina.

Chapter 1

Introduction and overview

Once upon a time there was a curious man. He knew nothing about the subject called "Information Technology." One day he visited a library and suddenly saw a book entitled *"Information Technology: The Art of Managing Information."* First, he thought: "This title is not for me. Let me ignore it and look at the next title." But then the curious man started thinking: "What is it? What is information technology? What is information? Why do I need to know how to manage it?" Since a curious man lives in all of us, it would be tempting to follow the sequence of his thoughts and try to answer these questions. Let us start with a simple question: What is information technology? This question is very important as far as this book or any other text related to information theory is concerned. The simplest answer to this question is already provided in the title of that book as: Information technology is the art of managing information. As soon as we accept this particular definition, the other two questions that appeared in his mind become extremely relevant. In this chapter, we will try to answer those two questions and develop a quantitative perception of classical and quantum information. Once a basic perception is built in the first part of this chapter, we will describe a short history of quantum computation and quantum communication at the end of this chapter.

1.1 What is information?

To a large extent, our general perception of information is qualitative. For example, often after a lecture we say, "this talk was quite informative" or "there was not much new information in this talk." This type of qualitative perception of information has been in existence from the beginning of human civilization, but a clear definition of information was not present until 1948. To begin with, we may define information as: *Information is something that we do not already know* [1]. Some simple examples may help us to develop a perception about the meaning of this simple notion of information. Suppose you are watching a football (soccer) match with your friends and you have seen that Ronaldo has scored a goal. Immediately after that, one of your friends shouts: "Oh it's a goal!" Here, when you see Ronaldo score, you gain some information, but you don't gain any information from your friend's shout because you already know that. So your friend's shout only provides some data to you, but no information. Thus *information is useful data for a particular analysis or decision task.* It helps us to choose reliably between alternatives. Let us give another example. "Sholay" is a popular Hindi movie. In this movie there are two characters called Veeru and Jai. In the movie Jai often tosses a coin, which has the same symbol on both the sides. Jai knows it, but Veeru does not know. Now whatever the call of Veeru, Jai never gains any information from the outcome of the toss since he already knows the result.

There are many technical definitions of information, but here we have opted for a simple definition which states that information is what we do not already know. However, with just a good definition we cannot compare the amount of information. Suppose I want to compare the capacity of your pendrive with that of mine, then the above definition of information will not help us to do the comparison. However, I can conclude that my pendrive is better than your pendrive if my pendrive can store 50 units of information, but your pendrive can store only 30 units of information. To do so, we need a quantitative measure of information. Claude Shannon introduced such a measure of information in 1948 [2]. The existence of a quantitative measure of information implies that information is a quantity and that leads to a fundamental question: Is information a physical quantity? If yes, then we may be able to construct some new physical laws for information and existing laws of the physical world must be applicable to information, too. Further, since the physical world is quantum mechanical the essential nature of information should be quantum mechanical. Thus before we start talking about quantum information, we need to establish that the information is physical.

1.1.1   Is information physical?

Different views about the nature of information have co-existed for centuries. One of those views is that information is not an abstract entity and it is always tied to a physical representation. This particular view was strongly established by Rolf Landauer in the later part of the last century [3]. Landauer argued that since information is always tied to a physical representation the limitations and possibilities of the real physical world would be applicable to information, too. We can obtain a stronger perception of this particular notion of information if we try to understand how information is really stored, transferred or processed in the real world. For example, consider a situation in which we are in an auditorium and I am delivering a lecture to convince you that information is physical. In this situation, how do you obtain information from me? The words spoken by me are conveyed by air pressure fluctuations which vibrate the membranes of your ears; nerves convert mechanical energy into electrical energy and finally the brain receives an electrical signal and you listen. So a physical process is involved in the communication of information. Similarly, writing on a piece of paper is essentially painting molecules of the paper in a certain meaningful fashion; in a magnetic hard disk we arrange magnetic dipoles in a certain meaningful fashion to store information. In brief, we cannot dissociate information from physical objects and consequently, information is not abstract and laws of physics are applicable to information. This fact that information is physical has a deeper meaning. It intrinsically implies that computer science is part of physics.

Since we need physical means to store, process and communicate information, the physical laws applicable to the physical resources used for the purpose of information processing, storage or communication would be applicable to information, too. A nice example is the following version of Einstein's postulate of the special theory of relativity: We cannot communicate information with velocity greater than that of light in vacuum. We know many things about the essential nature of physical observables. Let us list a few of them as examples and check whether these specific characteristics of physical observables are also observed in information or not.

- **Physical observable can be expressed in various ways without losing its essential nature:** For example, a cricket ball delivered by Kapil Dev and the sound coming out of a drum beaten by one of his excited fans can have the same energy. The same is true for information as it can also be expressed in various ways. For example, the following two statements: "I don't know where Malda is" and "I am completely unaware of the location of Malda" have something in common, although they share only one word in common. Loosely speaking the thing they have in common is their information content [4]. Essentially, the same information can be expressed in various ways, for example, you may substitute numbers for letters in a scheme such as a=1; b=2; c=3 and so on. The fact that information can be expressed in various ways without losing its essential nature is very useful in computation. This is so because it allows automatic manipulation of information. To be precise, it allows us to construct computing machines, which can process information by handling binary digits only [4].

- **Physical quantities can be transformed from one form to another:** For example, electrical energy can be converted to kinetic energy. The same is true for information because information is not sensitive to exactly how it is expressed and it can be easily transformed from one form to another. Let us elaborate this point with an example. Suppose I have a database of all the Nobel lectures delivered so far. If it is printed on paper, then the molecules of the paper are painted in a meaningful fashion with the molecules of the ink, but if it is kept on a magnetic hard disk, then the small magnetic dipoles are arranged in a meaningful fashion. Thus the same information is stored in two different forms and it is easy to visualize that we can transform it from one form to the other very easily. For example, it is an easy task to print the content of a file stored on the hard disk.

Rolf William Landauer was born in Stuttgart on February 4, 1927 into a Jewish family. To avoid Nazi pogrom, the family left Germany in 1938. They migrated to the United States, where Landauer obtained his undergraduate degree from Harvard University in 1945. Initially he joined the National Advisory Committee for Aeronautics (later known as NASA). After two years he joined IBM and there he became interested in the fundamental limits of computation. In 1961, he showed that computation itself does not require a minimum amount of energy [5]. However, erasing information requires a minimum amount of energy. In 1973 Charles Bennett showed that reversible computers can overcome this limitation [6].

Landauer was an excellent letter writer. In many of his letters he criticized optical computing and quantum computing. This is interesting because of the fact that his works on the physical nature of information and reversible computing played a very important role in the foundation of quantum computing. The depth of Landauer's contribution to the subject can be understood from the title of his obituary, written by Seth Lloyd in Nature [7]. The title was: "Head and heart of the physics of information."

Photo courtesy: C. H. Bennett. With permission.

Here we would like to suggest that readers try to find an example of information storage or information processing mechanism that does not need any physical representation. The harder we try, the more we will be convinced that we cannot dissociate information from physical representation.

We have already seen that information behaves like a physical quantity. However, if it has to appear in a physical law (like force and acceleration do), then it must have a quantitative measure. In 1948 Shannon published his classic work [2] in which he provided a meaningful quantitative measure of information. In the following subsection we introduce this particular quantitative measure of information, known as Shannon's entropy. The beauty and power of this particular measure of information and that of its quantum analogue (Von Neumann entropy) will be visible throughout this textbook.

1.1.2 Quantitative measure of information

Here we briefly discuss how information is measured. Before we do so it would be apt to get familiar with a set of terms/words that are frequently used in the context of measures of information. Let us start with *"event"*. Throwing a dice, tossing a coin, or going to meet a friend is an event. Every event has a set of possible outcomes. When we observe a particular outcome then we say that we have done a measurement. Thus measurement means observing the outcome of an event. Now suppose I do a measurement and tell you the outcome of the measurement. You may or may not gain any information from my disclosure. Actually, the amount of information that you gain from my disclosure will depend on your prior knowledge of the outcome. For example, if I tell you that you are male (female) then you don't gain any information, as you already know it. But if I throw a fair dice or toss a fair coin and tell you the outcome then you gain some information from my disclosure. In both cases you do not have any prior knowledge of the outcome, but in the case of fair dice you know that the outcome would be one of the six equally probable outcomes and in the case of a fair coin it would be one of only two equally probable outcomes. So your ignorance is more in the dice experiment. Consequently, it is expected that you gain more information in that case. Information is actually a measure of ignorance. The more ignorant we are the more information we gain from the outcome of an event. This is consistent with our simple definition of the information, which states that information is something that you already do not know. Now we can safely say that if the outcomes are equiprobable, then the more outcomes are possible in an experiment (i.e., the more chaotic the system is), the more information it is expected to contain. These are the intuitive (expected) properties of information, but these expected properties can yield an excellent measure of information. Let us see how.

Cartoon 1.1: Do you agree with Bob's argument?

First assume that I have an unfair coin that yields heads 90% of the time. Thus the probability of getting a head when the coin is tossed is $p_{head} = 0.9$ and consequently the probability of getting a tail is $p_{tail} = 0.1$. The information we gain from the result of a toss of this coin is expected to be less than the information obtained when we toss a fair coin. This is so because in case of this unfair coin we know that most of the time the result will be heads. This example indicates that the measure of information should be a function of the probability distribution of the possible outcomes. In other words, if H is a measure of information then H must be a function of p_i, where p_i is the probability of the outcome i. Now, suppose I rub one side of the unfair coin to change the probability distribution as $p_{head} = 0.91$ and $p_{tail} = 0.09$. This small change is not expected to considerably modify the amount of information $H(p)$ to be obtained from the outcome of a toss. Thus we may consider that a small change in the probability distribution of outcomes should only lead to a small change in the amount of information. Consequently, $H(p)$ is expected to be continuous in p.

Suppose you throw a fair dice and toss a fair coin and see the outcomes of both. You are expected to obtain some information from both events. Since the events are independent of each other, the total information obtained by you should be the sum of the information obtained by you in the individual events. This can be mathematically stated as: If x and y are drawn independently from the distributions $p(x)$ and $p(y)$ respectively, then

$$H(p(x), p(y)) = H(p(x)) + H(p(y)), \tag{1.1}$$

where $H(p(x), p(y))$ is the information associated with seeing a pair (x, y). Our task is to find out the functional form of $H(p)$ which satisfies (1.1). The task is easy because $H(p)$ is a continuous function of p. There exists only one smooth algebraic function which can satisfy (1.1). The solution of (1.1) is unique up to a constant factor and $H(p) = k \log p$, for some constant k. Value of k is arbitrary and it depends on the choice of unit and the base of the logarithm. Now, the average information gain when one of a mutually exclusive set of events with probabilities $p_1, \cdots, p_n$ occurs is $k \sum_i p_i \log p_i$. If the base of log is chosen as 2 then the unit of information is bit[1] and one bit is defined as the initial uncertainty in a binary situation of equal probability. Therefore,

$$1 \;=\; kp_1 \log_2 p_1 + kp_2 \log_2 p_2 = k\tfrac{1}{2}\log_2 \tfrac{1}{2} + k\tfrac{1}{2}\log_2 \tfrac{1}{2} = -k.$$

Or,

$$k = -1.$$

Thus we obtain a quantitative measure of information associated with a probability distribution as

$$H(p(x)) = -\sum_{x}^{N} p(x) \log_2 p(x). \tag{1.2}$$

Here the total number of possible outcomes is N and unit of information is bit. This particular quantitative measure of information is known as Shannon's entropy as it was introduced by Claude Shannon in 1948 [2] in analogy with Boltzmann entropy, which is a measure of disorder in a thermodynamic system. There is an interesting story behind this nomenclature. Shannon was concerned about the name. He thought of calling $H(p)$ "information" but the word was already in frequent use. Then he thought of calling it "uncertainty". Finally, he discussed the idea with Von Neumann, who suggested Shannon to call it "entropy" and justified the nomenclature as: "In the first place, a mathematical development very much like yours already exists in Boltzmann's statistical mechanics, and in the second place, no one understands entropy very well, so in any discussion you will be in a position of advantage."

In brief, if X denotes a random variable whose possible outcomes are random values x with probability $p(x)$ then the information content (or self-information) of X is $H(p(x)) = H(X)$. It is standard practice to use $H(X)$ for $H(p(x))$, but $H(X)$ is not a function of X, rather it is the information content of the variable X. Similarly, we write $H(p(y)) = H(Y)$. Now we may summarize the idea as follows: Shannon's entropy $H(p)$ = Amount of ignorance before the observation = Amount of uncertainty before the observation = Amount of information obtained after the observation.

Now we will provide a few simple examples to justify that $H(X)$ is a good measure of information content. Let us start with the second example of Section 1.1. Here Jai knows the outcome of the toss (say a head) so for Jai, X = head, and $p_{head} = 1$ and there are no other terms in the sum, leading to $H = 0$, so X has no information content. In all other situations, we gain some information from the outcome. To be precise, we always learn something unless we already know everything. Mathematically, this

---

[1]If the unit of logarithm is chosen as e then the unit of information is called "nat" for natural logarithm. $1\,\text{nat} \approx 1.44\,\text{bits}$. Further, the word "bit" originates from **bi**nary di**git**.

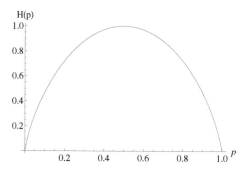

Figure 1.1: Variation of $H(p)$ with p where only two outcomes are possible and one of the outcomes has probability p.

implies that $H(p) \geq 0$, with $H(p) = 0$ iff[2] just one p_i is nonzero. This can be proved easily. When only one p_i is nonzero, then all other p_i's have to be zero and that particular nonzero $p_i = 1$. Therefore, $H(p) = -0 \log_2 0 - 0 \log_2 0 - \cdots - 0 \log_2 0 - 1 \log_2 1 - 0 \log_2 0 - \cdots - 0 \log_2 0 = 0$. When more than one outcome has nonzero probability then

$$0 < p < 1 \Rightarrow \log_2 p < 0 \Rightarrow -p \log_2 p > 0 \Rightarrow H(p) > 0.$$

We have seen that when the outcome of a toss is already known then the amount of information obtained from the observation of the outcome is zero. In case of a fair coin tossing $H = -2 \times \frac{1}{2} \log_2 \frac{1}{2} = 1$ bit. This example is important because it provides us justification of the choice of base of logarithm as 2. Clearly it provides us an excellent convention for unit of information. Thus the amount of ignorance we have before we know the outcome of an equally probable event having two possible outcomes or the information we gain after learning the outcome of the same event is called a bit. Let us see what happens when the coin is biased. Suppose $p_{\text{head}} = p$ and $p_{\text{tail}} = 1 - p$, then the entropy is $H(p) = -p \log_2 p - (1-p) \log_2(1-p)$. This is called binary entropy and often referred to as $H_2(p)$. Variation of this entropy is shown in Fig. 1.1. Fig. 1.1 clearly shows that from a coin-tossing experiment (or from any other experiment where only two outcomes are possible) we obtain maximum information when both the outcomes are equally probable. This is in accordance with our expectations. This observation can be further generalized to the events where more than two outcomes are possible. It is expected that when the system is unbiased, i.e., when all the outcomes are equally probable, then our ignorance before the observation is highest. It is also known that the more options there are the less we know what will happen next, and consequently our ignorance (entropy) is more. These two statements may be stated mathematically as $H(p) \leq C(X)$, where $C(X)$ is a constant that depends on the number of

---

[2]Iff means if and only if.

possible values of the random variable X, with $H(p) = C(X)$ for a uniform probability distribution, and $X' > X \Rightarrow C(X') > C(X)$. This can be proved as follows [1]:

The extreme values of a function $f(x_1, x_2, \cdots, x_N)$ under the constraints

$$
\begin{array}{rcl}
g_1(x_1, x_2, \cdots, x_N) & = & a_1 \\
g_2(x_1, x_2, \cdots, x_N) & = & a_2 \\
\vdots & & \vdots \quad \vdots \\
g_M(x_1, x_2, \cdots, x_N) & = & a_M
\end{array}
\tag{1.3}
$$

are usually obtained by solving M equations of Lagrange's undetermined multiplier λ_j as

$$
\frac{\partial f}{\partial x_i} - \sum_{j=1}^{M} \lambda_j \frac{\partial g_j}{\partial x_i} = 0.
\tag{1.4}
$$

In our case we have to maximize the entropy $H(p) = -\sum_{i=0}^{N} p_i \log_2 p_i$ with respect to the constraint condition $\sum_{i=0}^{N} p_i = 1$. Therefore, for this particular case equation (1.4) reduces to

$$
\frac{\partial}{\partial p_j}\left(-\sum_{i=0}^{N} p_i \log_2 p_i\right) = \lambda \frac{\partial}{\partial p_j} \sum_{i=0}^{N} p_i.
\tag{1.5}
$$

This equation can be easily solved for λ as

$$
\begin{array}{rcl}
-\log_2 p_j - 1 & = & \lambda \\
\Rightarrow \log_2 p_j & = & -(1+\lambda) \\
\Rightarrow p_j & = & 2^{(-(1+\lambda))}.
\end{array}
\tag{1.6}
$$

Now from the constraint condition $\sum_{i=0}^{N} p_i = 1$ we have

$$
\begin{array}{rcl}
\sum_{i=0}^{N} 2^{(-(1+\lambda))} & = & 1 \\
\Rightarrow N 2^{(-(1+\lambda))} & = & 1 \\
\Rightarrow N & = & 2^{(1+\lambda)} \\
\Rightarrow \log_2 N & = & (1+\lambda).
\end{array}
$$

Therefore,

$$
p_i = 2^{(-(1+\lambda))} = 2^{-\log_2 N} = 2^{\log_2 \frac{1}{N}} = \frac{1}{N}.
\tag{1.7}
$$

Thus the entropy maximizes when all the possible events are equally likely and the maximum entropy is

$$
H(p)|_{max} = \log_2 N.
\tag{1.8}
$$

Equation (1.8) clearly shows that $H(p)|_{max}$ monotonically increases with N, i.e., if the number of possible outcomes N is more then $H(p)|_{max} = \log_2 N$ is also more and it mathematically establishes the fact that the more options there are the less we know about what will happen next. For example, if X denotes the throw of a fair dice, then $H(X) = -6 \times \frac{1}{6} \log_2 \frac{1}{6} = \log_2 6 \simeq 2.58$ bits. But if X denotes the toss of a fair coin then $H(X) = 1$ bit. Thus we have seen that all the expected properties of a quantitative measure of information is satisfied by Shannon's entropy $H(p)$. The most basic problem in classical information theory is to obtain a measure information. Shannon's entropy is the key concept in this context.

Here it would be apt to note that Shannon's entropy can also be used to introduce the following set of related measures of information which relate to two random variables X and Y [8].

Relative entropy: Relative entropy is defined as

$$
\begin{aligned}
H(X||Y) &= -\sum_{x,y} p(x) \log_2 (p(y)) - H(x) \\
&= \sum_{x,y} p(x) \log_2 \left(\frac{p(x)}{p(y)} \right).
\end{aligned}
$$

This provides a measure of similarity between X and Y. If $X = Y$ then $H(X||Y) = 0$.

Joint entropy: Joint entropy is a measure of the combined information obtained by measuring two random variables X and Y. We have already mentioned that when X and Y are independent then the joint entropy $H(X,Y) = H(X) + H(Y)$. Now X and Y may or may not be independent of each other and the above equality does not hold when X and Y are not independent. In general, the joint entropy is

$$
H(X,Y) = -\sum_{x,y} p(x,y) \log_2(p(x,y)) \leq H(X) + H(Y). \qquad (1.9)
$$

The above type of inequalities are called subadditivity.

Conditional entropy: Conditional entropy is defined as:

$$
H(X|Y) = -\sum_{x,y} p(x|y) \log_2(p(x|y)),
$$

where $p(x|y) = \frac{p(x,y)}{p(x)}$. It provides us a measure of the information obtained from learning the outcome of X given that Y is known. The relation between conditional entropy and joint entropy is

$$
H(X,Y) = H(X) + H(Y|X). \qquad (1.10)
$$

Mutual information: This provides us a measure of correlation between two random variables. It is defined as

$$
I(X:Y) = H(X) - H(X|Y).
$$

Clearly it is the difference between the information gained from learning X (i.e., Shannon's entropy $H(X)$) and the information gained from learning X when Y is already known (i.e., conditional entropy $H(X|Y)$). This is a symmetric quantity and consequently

$$I(X:Y) = I(Y:X) = H(Y) - H(Y|X). \qquad (1.11)$$

Now substituting (1.10) into (1.11) we obtain

$$I(X:Y) = H(X) + H(Y) - H(X,Y).$$

Mutual information plays a very crucial role in our understanding of security and information leakage in the context of cryptography. Assume that Alice creates a random key X and sends it to Bob by some means. However, Bob receives Y, which is a noisy version of X. Eve is an unauthenticated party who tries to obtain the key and somehow obtains Z, which is another noisy version of X. Now the security of the key distribution protocol depends on mutual information, $I(X:Y)$, $I(X:Z)$ and $I(Y:Z)$. In an ideal situation we should have $I(X:Y) = H(X) = H(Y)$ and $I(X:Z) = I(Y:Z) = 0$, which implies that Bob has received a perfect copy of the key sent by Alice and Eve has failed to steal any information. In a realistic situation, Alice and Bob can establish a secret key[3] if and only if $I(X:Y) > \min\{I(X:Z), I(Y:Z)\}$. This is true for both classical and quantum cryptography. This idea will be clearer from the following two examples.

Example 1.1: Show that if $X = Y$ then $I(X:Y) = H(X)$.
Solution: As $X = Y$ so $H(X,Y) = H(X,X) = H(X)$ and consequently

$$I(X:Y) = H(X) + H(Y) - H(X,Y) = H(X) + H(X) - H(X) = H(X).$$

Example 1.2: Show that if X and Y are independent then the mutual information $I(X:Y) = 0$.
Solution: As X and Y are independent so $H(X,Y) = H(X) + H(Y)$ and consequently

$$I(X:Y) = H(X) + H(Y) - H(X,Y) = H(X) + H(X) - (H(X) + H(Y)) = 0.$$

We have already mentioned that the power and beauty of Shannon's entropy will be visible in the entire text. Just to give a flavor of its relevance, in the following subsection we have briefly stated Shannon's first coding theorem or Shannon's noiseless coding theorem, which provides a bound on data compression in absence of noise. Let us see how.

---

[3]Alice and Bob can establish the key using error correction and classical privacy amplification protocols. Here we have assumed that the privacy amplification protocol uses only one-way communication. The meaning of error correction and classical privacy amplification in the context of cryptography will be elaborated in Chapter 8.

1.1.3   Shannon's first coding theorem

Before we state Shannon's first coding theorem, we need to understand the meaning of coding. Coding is the art of expressing information in such a way that the average length of the entire message is minimized. This essentially exploits the fact that physical observable (information in this case) can be expressed in various ways without losing its essential nature. The idea of coding can be clarified through a simple example. Let us assume that an alphabet has four characters A, B, C, and D. We can encode this in a simple manner as A=00, B=01, C=10, and D=11. In this scheme, an N letter message will be encoded by $2N$ bits. Now if the letters appear with different probabilities, say A appears with probability $\frac{1}{2}$, B appears with probability $\frac{1}{4}$, C and D appear with probability $\frac{1}{8}$, then we can encode as A=0, B=10, C=110 and D=111. This simple encoding scheme is unambiguous, as a letter stops after every zero or three consecutive ones. In this case, the average length of an N letter long message is $N\left(\frac{1}{2}+\frac{1}{4}2+\frac{1}{8}3+\frac{1}{8}3\right) = \frac{7}{4}N < 2N$. Now we can easily see that if the outcomes are equally probable then the average length of the N letter long message in this scheme is $N\left(\frac{1}{4}+\frac{1}{4}\times 2+\frac{1}{4}\times 3+\frac{1}{4}\times 3\right) = \frac{9}{4}N \geq 2N$. It is clear from this example that the coding (data compression) is possible if and only if the outcomes are biased.

We may note that the possible coding schemes are not unique. If we have two possible encoding schemes as discussed in the previous example, then we can use the probability distribution to calculate the average length and conclude which of one of these two schemes is better. However, this does not tell us how much compression of the message is possible. Shannon's first coding theorem, which is also called source coding theorem or the noiseless coding theorem, really provides an answer to this question as samples drawn from a distribution $p(x)$ can on average be described by $H(p)$ bits rather than $\log_2 X$ bits.

In coding, we essentially exploit the difference between $H(p)$ and $\log_2 X$ to store or transmit a string of N samples drawn from a distribution $p(x)$ with $NH(p)$ bits. This is the essence of Shannon's first coding theorem. Now we can see that in the previous example, when the outcomes were biased, then $H(p) = -\frac{1}{2}\log_2\frac{1}{2}-\frac{1}{4}\log_2\frac{1}{4}-\frac{1}{8}\log_2\frac{1}{8}-\frac{1}{8}\log_2\frac{1}{8} = \frac{1}{2}\times 1+\frac{1}{4}\times 2+\frac{1}{8}\times 3+\frac{1}{8}\times 3 = \frac{7}{4}$ bits. Thus the coding scheme described above (i.e., A=0, B=10, C=110 and D=111) is optimal. Shannon's theorem provides us this bound.

Here we have briefly discussed this theorem for three reasons: firstly, because this is one of the foundation stones of the modern information theory; secondly, because it provides us an important and useful application of Shannon's entropy; and thirdly, because the quantum counterpart of this theorem, which has played a very important role in the development of quantum computing, was obtained in 1995. Now it would be apt to note that this kind of encoding will work only in the absence of noise. In the presence of noise we have to add some redundancy to the encoded message. The idea will be described in detail in Chapter 6.

Claude Shannon was born on 30th April, 1916 in Gaylord, Michigan, United States. He obtained his Ph.D. from Massachusetts Institute of Technology (MIT) in 1940 for his works on population genetics. At MIT he also worked on the differential analyzer, an early type of mechanical computer. In 1941 he joined Bell Laboratories and in 1948 he published his seminal work, "A mathematical theory of communication" in the *Bell System Technical Journal* as two papers [2]. These two papers founded the subject of information theory. In these papers he introduced the word "bit" and the basic idea of error correction. The importance of these two papers is so much that in 1990, *Scientific American* called these papers: "The Magna Carta of the Information Age."

He was fond of riding a unicycle, and he used to ride his unicycle in the corridors of Bell Laboratories and MIT. He also invented a two-seater version of the unicycle, but probably no one had ever shown enough courage to join him in the adventurous journey. Apart from conventional research he worked on many ideas which he found fascinating. For example, he was always interested in jugglers, and that led him to design a juggling machine and to formulate juggling theorem, which describes the relation between the position of the balls and the action of the hands. He died on 24th February, 2001 in Medford, Massachusetts. [a]. Photo credit: Reprinted with permission of Alcatel-Lucent USA Inc.

---

[a]More about Shannon can be found at:
1. http://www.nyu.edu/pages/linguistics/courses/v610003/shan.html
2. http://www.youtube.com/watch?v=sBHGzRxfeJY

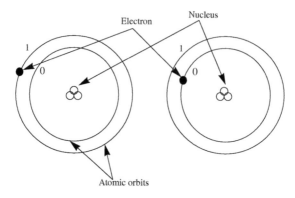

Figure 1.2: Physical realization of a bit.

1.1.4 More about the bit

We already have a definition of the bit (unit of information), and we already know that the information is always associated with some kind of physical system. So a bit must also be associated with some kind of physical system. In practice we use different two-state systems to represent a bit. We usually denote the outcomes of a measurement on a two-state system as "0" and "1". In fact, any two clearly distinguishable states may physically represent a bit. One of the states would denote "0" and the other one would represent "1". For example, we can use our hands to physically represent a bit. To visualize it assume that you are at one end of a big hall and I am at the other end. You can see me and we have pre-decided that I'll send you some information by raising up or lowering down my right hand. Say our chosen convention is such that if my right hand is up for 2 seconds then you note "1" and if it is down for 2 seconds then you note "0". Now if my right hand is up for 4 seconds, down for 6 seconds and again up for 2 seconds, then you receive 110001. The idea is that two distinguishable states (positions) of my hand (i.e., up and down) can be used to represent a bit. Another nice example is given in Fig. 1.2. Here a two level atom is considered as bit. We consider that there are only two allowed energy states of the atom and there is only one electron. When the electron is in the ground state then we call it "0" and when it is in the excited state we call it "1". These two clearly distinguishable atomic energy states form a bit. Similarly, a coin can form a bit.

1.2 Why do we need to know how to manage information?

Until now we have been discussing: What is information? Let us now try to answer the second question: Why do we need to know how to manage information? Let us start with the history of information starting from the the Vedic age. In Vedic age the amount of information available to a man was very limited because his world was small. He did not know anything outside his region. At that time, he could remember all the information available to him. That is why the Veda is also called Shruti[4]. Gradually, the amount of knowledge increased and it became impossible to store everything on our biological hard disk. So a need for some device to store information became important. Actually, it was the need of society that compelled different isolated civilizations to develop crude methods of writing almost at the same time. We learned to write, i.e., we learned to paint molecules of leaves or stones in a certain meaningful fashion and thus we learned to store information. Then we invented different instruments (pencils, papers, inks, colors, etc.) to store and share available information. The entire civilization stored a lot of information in the next few centuries and the socioeconomic conditions of that time demanded a technology to share this information. For example, think of the Roman Empire: if the Emperor wished to circulate an order to all the cities within his territory, then he needed multiple copies of the same order. So they needed an instrument that could produce multiple copies of the same text very fast. Once again due to the social need the printing machine was invented around 1440. The printing machine had a huge impact on the socioeconomic conditions of many countries like India. Indians got in touch with the ideas of modern science, democracy and independence through this device. So it is not right to think that information has started affecting our society and economy only in the recent past; earlier also it had great impact. But at that time the word "information" had not been used in the current form. Gradually we have learned to write things on magnetic tapes and other state-of-the-art materials. At present, we have excellent storage devices like hard disks, pendrives and blue ray disks. Bit density[5] in all these devices has considerably increased in recent years. This implies that we can now store a large amount of information in a small space. Processors have also been improved and processors with multicore technology have been introduced. When everything is going quite satisfactorily then why do we need to learn a new technique like quantum information processing? Part of the answer to this question lies in the challenges that are faced by toady's IT industry. For example, Moore's law states that the number

---

[4]Shruti means that which is heard.

[5]Bit density is defined as the number of bits stored per unit of length or area of the storage medium.

of transistors on a chip doubles every 18 months. This linear growth implies the decrease in feature size and that implies the decrease in operating voltage. On one hand, because of the decrease in feature size, the size of transistors is already approaching quantum domain where tunneling and other quantum phenomena emerge. On the other hand, operating voltage is approaching values less than 1 Volt and as a consequence of that the number of electrons present in a memory cell will become a very small integer number. Thus quantum behavior will appear. In fact around 2020 the continuous decrease in feature size is likely to cease and the emergence of quantum behavior is expected. Further, the cost of each new IC generation is increasing exponentially and the reduction of the minimum feature size is probably beyond the current limits of industrial lithographic techniques.

We still need to improve to cope with the needs of the society. For example, the total amount of information doubles every 18 months, so the bit density of storage devices has to increase. Apart from the above-stated problems there are other problems, too. All the existing techniques of information storage are limited by the physical laws or technological bounds [1]. For example, in magnetic hard disks magnetic wall energies and the head height limits the bit density. Similarly, in optical storage devices diffraction limit, which is proportional to the wavelength, restricts the bit density. Consequently, the present growth is not expected to continue forever and the onset of quantum effects below a threshold size is inevitable. In this situation, we could have taken a pessimistic approach and remained satisfied with the present technology. But scientists do not believe in pessimism, so they come up with many optimistic options. All these options essentially suggest that we move toward another paradigm of computing. One of the options is to exploit the quantum mechanical effects itself.

1.2.1   Which technology?

There are several new proposals such as DNA computing, optical computing, quantum computing, etc. It would be too early to say which one will rule future development, but at the moment quantum computing appears relatively more promising for several reasons. Here we list a few reasons:

1. Quantum cryptography is unconditionally secure and is already in use[6].

2. Several quantum algorithms, which are faster than the best known classical algorithms for the same tasks, are already proposed.

3. It has already been shown that certain tasks (like quantum teleportation and dense coding), which are impossible in the classical world, can be achieved in the quantum domain.

---

[6] For example, quantum cryptography was used by the organizers of the Soccer World Cup, 2010 in South Africa and some commercial products are already in the market.

4. Because of the quantum parallelism the storage capacity of the quantum register is exponentially higher than the classical register.

Keeping all these things in mind, it may not be a bad idea to learn the fundamentals of quantum communication and quantum computation. So let's learn quantum information.

We will gradually learn in detail about all the above-mentioned advantages of quantum communication and quantum computation, but at the beginning we have to learn about the qubits (or quantum bits) which are the quantum analogue of the classical bits.

1.2.2 The qubits

When the information processing task is done with the help of quantum mechanical systems then in analogy to classical computation and classical information processing we obtain quantum computation and quantum information processing. The analogy continues and there exists a building block of quantum information, which is the quantum analogue of a bit. Such an analogue is called a quantum bit or a qubit. The difference between a bit and a qubit lies in the fact that a bit is either in the state 0 or in 1, but the qubit can be in a superposition state, i.e., a qubit is allowed to exist simultaneously in the states 0 and 1. See Fig. 1.3a, where we show a two-level atom with an electron. The electron is shown by an electron cloud (which is a cloud of probability) that depicts that the electron can simultaneously exist in both the energy states. Thus the atom shown in Fig. 1.3a is a qubit which can simultaneously exist in two different states (say 0 and 1). Similarly we can think of a potential well where a particle is trapped and only two energy states are allowed. Then the particle trapped in the potential well represents a qubit. A quantum dot may be viewed as an example of such a potential well, and consequently we may use quantum dots to implement qubits.

Another example of a qubit is a single photon which encounters a beam splitter, as shown in Fig. 1.3b. A beam splitter transmits part of the incident light and reflects the rest. Now when the single photon encounters the beam splitter then it emerges in a superposition of the reflected path and the transmitted path. If we consider one path as 0, and the other as 1, then the photon is simultaneously in both the states and thus we have a qubit.

If we add two detectors along the two paths shown in Fig. 1.3b, then only one detector will be clicked at a time (remember that our input state is a single photon). Similarly, if we observe the state of the electron in the atom shown in Fig. 1.3a then we will observe it either in the ground state (0) or in the excited state (1). Thus measurement destroys the superposition, and after measurement a quantum state collapses to one of the possible states. This is an important postulate of quantum mechanics and is discussed in detail in Chapter 3.

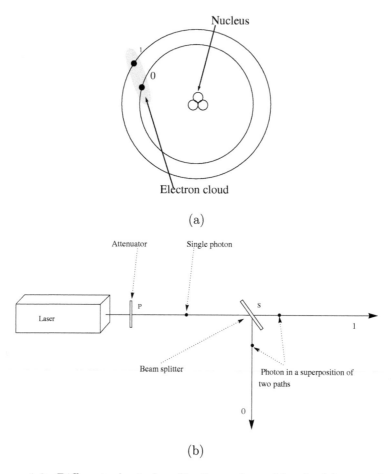

Figure 1.3: Different physical realizations of a qubit. In (a) a two-level atom is visualized as a qubit. Here the electron exists in superposition of two possible states. In (b) an attenuator reduces the intensity of the laser beam to create an approximate single photon source. Then the single photon encounters a beam splitter and the post beam splitter state of the photon is a superposition of two states, which is our qubit.

We can also think of using a photon as a qubit in a different manner. As it can have only two polarization states, it is an ideal two-level quantum system and we may use it as a qubit. Specifically, we may use the orthogonal polarization states of a photon to define 0 and 1. For example, if we choose to describe the state of a qubit as superposition of the linear polarization states in the horizontal and vertical directions, then we can set

$$\updownarrow \Rightarrow 0$$
$$\leftrightarrow \Rightarrow 1.$$

Similarly, we can also choose circular polarization states to describe the state of a qubit as superposition of the left circularly polarization state and right circularly polarization state. To be precise, we can set

$$\circlearrowleft \Rightarrow 0$$
$$\circlearrowright \Rightarrow 1.$$

Similarly, spin states of electron or nucleus may be used as a qubit. Spin states of the electrons are used as qubits in ion traps and spin states of the nucleus are used as qubits in NMR. In both cases it is a convention to use spin up state as 0 and spin down state as 1.

Now if we consider two qubits, we find that together they can simultaneously represent 4 alternative states (say binary numbers 00, 01, 10 and 11). The idea can easily be extended to n-qubit system where we can simultaneously store 2^n alternatives. The most important point is that a system of qubits can exist in a superposition of large number of possible alternative states. Superposition states allow many computations to be performed simultaneously, and that gives rise to what is known as *quantum parallelism*. The superposition can be viewed as the essential resource behind the advantages of the quantum computer. The idea will gradually become more clear in the following chapters.

So far we have provided an introduction to the basic building blocks of classical and quantum information. Now we would like to introduce you to the brief history of quantum computation and quantum communication in the next section. This will provide an overview of the subject, and in the subsequent chapters we will describe the ideas in detail.

1.3 A brief history of quantum computation and quantum communication

We have already seen that the information as well as all the devices used for information processing are physical. On the other hand, quantum mechanics is the best-known physical theory that can describe the properties of matter and field, and classical physics is only a special case of quantum physics. Since classical physics is only a subset of quantum physics, there may exist some quantum behavior that cannot be observed and explained in the classical domain. Although most of the information processing devices (computers) used today can adequately be described by the classical physics, we can still ask an interesting question: What happens when the physical basis for a computer is an explicitly quantum system which cannot be described by classical physics?

The attempt to answer this fundamental question led to the idea of quantum computing. Quantum computing is still in its childhood (if not in infancy), but even to reach this stage it owes a lot to various subjects. Several different lines of development have led to the present stage. No wonder it is difficult to narrate a story of chronological development of quantum computing and quantum communication. Further, we will intentionally ignore the history of classical information theory and quantum mechanics: the parents of quantum computing. To be precise, we will write a biography of quantum computing and quantum communication from their very origin and will not describe the life of their parents before their birth. It is really difficult to write such a biography because even the origin of quantum computing is not clear. Today, many physicists argue that quantum computing originated in 1964 when John Stewart Bell introduced Bell's inequality. We don't agree with this opinion because Bell's original work was not related to quantum communication or quantum computation. Its application in quantum communication is found only in 1991 when Ekert's protocol [9] of quantum key distribution was introduced. There is yet another group of scientists who believe that quantum computation started in 1982, when Feynman showed that classical computers cannot efficiently simulate a quantum mechanical system, but an appropriate class of quantum machines can simulate the quantum system [10]. However, we believe that quantum computing and quantum communication originated around 1970. Keeping this in mind, we will describe the chronological development of quantum computing and quantum communication starting from 1970. We hope this will provide a clear overview of the subject.

1970: In 1970 Stephan Wiesner wrote a seminal paper entitled "Conjugate Coding" [11]. The paper contained the root of many future developments of quantum information theory and quantum computing. The nocloning theorem, which states that arbitrary quantum states cannot be perfectly cloned (copied), was implicitly used in this paper and the basic idea of quantum cryptography was also introduced. This is an interesting paper, but the history of its publication is more interesting. In 1970, Wiesner submitted the paper in IEEE Transactions on Information Theory. The paper was immediately rejected because it did not use computer science jargon. The paper was later published in its original form in 1983 in the newsletter of ACM SIGACT [11].

1970-79: In this decade several other developments took place. Atomtraps were introduced and scanning tunneling microscopes were developed. All these developments directly helped us to obtain complete control over single quantum systems, and indirectly helped us to achieve the precision required for future development of quantum computing devices.

1973: Charles Bennett provided a model of the reversible Turing machine [6]. Bennett's reversible Turing machine was classical. However, it was an important step toward the works of Paul Benioff and David Deutsch which, in fact, led to the quantum Turing machine.

1980-82: Paul Benioff developed the idea of a "quantum mechanical Hamiltonian model of the Turing machine" that does not dissipate any energy [12, 13, 14, 15]. But his model of Turing machine was essentially classical and was equivalent to Bennett's reversible Turing machine.

1982 was one of the most eventful years in the history of quantum computing. On one hand, Benioff's work established that quantum machines can be used to efficiently simulate classical computers (classical Turing machines). On the other hand, Feynman asked the opposite question: Can classical computers efficiently simulate quantum mechanical systems? This question and its answer played a crucial role in the future development of quantum computing as a field. In the same year the nocloning theorem was also introduced in the present form.

1982: Richard Feynman showed that quantum mechanical systems cannot be simulated efficiently in a classical computer [10]. In other words, a quantum mechanical system can only be simulated by another quantum mechanical system. He also suggested a way to build a computer based on the laws of quantum mechanics. His conceptual model of quantum computing machine is now known as the Feynman quantum computer.

1982: W. K. Wootters and W. H. Zurek introduced the nocloning theorem in its present form [16]. Almost simultaneously and independently it was introduced by Dieks [17].

1984: Charles Bennett and Gilles Brassard [18] used the idea of Wiesner and nocloning theorem to introduce a complete protocol of quantum key distribution (QKD). The protocol is now known as BB84 protocol, after the inventors. Unconditional security of the key ensured by this protocol gives it a clear edge over any classical protocol of key distribution. Historically, it was a very important event as it led to the beginning of secure quantum communication.

1985: David Deutsch [19] introduced a notion of universal quantum computer. To be precise, he introduced a quantum Turing machine and provided a physical meaning to the Church-Turing hypothesis. He also designed a simple algorithm (now known as Deutsch's algorithm), which suggested that quantum computers may solve certain computational tasks faster than their classical counterparts.

1992: Charles Bennett and Stephan Wiesner introduced the idea of superdense coding [20], a process through which Alice can communicate two classical bits of information by sending one qubit of information to Bob, provided they share an entangled state. This is a phenomenon of quantum paradigm. It does not have any classical counterpart.

1993: Charles Bennett, Gilles Brassard, Claude Crépeau, Richard Jozsa, Asher Peres, and William K. Wootters introduced the notion of quantum teleportation [21], another phenomenon, which can be seen in quantum domain only. Teleportation is a quantum task in which an unknown quantum state is transmitted from a sender (Alice) to a spatially separated receiver (Bob) via an entangled quantum channel and with the help of some classical communications.

1994: Deutsch's algorithm indicated that quantum algorithms may perform certain tasks faster than their classical counterparts. But we had to wait nine years to obtain a really interesting quantum algorithm. In 1994, Peter Shor introduced a quantum algorithm for finding the prime factors of an integer and another for discrete logarithm problem [22]. He showed that these problems can be solved efficiently[7] in a quantum computer. These two algorithms created huge excitement, optimism and interest among physicists and computer scientists because there did not exist any efficient classical algorithm for these problems. It strongly indicated that the quantum computers may be more powerful than the classical computers. The factorization algorithm, which is now known as Shor's algorithm, got more attention because it threatened the existence of classical cryptography. Most of the classical secure communication protocols that are used today depend on the RSA cryptographic protocol, which depends on the inefficiency of the classical computer to factorize a large integer in polynomial time.

---

[7]An algorithm is said to be efficient if the time required by it to solve a problem is polynomial of the size of the problem. Similarly, an algorithm is said to be inefficient if the time required by it to solve a problem is super-polynomial (usually exponential) of the size of the problem. This idea is discussed in detail in Chapter 2.

Richard Feynman was born on 11th May, 1918 in New York. He was actively involved in the development of the first atomic bomb. He is well known for his works on quantum electrodynamics, path integral formulation of quantum mechanics, superfluidity, theory of weak interaction, etc. In 1965 he received the Nobel Prize for his work on quantum electrodynamics. He found the exact cause of the explosion of the Challenger space shuttle. He was a prolific teacher and his talks often initiated new fields of research. For example, pioneering ideas of quantum computing and nanotechnology originated from his talks.

He was the most interesting character of 20th-century physics. He died on February 15, 1988. In the seventy years of his life, he was involved in many interesting activities. To name a few, he studied Maya hieroglyphs; he was a prankster, a juggler, a safe-cracker, a bongo player, a painter and, of course, a physicist. The interesting and adventurous nature of his character is perfectly depicted in two very interesting and humorous autobiographical books: *Surely you're joking, Mr. Feynman!* and *What do you care what other people think?*

Photo credit: Linn Duncan/University of Rochester, courtesy AIP Emilio Segre Visual Archives. With permission.

1995: Benjamin Schumacher [23] provided a quantum analogue of Shannon's noiseless channel coding theorem, which quantifies the physical resources required to store or communicate the output from an information source. In the process of developing a quantum analogue of Shannon's noiseless channel coding theorem, Schumacher also defined "quantum bit" or qubit as quantum analogue of classical bit.

1995: Peter Shor introduced a 9-qubit quantum error correcting code [24] which can correct one error of general type.

1996: Raymond Laflamme, Cesar Miquel, Juan Pablo Paz, and Wojciech Hubert Zurek [25] improved Shor's original idea of quantum error correction and showed that a 5-qubit code is sufficient for correction of one quantum error of general type.

1996: Peter Shor introduced the idea of fault-tolerant quantum computation [26].

1997: Lov Grover [27] gave another example, where a quantum algorithm is faster than the classical counterpart. He developed a quantum algorithm for the search of an unstructured database. The algorithm is now known as Grover's algorithm.

1997: The first ever experimental demonstration [28] of the quantum teleportation phenomenon was reported by Dik Bouwmeester, Jian-Wei Pan, Klaus Mattle, Manfred Eibl, Harald Weinfurter and Anton Zeilinger, at University of Wien, Austria. They had used polarization of photon as a qubit.

1998: Isac Chuang, Neil Gershenfeld and Mark Kubinec [29] demonstrated the world's first 2-qubit quantum computer with a solution of chloroform molecules. They implemented Grover's search algorithm for a system with four states (i.e., two qubits). Their quantum computer used nuclear magnetic resonance (NMR) to manipulate the atomic nuclei of the chloroform molecule.

1999: L. M. K. Vandersypen, M. Steffen, M. H. Sherwood, C. S. Yannoni, G. Breyta, and I. L. Chuang implemented Grover's database-search algorithm using a 3-qubit quantum computer [30].

2000: D. P. DiVincenzo proposed a list of the requirements [31] that every potential technology proposed for quantum computing must satisfy for the successful physical implementation of quantum computation. The list contains five basic criteria for quantum computation and two additional criteria for quantum communication. These criteria are known as the DiVincenzo criteria and are extensively used to test and compare the feasibility of proposed technologies.

2000: L. M. K. Vandersypen, M. Steffen, G. Breyta, C. S. Yannoni, R. Cleve, and I. L. Chuang at IBM, Almaden, California implemented the order-finding algorithm using a 5-qubit NMR quantum computer [32]. The order-finding is a single-step task in a quantum computer, but it requires repeated cycles in a classical computer[8].

2001: L. M. K. Vandersypen, M. Steffen, G. Breyta, C. S. Yannoni, M. H. Sherwood and I. L. Chuang successfully demonstrated Shor's algorithm on a 7-qubit NMR quantum computer [33]. The computer correctly deduced the prime factors of 15.

---

[8]Discussed in detail in Chapter 5.

2003: Z. Yuan, C. Gobby, and A. J. Shields from Toshiba Research Europe, UK, demonstrated quantum cryptography over fibers longer than 100 km [34]. This opened up the possibility of commercial use of quantum cryptography.

2005: H. Häffner *et al.* [35] succeed in creating the first qubyte (quantum byte), which is a series of eight qubits. They used ion traps to construct an 8-qubit W state.

2010: Quantum cryptography was used for secure communication during the Soccer World Cup held in South Africa [36]. This is probably the first occasion when quantum cryptography was used for security of such an important event.

2010: Free space quantum teleportation over 10 miles (16 km) was achieved by Xian-Min Jin *et al.* in China [37].

2012: Free space quantum teleportation and entanglement distribution over 100-kilometer was achieved by Juan Yin *et al.* in China [38].

1.4   Solved examples

1. Durjadhan, Judhustir and Bidur are playing a board game called *pasha*. The probability of victory for Durjadhan, Judhustir and Bidur is 0.5, 0.25 and 0.25, respectively. How much information do you gain when you learn the result of the game?
   Solution: I obtain $H(X) = -\frac{1}{2}\log_2\frac{1}{2} - \frac{1}{4}\log_2\frac{1}{4} - \frac{1}{4}\log_2\frac{1}{4} = \frac{1}{2}\log_2 2 + \frac{1}{4}\log_2 4 + \frac{1}{4}\log_2 4 = \frac{1}{2} + \frac{1}{2} + \frac{1}{2} = \frac{3}{2}$ bits of information.

2. Let
$$X = \begin{cases} a \text{ with } p(a) = \frac{1}{2} \\ b \text{ with } p(b) = \frac{1}{4} \\ c \text{ with } p(c) = \frac{1}{8} \\ d \text{ with } p(d) = \frac{1}{8} \end{cases}.$$

   Find $H(X)$. Notations are the same as those used in the present chapter.
   Solution: $H(X) = -\frac{1}{2}\log_2\frac{1}{2} - \frac{1}{4}\log_2\frac{1}{4} - \frac{1}{8}\log_2\frac{1}{8} - \frac{1}{8}\log_2\frac{1}{8} = \frac{7}{4}$ bits.

3. Consider a random variable that has a uniform distribution of 8 outcomes. What is your amount of ignorance before you know the outcome?
   Solution: The amount of my ignorance before I know the outcome is $H(X) = -\sum_{i=0}^{8}\frac{1}{8}\log_2\frac{1}{8} = \log_2 8 = 3$ bits.

4. Show that Shannon's entropy satisfies (1.1).
   Solution:

$$
\begin{aligned}
H(p(x), p(y)) = H(X, Y) &= -\sum_{x,y} p(x)p(y) \log\left(p(x)p(y)\right) \\
&= -\sum_{x,y} p(x)p(y) \log p(x) \\
&\quad - \sum_{x,y} p(x)p(y) \log p(y) \\
&= -\sum_x p(x) \log p(x) - \sum_y p(y) \log p(y) \\
&= H(p(x)) + H(p(y)) \\
&= H(X) + H(Y).
\end{aligned}
$$

Note that the events are mutually independent.

5. Assume that we are playing a game in which I ask you to choose a number between 1 and 16, and keep it secret. Then, I will ask you a few questions. You have to answer them yes or no. Now use Shannon's entropy to compute the minimum number of questions that I need to ask you to know the number you have chosen, and also use it to construct the questions.
   Solution: Initially there are 16 numbers and you can choose any of them with equal probability. Therefore, my ignorance is $\log_2 16 = 4$ bits. Now when I ask you a question, you answer either yes or no. So there are two outcomes, I get the maximum information only when the outcomes occur with equal probability. So after each answer I get maximum $-\frac{1}{2}\log_2\left(\frac{1}{2}\right) - \frac{1}{2}\log_2\left(\frac{1}{2}\right) = 1$ bit of information. The amount of my ignorance is 4 bits and from each answer I can obtain maximum 1 bit of information. Therefore, I would have to ask you at least four questions.
   What are these questions? Shannon's entropy can help us to construct them, too. My first question should be such that it reduces my ignorance from 4 bits to 3 bits. So I have to ask you a question so that after your answer, the total number of possible alternatives becomes 8. So I may ask you whether the number is 8 or less (or is it even/odd?). If you say yes, my next question will be: Is it less than or equal to 4? And if you say no, I'll ask you: Is it less than or equal to 12? How I finally obtain the number is shown in Fig. 1.4.

6. A jeweler has made nine gold coins. All look similar, but one of them is light in weight. If you have to trace the faulty coin with the help of a weighing pan, what is the minimum number of measurements that you have to do? Use Shannon's entropy to obtain the answer.
   Solution: As any of the nine coins can be faulty and all are equally probable so the initial uncertainty is $\log_2 9 = \log_2 3^2 = 2\log_2 3$. Now we perform the measurement in such a way that we put $\frac{1}{3}$ of the coins in the left pan and an equal number of coins in the right pan. Three outcomes are possible: left and right pans are equal, left pan is heavier, right pan is heavier. The experiment is devised in such

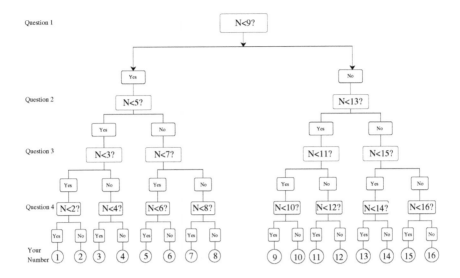

Figure 1.4: The diagram shows how to find the number you have chosen. Every question reduces my ignorance by 1 bit.

a way that each of the outcomes has the same probability (i.e., $\frac{1}{3}$). Thus the information obtained from the result of measurement is $-3 \times \frac{1}{3} \log_2 \frac{1}{3} = \log_2 3$. So we need at least two measurements to locate the faulty coin.

Strategy: As all the three outcomes are to be made equally probable, divide nine coins into three sets so that each set has three coins (let us call them set A, set B and set C). Now we put set A in the left pan and set B in the right pan and set C is kept aside. Since the probability of the faulty coin being in any of the three sets is equal, the outcomes are equally probable. If the weights are same, then the faulty coin is in set C, if the left (right) pan is heavier, then it is in B (A) set. Now the uncertainty is reduced, we have to find it from three coins only. So we put one coin in the left pan and one in the right pan and the third coin is kept aside. If the pans are balanced then the third coin is faulty, otherwise whichever pan is lighter contains the faulty coin.

This is known as Rényi's fake coin problem.

7. A source emits two symbols, a and b. What probability distribution of these events maximizes the source entropy?

    Solution: To maximize the entropy both the events must have equal probability so the probability distribution will be $p(a) = \frac{1}{2}$, $p(b) = \frac{1}{2}$.

8. Our DNA is made up of four types of nucleotides, namely adenine

(A), guanine (G), cytosine (C), thymine (T). Assume that the probabilities of having each nucleotide in a DNA strand comprised of N nucleotides are: $p(A) = \frac{1}{2}, p(G) = \frac{1}{4}, p(C) = \frac{1}{8}, p(T) = \frac{1}{8}$. How much information do you gain when you learn the DNA sequence (i.e., the sequence of nucleotides in that particular DNA strand)? Usually billions of nucleotides form a strand of DNA. Just consider $N = 10^9$ and compare the information content of this DNA strand with the capacity of a typical hard disk.

Solution: The information content of the DNA strand is $NH(p) = N\left(-\frac{1}{2}\log_2\frac{1}{2} - \frac{1}{4}\log_2\frac{1}{4} - \frac{1}{8}\log_2\frac{1}{8} - \frac{1}{8}\log_2\frac{1}{8}\right) = 1.75N$ bits. If $N = 10^9$ then the DNA strand contains 1.75×10^9 bits $\approx$ 2Gb of information. This is a large amount of information, but much less compared to the capacity of the modern hard disks (say 500 Gb). But it is important to note that a longer DNA strand (which is a single macromolecule) may contain more information than the capacity of a typical hard disk.

9. If $\{p_i\}$ and $\{q_i\}$ are two arbitrary probability distributions, which are normalized (i.e., $\sum_i p_i = \sum_i q_i = 1$), then prove that $\sum_i p_i \log p_i \geq \sum_i p_i \log q_i$.

Solution: Using $\log x \leq 1 - x$, we can write

$$\sum_i p_i \log \frac{q_i}{p_i} \leq \sum_i p_i \left(1 - \frac{q_i}{p_i}\right) = \sum_i p_i - \sum_i q_i = 0,$$

which implies

$$\sum_i p_i \log p_i \geq \sum_i p_i \log q_i.$$

10. Show that $2H(X) \geq H(X|Y) - H(Y|X) + H(X,Y)$.

Solution: As $H(X,Y) = H(Y,X)$, therefore

$$\begin{aligned} H(X) + H(Y|X) &= H(Y) + H(X|Y) \\ H(X) - H(Y) &= H(X|Y) - H(Y|X). \end{aligned}$$

But from subadditivity we know that

$$H(X) + H(Y) \geq H(X,Y).$$

Adding the last two equations we obtain

$$2H(X) \geq H(X|Y) - H(Y|X) + H(X,Y).$$

1.5   Further reading

1. An excellent introduction to the subject can be obtained from Quantum computing, A. M. Steane, Reports on Progress in Physics, **61**, (1998) 117-173, quant-ph/9708022.

2. Shannon's classic works were published as two papers in The Bell System Technical Journal, **27** (1948) 379-423, and **27** (1948) 623-656. Now it is available for free in a combined format at http://cm.bell-labs.com/cm/ms/what/shannonday/shannon1948.pdf

3. Landauer's paper on physical nature of information: R. Landauer, The physical nature of information, Phys. Lett. A **217** (1996) 188-193. Can be read for free at http://www.uni-leipzig.de/~biophy09/ Biophysik-Vorlesung_2009-2010_DATA/QUELLEN/LIT/A/B/3/ Landauer_1996_physical_nature_information.pdf

4. Richard Feynman's classic paper, Simulating physics with computers, published in Int. J. Theo. Phys. **21** (1982) 467-488, can be read for free at https://www.cs.berkeley.edu/~christos/classics/Feynman.pdf.

5. Chapter 4 of N. Gershenfeld, The physics of information technology, Cambridge University Press, Cambridge, UK (2002).

6. R. P. Feynman, Feynman lectures on computation (T. Het and R. W. Allen (eds.)), Perseus Books, Massachusetts, United States (1996).

7. Most of the papers cited in this chapter played a very important role in the development of the subject. We recommend them for further reading. There is great fun in learning things from the original papers.

8. A set of lucid introductory articles written by well-known experts of the field are available. A few of them are cited here [39, 40, 41, 42, 43]. We believe that readers will enjoy reading this set of articles and other similar articles. Further, there exists a set of excellent course materials [44, 45, 46, 47, 48]. The relevant parts of these articles and lecture notes are in general recommended for further reading in every chapter. To avoid repetition we will not mention them in the "further reading" sections of the other chapters.

9. A complete list of papers on quantum information can be found at [49]. It has not been updated recently but is still a good source of necessary information.

10. We have mostly elaborated Landauer's classical argument that information is physical. A quantum extension of the idea is elaborated upon in an excellent article: D. P. DiVincenzo and D. P. Loss, Quantum information is physical, cond-mat/9710259.

11. An excellent book on quantum information is M. A. Nielsen and I. L. Chuang, Quantum computation and quantum information, Cambridge University Press, New Delhi, India (2008). The relevant parts of this book are in general recommended for further reading in every chapter. To avoid repetition we will not mention it in the "further reading" sections of the other chapters.

1.6   Exercises

1. Name 10 ways by which you can either communicate or store information and identify the fundamental physical process involved in them. What do you conclude? Do you think that the information is always associated with a physical entity?

2. Assume that you are playing KBC-2 with Amitabh Bachhan[9]. In the program when he asks a particular question to a participant, he provides four alternative answers. One of them is correct. He has asked you a question, and has provided four possible answers. You don't have any idea about the correct answer. Therefore, all the options are equiprobable for you. Now you decide to use a lifeline called 50-50 (which removes two incorrect options at random). So now you are left with two options only, but still you don't know the answer. Can you quantitatively measure the amount of your ignorance before you use the lifeline, and by what amount is the ignorance reduced after you use the lifeline?

3. Suppose Jai tosses an unfair coin. He knows that the same symbol is embossed on both sides of the coin. Veeru is completely unaware of this fact. Do you think Veeru earns some information from the outcome of the toss? Justify your answer.

4. Mathematically prove that you always learn something unless you already know everything.

5. A source emits four symbols, 0, 1, 2 and 3 with probabilities 0.2, 0.2, 0.2 and 0.4. On average, at least how many bits per symbol are required to faithfully represent messages from this source?

6. Suppose that eight sprinters are participating in the final of the 100-meter sprint event of the Olympic games. The probabilities of winning for the eight sprinters are $(\frac{1}{2}, \frac{1}{4}, \frac{1}{8}, \frac{1}{16}, \frac{1}{64}, \frac{1}{64}, \frac{1}{64}, \frac{1}{64})$. Calculate the entropy of the sprint event.

7. We have a biased dice having probability of different outcomes as $(\frac{1}{6}, \frac{1}{6}, \frac{1}{6}, \frac{1}{6}, \frac{1}{8}, \frac{5}{24})$. Calculate the amount of ignorance that you have before you observe the outcome of the throw of the dice. If you decide to use a code to communicate the results of subsequent N throws of this dice, then calculate the minimum average length of your code.

8. You have a fair dice and I have an unfair dice. Who is more ignorant about the outcome of his throw? Assume that you throw your dice and I throw my dice.

---

[9]Amitabh Bachhan is one of the most popular film stars of India. He hosts a one-to-one quiz program on Indian television, entitled KBC (Kaun Banega Crorepati, which can be translated as: Who Wants to Be a Millionaire?).

9. Prove that $H(p)$ is continuous. [Hint: since the sum of continuous functions is also continuous, it would be sufficient to show that $p \log p$ is continuous in p.]

10. How will you solve Rényi's fake coin problem discussed in Solved Example 6, when number of gold coins prepared by the jeweler is 27 and only one of them is faulty.

11. Read Shannon's original paper entitled "A Mathematical Theory of Communication", which is available for free at http://cm.bell-labs.com/cm/ms/what/shannonday/shannon1948.pdf. Write a short note on this work. List all the points that you understand and also make a list of points/topics that you cannot understand.

Chapter 2

Basic ideas of classical and quantum computational models and complexity classes

In this chapter we briefly describe the basic ideas of classical and quantum computational models. To be precise, we will provide an overview of the classical (irreversible, probabilistic and reversible) and quantum Turing machines, the need for error correction and the quantum circuit model of computation. We also describe a few complexity classes that are relevant for the study of quantum computing and quantum cryptography.

2.1 Elementary idea of complexity of an algorithm

Classical theory of computation basically considers two questions: (1) What is computable? (2) What resources are required by a computing machine to solve a problem? The fundamental resources required for computing are means to store and manipulate symbols. Thus the most important resources are time (the amount of time required by the computing machine to complete a specific computational task) and space (the amount of memory required by the computing machine to perform the computation). Resource requirements are measured in terms of the length (size) of the problem. Let us think about an algorithm that can find out the square of x provided x is given. To specify the problem we need to provide $L = \log_2(x)$ amount of information ($L = \log_2(x)$ is the number of bits needed to store the value of

x) to the computing machine. This L can be visualized as the size of the problem. The computational complexity of the algorithm is determined by the number of steps s required by a computing machine to solve the problem of size L by using that particular algorithm. It is expected that the number of steps s required by the computing machine would depend on the size of the problem L. Thus s is expected to be a function of L. For example, we can think of the following cases: $s_1 = A(L^3 + L)$, $s_2 = A(L^2 + L)$ and $s_3 = A2^L = Ax$, where A is a constant. In the first two cases, the problems are considered tractable or computable, and we say that these problems can be *efficiently* solved in the computing machine. In the last case, the computational task is considered to be hard or uncomputable.

Let us make the notion a bit more compact. To do so, we introduce a coarse measure of complexity. To describe this measure we need to introduce the following two notations:

1. O (pronounced as "big oh") notation which provides an upper bound on the growth rate of a function. The idea will be clear from the following examples: In this notation $O(s_1) = O\left(L^3\right)$, $O(s_2) = O\left(L^2\right)$ and $O(s_3) = O\left(2^L\right)$. Thus in this notation we suppress the constants and lower order terms to obtain a coarse upper bound. In this notation $O(s(L))$ is used to denote the upper bound on the running time of the algorithm.

2. Similarly, we may introduce a notation to denote the lower bound. Such a notation is usually called "big omega" notation and denoted as Ω.

In the above examples, we have considered time as the computational resource, but these ideas are equally valid for any other choice of resource; say, space.

In general an algorithm is said to be efficient with respect to a particular resource if the amount of that resource used to implement the algorithm to solve the corresponding computational task is $O\left(L^k\right)$ for some k. In this case we say that the algorithm is polynomial with respect to that resource. Similarly, if an algorithm runs in $O(L)$ then we call it linear, and if it runs in $O(\log L)$ then we call it logarithmic. We know that linear and logarithmic functions cannot grow faster than the polynomial function. Consequently, these algorithms are also called efficient. On the other hand, if an algorithm requires $\Omega(c^L)$ resources (similar to our third example where $s_3 = A2^L$) where $c > 1$ is a constant then the algorithm is called exponential. It may be noted that the bound mentioned here is the lower bound. Precisely, if the resource requirement of an algorithm cannot be bounded above by any polynomial, then we call it super-polynomial (sometimes we loosely call it exponential, too) and the algorithm is considered to be inefficient and the corresponding task (problem) is considered as hard.

Here we have given the basic idea of complexity, but have not discussed the idea of the complexity classes. That will be done in detail in Section 2.4. Before that we need to know about the Turing machine. To understand the beauty and power of the Turing machine we need the concept of efficient solution, which has already been discussed. Thus we have obtained the adequate background to learn about the Turing machine.

2.2 Turing machine

The history behind the introduction of the Turing machine is fascinating. To feel the importance and the logic behind the development of the Turing machine we have to understand the situation of science and technology at the beginning of the 20th century.

Alan Turing was born on 23rd June, 1912 in London[1]. Almost twelve years before his birth the famous German mathematician David Hilbert proposed an interesting problem, which may be lucidly stated as: Is it possible to construct a universal mechanical (mathematical) procedure by which the truth or falsity of any mathematical conjecture can be proved? Subsequently, Gödel established the existence of mathematical conjectures which cannot be proved or disproved, and these are called undecidable. Thus Gödel's work modified the original question of Hilbert as: Is it possible to design a mechanical (mathematical) procedure by which the decidability or undecidability of any mathematical conjecture (i.e., whether the mathematical conjecture can in principle be proved or not) can be proved? Turing had addressed this question, which he had heard during a course of lectures that he attended at Cambridge University, England. Probably, by mechanical procedure, Hilbert indicated an algorithm; but Turing took it in its literal meaning and tried to model the thought process of a mathematician. Turing's arguments may be lucidly explained through a crude model of proof process. All of us have proved some theorems and have seen some mathematicians (maybe our mathematics professors) proving some theorems on the board or in a set of papers. So we have some perception about how a mathematician proves something. But to make the observation more compact and to transform it into a mechanical procedure, let us think of a Lilliput standing on the shoulder of the mathematician and looking inside the mathematician's head and on the paper on which the mathematician is writing the proof.

The Lilliput observes that the mathematician is writing on a paper in the following manner: First a line of calculation, then another, then another until he finishes the page. Once a page is finished he writes on another page, and it continues until the proof is completed. Does the mathematician really require a two dimensional paper? The answer is no, as he can attach

---

[1]To learn about the interesting life of Alan Turing we recommend the readers to explore the web-site: http://www.turing.org.uk/turing/

the start of the second line to the end of the first and so on. This allows us to replace the two dimensional paper by a long continuous strip of paper. The strip is called tape. A question that should immediately arise here is: How long should the strip be? It should be unlimited (infinite). This is because while the mathematician proves a new theorem or solves an unknown problem, he does not know in advance how many pages it will require. He can keep a very large number of sheets of paper to ensure that he never runs short of paper[2]. Thus if the mathematician can do a computational task (or prove a theorem) on paper or board, then the unlimited tape must be sufficient for the task.

The Lilliput also observes that the mathematician often goes back and forth on the computational steps. To replicate this activity of the mathematician there must be a "read/write" head in our machine which is allowed to go back and forth along the tape. Now we can visualize that if we wish to realize a machine which simulates a mathematician then one dimensional tape simplifies the motion of the read/write head as it is required to move along a line only. It would be sufficient to enable the read/write head to move one step forward and one step backward as that would allow the head to move from its existing position to any other position on the tape in multiple steps.

Now the mathematician uses a finite alphabet (consisting of mathematical symbols, alphanumeric[3], parentheses, etc.) to write his proof. This can be simulated in a machine by a simpler alphabet which consists of blank space and two symbols, say 0 and 1. This is allowed because a one-to-one correspondence between a string of zeroes and ones can be made with every symbol present in any larger set of symbols.

We have an unlimited tape, a read/write head capable of moving back and forth, and an alphabet consisting of blank space and two symbols, but this is not sufficient to perform the computational task. The Lilliput standing on the shoulder of the mathematician and looking inside the head of the mathematician observes that to prove something, the mathematician uses some logical rules, some earlier conclusions and some axioms. Consequently, if we wish to simulate the mathematician then our computing machine must be equipped with these rules, axioms and conclusions. Assume that the read/write head or an additional component attached to the read/write head is equipped with these features.

Now we can simulate the mathematician by an abstract mechanical procedure (an abstract machine) as we have everything that the mathematician uses for his proof. But we can still make the life of the machine simpler by dividing the long tape into a sequence of identical cells (which

---

[2]In reality the number of pages cannot be infinite as the universe itself is finite. However, it is fine to consider the tape of the abstract Turing machine as infinite. This is a kind of idealization that helps us to visualize the process of computation.

[3]Alphanumeric is the set of alphabets and numbers.   Example, $\{A, B, \cdots, Y, Z, 0, 1, \cdots, 9\}$.

are marked off) and allowing the machine to write only one symbol (i.e., 0 or 1) or to keep blank space inside each cell. This is sufficient because our machine is using an alphabet which has only two symbols. Depending upon the set of instructions present in the head of the machine, it will either change the symbol or leave it unchanged and then the head will move one step forward or backward. Thus our abstract machine can do everything that a mathematician can do, and consequently if any mathematician can compute a function (i.e., if he can do a computational task), then our machine can also compute it[4]. Thus any mathematical conjecture which can be proved using any procedure can also be proved in the Turing machine. In other words, if a mathematical conjecture cannot be proved in the Turing machine then it cannot be proved by any other means. This answers modified Hilbert's problem. The only question that remains unanswered is whether the machine can do it efficiently. This abstract machine is called the Turing machine. A schematic diagram of the Turing machine is provided in Fig. 2.1.

This remarkable idea of the abstract computing machine was proposed by Turing in a paper in 1936 [50]. Almost simultaneously and independently Alonzo Church [51] obtained a similar answer to the modified Hilbert's problem by introducing λ-calculus. In his seminal paper Turing showed that there exists a "universal Turing machine" that can be used to simulate any other Turing machine. Further, he claimed that the Turing machine completely captures the algorithmic process of performing a computational task. This is qualitatively seen in the analogy of the mathematician used here to develop the idea of the Turing machine. The works of Church and Turing implied that *if there exists an algorithm which can be performed on any piece of hardware, then there is an equivalent algorithm for a universal Turing machine which can also perform the same task.* This is known as the Church-Turing hypothesis. Note that it does not say anything about the efficiency of computation.

2.2.1 Deterministic Turing machine

The systematic effort of modeling a mathematician's thought process led to the idea of the deterministic Turing machine. We can summarize the idea as follows. The Turing machine is an abstract machine consisting of the following three parts:

1. An unlimited tape T, which is divided into cells at positions $\cdots, -1, 0, 1, \cdots$. Any cell of T contains a symbol q from a finite alphabet Q which includes the blank. The restriction on T is that there can be only finitely many non-blank cells on the tape. A binary choice of states, 0 or 1, is sufficient. However, a larger alphabet is also allowed.

---

[4]The analogy between the mathematician and the abstract machine is very lucid and it provides only a crude model of the Turing machine. However, it provides a strong perception of the intuition that led to the Church-Turing hypothesis.

Tape (T)

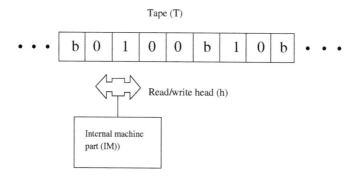

Figure 2.1: Schematic diagram of the Turing machine. The symbol b denotes a blank cell.

2. A read/write head h, capable of moving back and forth scanning the content of each cell. The head h can scan one cell at a time and can move one step forward or backward. In a particular computational step, h may change q to $q' \in Q$ or keep it unchanged.

3. An internal machine part IM which is capable of assuming any one of a finite set of internal states. IM also contains a set of rules that specify, given the current internal state, how the state must change, given the content of the cell currently being read under the head.

Modern computers are not Turing machines. However, a similarity between the Turing machine and a modern computer can be easily visualized. The head may be visualized as a small processor or a logic block, whose inputs are its existing internal memory state and the bit value at the current tape position.

We have already mentioned that the Turing machine is an abstract machine introduced in 1936. Very soon after, Von Neumann developed a theoretical model for a computer which is as capable as a universal Turing machine. In 1947 Bardeen, Brattain and Shockley developed the transistor, and the hardware development started. Since that time hardware has been considerably improved, but the dream journey is not expected to continue forever as the size of transistors is approaching quantum limit (see Section 1.2). Despite all the developments of modern hardware and increase in computational power, the interest in the Turing machine remains. Here we would like to note that the Turing machine model is not the only model of computation. Several other models have also been developed and compared. The results of these studies indicate that if a computation can be efficiently performed on any computational model, it can also be efficiently performed on the Turing machine. The inclusion of the word "efficient" strengthened the Church-Turing hypothesis, and it led to the strong Church-Turing hypothesis: *Any algorithmic process can be simulated efficiently using a Turing machine.*

Alan Turing is one of the founding fathers of modern computer science. He was born on June 23rd, 1912. He was a man far ahead of his time. He was a mathematician, a philosopher, a code breaker, and was gay. He was also interested in fundamental issues related to biology. The end of his interesting life was very tragic. At the age of 39, he was home-arrested by the Manchester police for having a homosexual relationship, which at that time in England was a crime. After that he was forced to stop his work in Cold War Intelligence at Bletchley Park. He was also compelled to take hormone injections to counter his homosexual tendencies. He tolerated unwanted and proactive police interventions and social discrimination for about two years. Finally, in June, 1954, he committed suicide by eating an apple poisoned with cyanide[a].

The photo shows the Alan Turing memorial statue in Sackville Park, Manchester. It is adopted from: http://en.wikipedia.org/wiki/Alan_Turing, photo taken by Lmno.

[a]More on the interesting life of Turing can be found in (a) A. Nemeth, Alan Turing: A man ahead of his time, http://www.ciampini.info/file/Alan%20Turing.pdf and (b) S. Turing, "Alan M. Turing: Centenary Edition", Cambridge University Press, Cambridge, UK (2012).

2.2.2 Probabilistic Turing machine

Since its introduction, the strong Church-Turing hypothesis has faced several challenges. Initially challenges came from the analog computers. Theoretically, it can be shown that analog computers can efficiently solve certain problems which do not have any known efficient solution on a Turing machine. But this analog machine will not work successfully in realistic situations. This is so because in a realistic situation there would be noise, and that would lead to errors which cannot be corrected efficiently in analog computers. Since the strong Church-Turing hypothesis had survived many challenges, people started widely accepting it.

Let us visualize the difficulty of error correction in analog computing

Consider a message u sent by Alice to Bob. An error e modifies the message as $u \longrightarrow u_1 = u+e$, where u_1 is the modified (noisy) message received by Bob. For example, if $u = 1001$ and $u_1 = 1101$ then $e = 0100$. This error can be corrected iff there is a one-to-one correspondence between u and u_1. In other words, it may be corrected iff there exists a set of messages C and a set of errors E such that $u + e \neq v + f \, \forall \, u, v \in C : u \neq v$ and $e, f \in E$. Now if this condition is satisfied, then the error can be corrected by Bob because he can uniquely deduce u from $u+e$. However, this model of error correction cannot be applied to analog computing because in the case of analog computing every u_1 is also a message. In brief, the possibility of infinitely many states makes it impossible to correct errors in analog computing machines. Apparently, the same logic is applicable to quantum states. This is because of the fact that the qubit is a two level quantum system, which can have continuum of possible states specified by $|\psi\rangle = \alpha|0\rangle + \beta|1\rangle : |\alpha|^2 + |\beta|^2 = 1$. Fortunately, it is found that the quantum errors can be digitalized. The quantum error correction code and fault-tolerant quantum computing are now well developed subjects and they will be discussed with appropriate importance in Chapter 6.

The strong Church-Turing hypothesis faced a real challenge when the Solovay-Strassen primality test algorithm appeared in 1977. This is a randomized algorithm designed by Robert Solovay and Volker Strassen [52] to find out whether an integer is a prime or not. It does not perform the task with certainty, rather it determines that the integer is probably prime or else a composite with certainty. The meaning of the last sentence may be made clearer by a simple primality test algorithm. Suppose I have to check whether n is prime or not. I choose a prime number m between 3 and $\sqrt{n}$ at random and divide n by that number. If n is divisible by m then we know with certainty that n is composite, but if not then we are not sure whether it's prime or not as it can have a factor other than m.

This was a crude example; the Solovay-Strassen algorithm is much more sophisticated and after k repetition of the Solovay-Strassen primality test, the probability of failure of the test is only $\frac{1}{2^k}$. Thus after a few repetitions of this randomized algorithm one can say with near certainty whether a number is prime or not. But this problem cannot be solved efficiently by a deterministic Turing machine. The Solovay-Strassen algorithm created huge interest among computer scientists and mathematicians. As a result, several randomized algorithms appeared that can efficiently solve certain problems (with a bounded failure probability) which cannot be solved efficiently in a deterministic Turing machine.

To circumvent this challenge, the deterministic Turing machine was generalized to a probabilistic Turing machine. A probabilistic Turing machine is capable of making a random binary choice at each step, and the state transition rules are modified to account for these random bits. Thus the probabilistic Turing machines are Turing machines with a built-in random number generator. Computation of "square root modulo a prime" is another example of such a problem, which has an efficient solution in the probabilistic Turing machine but has no efficient solution in the deterministic Turing machine. Now the generalization of the Turing machine leads to a modification of the strong Church-Turing hypothesis as: *Any algorithmic process can be simulated efficiently using a probabilistic Turing machine.*

The examples above tend to indicate that a probabilistic Turing machine may be more powerful than a deterministic one, but so far we do not have a proof whether a deterministic Turing machine and a probabilistic Turing machine are equivalent. Now we would like to note that the above modification of the strong Church-Turing hypothesis was ad hoc, and tomorrow we may find a new computational model in which one can efficiently solve problems that are not soluble even in a probabilistic Turing machine. It is not expected that we will keep on changing the strong Church-Turing hypothesis on every such occasion. This observation leads to a question: Is it possible to design a computational model that can simulate any other computational model? In 1985, David Deutsch tried to answer this question from a physical perspective. Since the best model of nature known at the moment is quantum mechanics, Deutsch considered a computing device based on the laws of quantum mechanics and proposed a quantum analogue of the Turing machine which is known as the quantum Turing machine [19]. This led to the final modification of the strong Church-Turing hypothesis, and now we have a quantum strong Church-Turing hypothesis, which is: *A quantum Turing machine can efficiently simulate any realistic model of computation.* Turing's universal computer was mathematical, but Deutsch's universal computer is physical, and consequently the quantum strong Church-Turing hypothesis is often called the physical Church-Turing hypothesis.

Credit for the introduction of the quantum Turing machine goes to Deutsch [19], but an effort in that direction was actually initiated by Paul

Benioff in 1980. Here we briefly describe the Deutsch's idea, but before that it would be apt to mention that in 1985 Deutsch also designed a simple algorithm (now known as Deutsch's algorithm), which suggested that the quantum computers may solve certain computational tasks faster than their classical counterparts. This indicated that quantum computers might have more computational powers than their classical counterparts. This indicative idea got more support from the introduction of Shor's algorithm for finding the prime factors of an integer in 1994 [22] and Grover's algorithm for unstructured database search in 1997 [27]. The speed-up achieved in these algorithms strongly indicated that a quantum computer may be more powerful than a classical computer. However, these are only indications, not proof. We discuss this point once again when we will discuss classical and quantum complexity classes in Section 2.4, but before that we need to describe Bennett's idea of the reversible Turing machine and Benioff's and Deutsch's idea of the quantum Turing machine.

2.2.2.1   Reversible Turing machine

The Turing machines described so far are not in general reversible. In 1973, Bennett established that each Turing machine can be generalized to a reversible Turing machine [6]. Bennett's work was motivated by Landauer's principle introduced in 1961 [5]. According to Landauer's principle, a computing device must dissipate at least $kT \ln 2$ amount of energy[5] for each bit of information it erases or throws away. A usual computer or a usual Turing machine is irreversible because these machines often throw away, overwrite and erase information[6]. There exists a simple way to construct a reversible Turing machine from an irreversible Turing machine. Just add an extra tape that is initially blank. The tape may be called the history tape, and after every operation it may be used to store all the information required to uniquely construct the last recorded state on the tape. Landauer pointed out a problem in this simple minded model. According to him, this will only postpone the process of erasure of information because the information on the tape has to be erased before the tape is reused. This is a problem, but we have an interesting trick to circumvent this problem. Let us start a backward computation from the end of the original computation by using the inverse of the original transition function. Since the forward computation is reversible, the backward computation will also be reversible, and the machine will return to its initial condition where the tape was blank. Thus the history tape may be cleaned without any dissipation of energy. But again we have a problem. The output of the forward

---

[5]k is Boltzmann constant which is equal to 1.38×10^{-23} Joules/Kelvin. So at room temperature the minimum amount of dissipated energy is $kT \ln 2 = 1.38 \times 10^{-23} \times 300 \times 0.693$ Joules $= 2.869 \times 10^{-21}$ Joules.

[6]The usual computers are made up of irreversible gates so they are essentially irreversible.

computation is now transformed back to the initial state. So we have not computed anything. Bennett invented an excellent trick [6] to circumvent this problem. He introduced a third tape. An extra copy of the output of the forward calculation is reversibly copied on this tape before the starting of the backward computation. Precisely, the final result of the computation is reversibly copied on the third tape. Now we can run the backward computation and clean the history tape reversibly. This solves all our problems and we have a three-tape reversible Turing machine. This was introduced by Bennett and he showed that the reversible Turing machine is universal. To emulate a v step irreversible computation we require $2v$ steps in the reversible computer (v steps for forward calculation and v steps for backward calculation; if v is large then we can ignore the steps required to copy the final result to the third tape). The disadvantage of the reversible Turing machine lies in the fact that we need a large history tape, and the advantage is that *any algorithmic process can be simulated efficiently using a reversible Turing machine.*

Bennett's universal Turing machine is classical. In 1980 Paul Benioff tried to extend Bennett's idea and provided a quantum mechanical Hamiltonian model of the Turing machine. Between 1980 to 1982 Benioff published a few papers [12, 13, 14, 15] on this topic and showed that it is possible to construct a quantum mechanical Hamiltonian model of the Turing machine that does not dissipate any energy (as it is reversible) and operates at the quantum limit imposed by the time energy uncertainty relation. His seminal works generalized Bennett's idea and showed the possibility of a quantum Turing machine. But it was not aimed to establish the idea of a universal quantum Turing machine or to challenge the classical strong Church-Turing hypothesis. Benioff's idea was just to replace the tape in the conventional Turing machine by a sequence of qubits and the read/write head of the conventional Turing machine by a quantum mechanical interaction. It was an important step toward the development of the quantum computer, but Benioff's pre-1985 proposals of Turing machines were essentially classical because in this model at the completion of each computational step the head was used to measure the state of the tape. Consequently, any superposition state present in the tape would collapse and the advantage of superposition will be lost. This point was noted by David Deutsch and he argued that Benioff's machine is classical because of the following two reasons [19]: (1) Complete specification of the state of the Turing machine at any instant is equivalent to the specification of a set of numbers, all of which are in principle measurable. This is not in accordance with the quantum mechanics. (2) Benioff's model of computation uses quantum kinematics and dynamics, but at the end of each elementary computational step, no characteristically quantum property of the model (such as interference, non-separability, or indeterminism) can be detected. Consequently, its computations can be perfectly simulated by a classical Turing machine.

Although Benioff's pre-1985 works [12, 13, 14, 15] did not provide us the quantum Turing machine in the true sense, it was an important step toward that. Feynman's 1982 work on the "universal quantum simulator" [10] was yet another milestone, although from a different perspective. Finally, a clear idea of a quantum Turing machine appeared in 1985, when Deutsch clearly established that a quantum Turing machine is possible, and probably it is more powerful than the classical probabilistic Turing machine. After the work of Deutsch, many people including Benioff have immensely contributed to the study of the quantum Turing machine. It is not the purpose of the present book to review them. We will just provide a brief perception of the quantum Turing machine.

2.2.2.2   Quantum Turing machine

To appreciate Deutsch's idea of the quantum Turing machine, it is important to understand that before Deutsch, the Turing machine was an abstract mathematical object. Deutsch considered it as a physical object. To do so, he first reinterpreted the meaning of computable functions. Here we would like to note that there are many equivalent but different statements of the Church-Turing hypothesis. One way to state it is: *Every function which would naturally be regarded as computable can be computed by the universal Turing machine.* Now it is difficult to imagine that a function is naturally computable, but could not be computed by any physical means or vice versa. Extending this argument, Deutsch considered that every computable function could be computed in a real physical system and reinterpreted Church-Turing thesis as: *Every function which may in principle be computed by a real physical system can be computed by the universal Turing machine.* This simple consideration had a profound effect on our understanding of computation as it provided a physical meaning to the abstract notion of computing.

In the next step Deutsch introduced a physical notion of perfect simulation. To understand that let us consider a computing machine M, which can perfectly simulate (without any approximation) a physical system S. To simulate S using M we have to use a set of labels as input and output. These labels are supposed to completely describe the state of S at a particular instant. Now if there exists a program $P(S)$ which can create these labels then the machine M can perfectly simulate the physical system S. In other words, $P(S)$ converts M into a black box, which is functionally indistinguishable from S. This consideration led to the physical version of the Church-Turing hypothesis which is: *Every finitely realizable physical system can be perfectly simulated by a universal model computing machine operating by finite means.*

It is interesting to note that neither classical physics nor the classical universal Turing machine obey this strong physical version of the Church-Turing hypothesis because classical physics is continuous and the classical universal Turing machine is discrete. This was one of the motivations that led to a quantum Turing machine. But what is significant is that the best model of the nature known so far is quantum mechanics, and since under this consideration the computing machine is physical, it is natural to look for a truly quantum model. There is no sense in looking into classical physics, which is just a special case of quantum physics.

We have described the essential idea behind the quantum Turing machine. Now let us describe the quantum Turing machine itself. The quantum Turing machine is essentially reversible. In the last section we have described a three-tape reversible Turing machine. Here we would like to note that a one-tape reversible machine is also possible, but in that case the tape has to be divided into the working region, the history region and the result region. The cost of shifting frequently between working region and history region would require v^2 steps to emulate a v step irreversible computation. It shows that the transformation from a one-tape machine to a many-tape machine and the opposite is straightforward and just a matter of convenience. Keeping this in mind, here we will describe a one-tape quantum Turing machine.

To a large extent the model is similar to the usual Turing machine. The quantum Turing machine consists of a finite processor (head) and an infinite memory (tape). Only a finite portion of the memory is ever used. The infinite memory may be physically visualized as a two-way infinite one dimensional lattice of qubits. This is analogous to the tape (with the states) in the usual Turing machine. The lattice positions are analogous to the cells in the tape of a classical Turing machine and the state of the qubit is analogous to the symbol stored in a particular cell. The head (processor) can exist in any one of a finite number of orthogonal states $|l\rangle$ with $l = 1, 2, \cdots, L$ and it moves along the lattice interacting with the qubits at or next to its location on the lattice. Now the elementary quantum Turing machine requires to perform one or more of the following operations: (1) head moves one lattice site to the right or left, (2) change the state of the qubit scanned or read by the head in accordance with the state of the head and that of the scanned qubit, (3) change the head state $|l\rangle$. The computation proceeds in steps of fixed time interval ΔT, and during each step only the head and a finite part of the memory (tape) interact, the rest of the memory remains static. The basic difference between Deutsch's idea and Benioff's idea lies in the fact that in Deutsch's model the tape (or the quantum memory register) stays in a superposition of computational states throughout the entire operations of the Turing machine, but in Benioff's model the tape is measured after every computational step.

The way we have visualized the Turing machine, it looks quite similar to the usual Turing machine, but conceptually it is quite different. There exist several models of the quantum Turing machine. The models are characterized by the time step operator T which corresponds to a single step of computation. Further, there exist several different but equivalent models of computation. The circuit model of computation is one of them. In quantum computing, the quantum circuit model is more frequently used than the quantum Turing machine model. We briefly describe the quantum circuit model of computation in the next subsection.

2.3   Circuit model of computation

There exists a set of underlying assumptions in the quantum circuit model of computation. We are not yet ready to describe them, as the corresponding algebra and notations have not yet been introduced. Here we provide only a brief introduction to the circuit model of computation.

Let us change our focus from an abstract classical Turing machine, and look into a real computer (maybe your PC or laptop). How does a real classical computer compute? It computes with the help of several integrated circuits, and these integrated circuits are constructed by making arrays or networks of gates. This is interesting; it seems that gates can be used to construct the model of a computing machine. To visualize this point let us look at a simple circuit. For example, consider a full adder circuit. It is easy to recognize that a full adder circuit does a particular computational task with the help of a few gates. Now we may ask: What does a gate do in a circuit? Each gate performs a small computational task. To be precise, a gate computes a function which maps the input states to the output states. If this mapping is bijective then the gate is reversible; otherwise the gate is irreversible. For example, NOT is a reversible gate because we can uniquely construct the input state from the output state, but NAND is irreversible because we cannot uniquely construct the input state from the output state.

In summary, we can say that every gate does a small computational task and they may be joined together in a meaningful fashion to make a computing machine. If we can show that a finite set of different gates can be combined to construct a circuit for performing any computation, then in principle, we have a universal computing machine. Without going into the mathematical detail, we can say that such sets exist and that's what we mean by universality or universal gate library. We may define a universal gate library for classical computation as a set of gates that is sufficient to construct a circuit for computing $f : \{0,1\}^n \rightarrow \{0,1\}^m$ for any positive integer n and m. There exist several such universal classical gate libraries, such as {NAND}, {NOR}, {Toffoli},[7] etc. Thus we can construct circuits to compute any function, and consequently anything that is computable can be computed in our computing machine.

This classical idea can be extended to quantum domain, if we can construct a universal gate library of quantum gates. Here we have to be cau-

---

[7]Toffoli, Deutsch, Hadamard and CNOT gates will be properly described in Chapter 4.

tious as the functions are bijective and the gates are probabilistic. But there exist several such gate libraries and we can join gates from them to construct a circuit for any allowed f. For example, {CNOT, all one qubit gates}, {Deutsch}, {Toffoli, Hadamard} [53] are universal quantum gate sets (gate libraries). Consequently, we have a quantum circuit model of computation.

The quantum circuit model of the computation is equivalent to the quantum Turing machine model of computation and other well known models of quantum computation. The equivalence of two computational models implies that the resources required to solve a particular problem are the same in both the models. Let us try to understand the resource requirement in the quantum circuit model of computation.

To measure the quality of quantum circuits we use several different cost metrics. One of them is the gate count or circuit complexity. Gate count is simply the number of quantum gates present in the quantum circuit. Since the quantum gates are unitary, their product is also unitary, but we are not allowed to use the product as a new gate that reduces the circuit complexity. The gates used to measure circuit complexity are to be collected from a well defined universal gate library. Now, consider that we have a computational task which can be computed by a quantum circuit. If the gate count of the circuit is polynomially bounded in the number of input qubits (bits) n, then the circuit is called efficient. For example, if we require n^4 gates to compute a function then the circuit is efficient for the task. But if we require at least 2^n gates to compute a function, then the circuit is not efficient for the purpose. The analogy between the circuit complexity and time complexity would be clearer if we consider that the circuit is divided into a sequence of discrete time-slices (time steps), where the application of a single gate (or the gates that can be applied in parallel) requires a time-slice or unit time [54]. The depth T_d of the circuit is defined as the total number of such time slices. $T_d \leq$ total number of gates, as gates that act on disjoint qubit lines can be applied in parallel (i.e., at the same time). Another measure of complexity of circuit is total number of qubit lines present in the circuit. This is called width or space of the circuit, and it is analogous to space in the Turing machine model. If $T_d = O(n^k)$ for some k for a particular computational task, then the circuit is efficient for that task and there must exist an algorithm which can perform the same computational task in a Turing machine in polynomial time.

2.4 Computational complexity and related issues

Computational complexity theory aims to classify computational problems according to their inherent difficulty (complexity). The computational complexity theory, which is a branch of theory of computation, is quite well developed. But still there are several open questions in this theory and many commonly believed assertions of this theory are not proved. They are believed to be true because the counter examples are not found. We will provide several examples of such unproven beliefs in computational complexity theory, but before we provide such examples, we need to have a perception of this theory. The fundamental building block of this theory is a complexity class. A complexity class is a set of problems of similar complexity (difficulty level). Simply stated, if we have a set of problems, that are solvable in a computer with equal level of difficulty, then the set forms a complexity class. By equal amount of difficulty we mean that to solve these problems we require a similar amount of computational resources.

We have already provided the definition of complexity class. Now we wish to classify the complexity classes according to the amount of complexity. P (polynomial time) and NP (nondeterministic polynomial time) are the most popular complexity classes. Simply put, the class P contains all the computational problems which can be efficiently (quickly) solved on a classical computer. By an efficiently solvable problem we mean that the problem can be solved in polynomial time or equivalently by using polynomial number of gates. On the other hand, the class NP contains all those problems whose solutions can be quickly verified on a classical computer. Since solutions of all problems in P can be verified efficiently, the problems that are in P are also in NP. However, the class NP may contain problems, whose solutions can be verified efficiently, but the problem itself cannot be solved efficiently. Existence of any such problem implies that P is a proper (strict) subset of NP, i.e., $P \subsetneq NP$. A probable example of such a problem is prime factorization. Assume that we have to find out prime factors of a large integer n. Until now there does not exist any efficient algorithm which can be implemented on a classical computer to quickly find out the prime factors. The best known method is the number field sieve[8], which requires a considerably large number of computational steps (s) to find the prime factors of a large number n. To be precise, if n is a 600 bit number then $s \approx 10^{31}$ which indicates that to factorize such a large number we would require about a million years with the present technology. Thus the problem is intractable in the classical computer at present, and it does not belong to P at the moment. Further, if we have been provided with a solution p of this problem (i.e., p is told to be a factor of n), then we can quickly verify the correctness of the solution by dividing n by p. This indicates that factorization is a computational problem which does not belong to P and belongs to NP. This suggests that $P \subsetneq NP$. We reach this conclusion by considering the number field sieve algorithm as the best algorithm. However, we cannot exclude the possibility that tomorrow someone may invent an efficient classical algorithm for factorization.

---

[8]This method was developed by Polard. This powerful method of factorization of an arbitrary number has complexity $O\left(\exp\left[c(\log n)^{\frac{1}{3}}(\log\log n)^{\frac{2}{3}}\right]\right)$, where in general case $c = \left(\frac{64}{9}\right)^{\frac{1}{3}} = 1.92$.

Most of us believe that there exist problems in NP which do not exist in P, but this has not been proved. At present we do not know whether $P = NP$. This is one of the most important unsolved problems of computer science. The importance of this open problem would be clearer if we consider the NP-complete problems. NP-complete is a subclass of NP which contains the hardest problems of NP, in the sense that if you can efficiently solve one NP-complete problem, then you can efficiently solve all NP problems. Thus if one can design an algorithm to solve an NP-complete problem in polynomial time then all the other problems which belong to NP would be solved in polynomial time[9]. On the other hand, if someone can prove that $P \neq NP$, that would imply that none of the NP-complete problems can be solved quickly in the classical computer. A large number of NP-complete problems are reported so far and many of them are very important. The examples of important NP-complete problems are scheduling problems, satisfiability problems, traveling salesman problem, etc. Here we would like to draw the attention of the readers toward the fact that the complexity depends on the algorithm. Consequently, invention of a new and efficient algorithm may shift a computational problem from one computational class to the other.

We can have a better perception of NP-complete problems if we understand the meaning of reduction in the context of complexity theory. To do so, let us assume that we have an algorithm for multiplication. Thus if we provide two numbers p and q as inputs then the algorithm provides us pq as solution. Now if $p = q$, then the same algorithm will give us square of p. This implies that an algorithm that can do multiplication can also do squaring. In other words, squaring cannot be more difficult than multiplication, and squaring can be reduced to multiplication. In general, if a problem X can be solved using an algorithm designed for solving problem Y, then X cannot be more difficult than Y, and we say that X reduces to Y. This notion of reduction is very useful.

Consider an arbitrary complexity class C. If all problems of C can be efficiently (i.e., with a polynomial-time overhead) reduced to a particular problem Y then that implies that no problem of C is much harder than the problem Y. In other words, if we can efficiently solve Y, we can efficiently solve all problems of C. If $Y \in C$, it is said to be complete for C. Thus Y is the hardest problem in C. There may be more than one problem in C having same hardness as that of Y. All problems of C can be solved, if any of these hardest problems can be solved. Consequently, efficient solution of one of the hardest problems in C would imply efficient solutions of all other hardest problems in C, too. Now, when $C = NP$, then the set of all the hardest problems of C forms a subset of NP which is known as NP-complete. Thus NP-complete problems are the hardest problems in

---

[9]If it ever happens then the challenge and fun of algorithm designing would reduce to a large extent.

class NP.

Cartoon 2.1: Does complexity class of a problem depend on our knowledge of algorithm?

In Subsection 2.2.2, we briefly mentioned the Solovay-Strassen primality test algorithm, which tells us that a given number is composite with certainty or prime with a bounded error. Thus if we see it as a yes/no question, the algorithm answers a question: Is the given number composite? If the answer is "no" (i.e., if the given number is prime), the algorithm always gives the correct result, but if the answer is "yes", the algorithm provides us a correct answer with bounded probability of failure. Say, we fix the bound at $\frac{1}{2}$. The Solovay-Strassen primality test algorithm is just a particular example of a yes/no problem for which an efficient solution in a probabilistic Turing machine exists with the properties that if the correct answer is "no", it always returns "no", but if the correct answer is "yes", it returns "yes" with probability at least $\frac{1}{2}$. A set of all such problems form a class called RP which stands for randomized polynomial time. Now note that in the Solovay-Strassen primality test, if we find a factor of the given number, it is definitely composite; thus the cases where we obtain "yes" as answers are always correct, and the cases where we obtain "no" as answers are mostly correct. This is the general property of RP. Thus the Solovay-Strassen primality test is in RP.

Analogously, we may define another class of problems, such that if the correct answer is "yes", it always returns "yes", but if the correct answer is "no", it returns "no" with probability greater than $\frac{1}{2}$. Such a class is called $co - RP$. The intersection of these two sets is called ZPP (zero-error probabilistic polynomial time), which contains problems for which both yes and no answers are always correct, but there may be instances of "I don't know".

Quantum complexity class analogous to RP is called RQP and is defined as the class of yes/no kind of problems that can be efficiently answered by a quantum Turing machine with always correct "no" answer and the correct "yes" answer with probability greater than $\frac{1}{2}$. Similarly, quantum analogue of $co - RP$ and ZPP are called $co - RQP$ and ZQP respectively. It is interesting to note that the problems in ZQP can be solved by a quantum Turing machine without errors in polynomial time.

The factorization problem can be efficiently solved in a quantum computer using Shor's algorithm. This is an interesting development since the factorization problem is the basis of RSA cryptography, which is extensively used today. But unfortunately the factorization problem does not belong to NP-complete. Consequently, Shor's algorithm cannot be used to solve a large class of problems, but the success of Shor's algorithm made many researchers optimistic about solving NP-complete problems using quantum algorithm or to verify whether $P = NP$ or not. However, so far there is no success in that regard and the initial optimism is gradually disappearing. Here it would be apt to note that the problems of NP that are not in NP-complete form a class known as NPI (NP intermediate). Since factorization is not proven to be in NP-complete, at present it is believed to be in NPI. As Shor's algorithm successfully solved this problem, and as no efficient quantum algorithm has so far been obtained for solving any NP-complete problem, many people believe that the quantum Turing machine can only solve problems in NPI in polynomial time, and it cannot solve problems in NP-complete in polynomial time. However, this is only a belief, and it does not have any proof until now.

At present, there exist a handful of quantum algorithms that perform better than the best known classical algorithms for the same purposes. These algorithms belong to a particular complexity class called BQP (bounded error quantum polynomial time). For example, Shor's algorithm belongs to BQP. The BQP is defined as the complexity class which contains all the computational problems that can be solved efficiently using quantum algorithms provided a bounded probability of error is allowed. The upper bound on the probability of obtaining a wrong solution is usually fixed at $\frac{1}{3}$ (any other fixed number smaller than $\frac{1}{2}$ will do). Now it is straightforward to see that $RQP \subset BQP$ and $co - RQP \subset BQP$. To understand the power of quantum computers we need to understand this computational class. To do so, we need to understand two more complexity classes, namely, $PSPACE$ and BPP.

$PSPACE$ is an important complexity class. It contains all those problems which can be solved in a small computer, but not essentially in small computational time. Both P and NP are clearly subsets of $PSPACE$. But once again, it is not known whether this inclusion is strict (i.e., whether $P \subsetneq PSPACE$ and $NP \subsetneq PSPACE$). BPP (bounded-error probabilistic polynomial time) is another class of computational problems that can be quickly solved using randomized algorithms provided a bounded probability of error is allowed. BPP contains both RP and $co-RP$. It is clear that BQP is the quantum analog of BPP and it includes BPP. Similarly, a quantum complexity class analogous to P is called QP (quantum polynomial time). The computational problems in QP can be solved with certainty in polynomial time by a quantum computer. Until now the exact relations between P, QP, NP, $PSPACE$ and BQP are not known, but it is known that the quantum computer can efficiently solve all the computational problems in P and it cannot efficiently solve any computational problem outside $PSPACE$. These facts indicate that BQP lies somewhere between P and $PSPACE$. To be precise, currently we know $P \subset BPP \subset BQP \subset PSPACE$ and $P \subset NP \subset PSPACE$, but so far there is no proof, which indicates that these relations are strict. At present we have some examples, where quantum algorithm performs better than the best known classical algorithm for the same task, but it does not strictly prove that the quantum computer is more powerful than the classical computer. In order to prove that the quantum computer is more powerful than the classical one, we have to show that it can efficiently solve at least one problem which exists inside $PSPACE$ but outside P. The existence of such an algorithm would implicitly mean that $P \neq PSPACE$. Such a proof has not yet been reported.

2.5   Solved examples

1. Use Landauer's principle to establish that in an ideal situation 3.5×10^{20} operations per second can be done in an irreversible computer by using 1 Watt of power. Compare this rate with the existing rate and write your comments.

   Solution: As in every irreversible operation we erase at least one bit of information so according to Landauer's principle we spent at least $kT \ln 2$ amount of energy for each irreversible operation. Here k is Boltzmann constant= 1.38×10^{-23} Joules/Kelvin and at room temperature $T = 300$ Kelvin.

   Therefore, we need at least $1.38 \times 10^{-23} \times 300 \times \ln 2 = 2.8696 \times 10^{-21}$ Joules for every operation.

   Now 1 Watt of power means expense of 1 Joule of energy per second which means that the number of operations that can ideally be done per second in an irreversible computer is

   $$\frac{1}{2.8696 \times 10^{-21}} = 3.48 \times 10^{20} \approx 3.5 \times 10^{20} \text{ operations.}$$

   At present, the speed of our PC's are in few GHz (say, 3.5 GHz i.e., 3.5×10^9 operations/second) so in an ideal situation it can be faster by a factor of 10^{11} (i.e., a hundred billion times faster than your present PC. Just imagine it!!).

2. Use Landauer's principle and calculate the maximum number of operations possible in an irreversible logic gates based computer, when it operates at a temperature 30°C and consumes 40 Watts of power. Compare the speed of the above mentioned maximally efficient reversible computer with that of the existing computers and write your comments.

   Solution: Follow the solution of the previous problem.

3. Consider that quantum algorithms perform different computational tasks in (a) $O\left(n^3\right)$ steps with certainty, (b) $O(n \log n)$ steps with certainty, and (c) $O(n \log n)$ steps with probability of error $\frac{1}{64}$. Identify the complexity class in each case.

   Solution: (a) Quantum algorithm solves the problem in polynomial time with certainty. Therefore the corresponding task belongs to computational class QP.

   (b) Quantum algorithm solves the problem in polynomial time with certainty. Therefore the corresponding task belongs to computational class QP.

   (c) Quantum algorithm solves the problem probabilistically in polynomial time with failure probability $\frac{1}{64} < \frac{1}{3}$. Therefore the corresponding task belongs to the computational class BQP.

4. There exists a quantum gate named Toffoli gate which can be used to design a NAND gate. Further, there exists another quantum gate called Hadamard gate which can be used to simulate a coin flip. How these two facts imply $BPP \subseteq BQP$?

   Solution: Since NAND gate is the universal classical gate, the fact that Toffoli gate can be used to design aNAND gate implies that all computational tasks that are possible using classical resources can also be done with quantum resources. Further, since the Hadamard gate can be used to simulate coin flip, a probabilistic classical Turing machine can be simulated by a quantum Turing machine. In other words, all the problems that can be efficiently done (with bounded error) in a classical probabilistic Turing machine can also be done in a quantum Turing machine (with bounded error). But the converse is not known. So either $BPP \subset BQP$ or $BPP = BQP$. This is how the given facts imply $BPP \subseteq BQP$.

5. Consider a single Toffoli gate as a circuit and tell us the width and depth of this simple circuit.

   Solution: Since Toffoli is a three qubit gate, the width of the circuit is 3, and since there is only one gate in the circuit, the depth of it is 1.

6. Classify the following statements into three groups (a) correct, (b) wrong and (c) not known with certainty?
(i) $P \subset NP$ (with strict inclusion).
(ii) The computational problems in QP can be solved with certainty in polynomial time by a quantum computer.
(iii) $P \neq PSPACE$.
(iv) After 100 repetitions of Solovay-Strassen primality test algorithm we could not find any prime factor of n. So n is prime.
(v) Alphanumeric is the set of alphabets and numbers.
(vi) Every function which may in principle be computed by a real physical system can be computed by the universal Turing machine.
Solution: Statements (i), (iii) and (iv) are not known with certainty. Statements (ii), (v) and (vi) are correct. So none of the statements are wrong.

2.6   Further reading

1. Hilbert's original paper: D. Hilbert, Weiterführung der Methoden der Variationsrechnung, Akad. Wiss. Göttingen (1900) 291-296, is available in German at
http://www.mathematik.uni-bielefeld.de/~kersten/hilbert/23.pdf and is also available in English (translation) at
http://aleph0.clarku.edu/~djoyce/hilbert/problems.html#prob23.

2. Turing's original work on Turing machine which was published in Proceedings of the London Mathematical Society, ser. 2. **42** (1936-7) 230-265; corrections, Ibid, **43** (1937) 544-546 is now available at http://www.cs.virginia.edu/~robins/Turing_Paper_1936.pdf.

3. Bennett's seminal paper on reversible Turing machine, entitled Logical reversibility of computation, was published in IBM J. Res. Dev. **17** (1973) 525. The article may be read for free at
http://webout.weizmann.ac.il/complex/tlusty/courses/InfoInBio/Papers/Bennett1973.pdf.

4. Deutsch's original paper: D. Deutsch, Quantum theory, the Church-Turing principle and the universal quantum computer, Proceedings of the Royal Society of London A, **400** (1985) 97-117. It is available at: http://www.cs.berkeley.edu/~christos/classics/Deutsch_quantum_theory.pdf

5. Here we have discussed only a few computational classes relevant to us. Interested readers may visit the complexity zoo site:
https://complexityzoo.uwaterloo.ca/Complexity_Zoo for a comprehensive list of complexity classes.

2.7 Exercises

1. Bounds on the best known classical algorithms for some specific computational tasks are given below. Try to identify the complexity classes in each of the following cases.
   (a) $O(n^2)$, (b) $O(n^n)$, (c) $O(2^n)$, (d) $O(3^n)$, but the validity of the solution can be checked in $O(n^3)$, (e) $\log(n)$.

2. Assume that two irreversible computers A and B are performing Boolean logic operations. Operating temperature of A is -4°F and that of B is 0°C. Find out the difference in minimum amount of energy required for deletion of one bit of information in A and B.

3. Is it correct to say $QP \subseteq BQP$?

4. A reversible three-tape Turing machine requires 2000 steps to finish a computational task. Approximately how many steps are required by (a) an irreversible Turing machine and (b) a single tape reversible Turing machine?

5. Provide examples of computational tasks that belong to the following complexity classes: (a) $co-RP$, (b) RP, (c) NP-complete, (d) NPI, (e) BQP, (f) QP, (g) ZQP.

Chapter 3

Mathematical tools and simple quantum mechanics required for quantum computing

It is not in the scope of this elementary textbook to introduce quantum mechanics in detail or to provide an elaborate mathematical background required for quantum computing. However, we will provide some basic ideas that are essential for us to proceed further. To be precise, here we will briefly describe vector space, different types of operators, Gram-Schmidt procedure, postulates of quantum mechanics, POVM, density operator, entanglement and its measures, partial trace, Schmidt decomposition, Bloch sphere, nocloning theorem, etc. Let us start our journey toward the world of quantum computation by reviewing some simple ideas of algebra.

3.1 A little bit of algebra required for quantum computing

As far as the algebra of quantum computing is concerned, the most important concept is Hilbert space, which is a complex vector space. To understand that, we need to understand, vector space first. In order to understand vector space, in general, we may recall vectors in conventional three dimensional space. In a three dimensional coordinate space we need three numbers to describe a vector. These three numbers are essentially projections along three directions (or three axes) specified by three orthonormal unit vectors. So in three dimensional space we need a set of three orthonormal unit vectors $\hat{i}$, $\hat{j}$ and $\hat{k}$ to describe any other vector $\vec{P}$ (in terms of these unit vectors) as $\vec{P} = x\hat{i} + y\hat{j} + z\hat{k}$. Now, if we generalize this idea and do not restrict ourselves to three dimensions and allow the scalar projections $(x, y, z,$ etc.$)$ to be complex, then we obtain a vector space. Thus we may lucidly describe a vector space in general as a generalization of familiar three dimensional space into an n-dimensional space where complex projections are also allowed. A formal definition of the vector space is provided in the following subsection.

3.1.1   Vector space

A vector space is a set of elements called vectors $(\vec{\alpha}, \vec{\beta}, \vec{\gamma}, \cdots)$, together with a set of scalars $(a, b, c, \cdots)$, which is closed under vector addition and scalar multiplication. Thus a set of objects constitutes a vector space if it obeys the following rules:

1. Closure: Addition of two objects of the set gives another object of the same set (i.e., addition of two vectors gives a vector). Therefore, a well defined operation will never take us outside the vector space.

2. Has a zero: For every object $\vec{V}$ there exists another object $\vec{0}$ such that $\vec{V} + \vec{0} = \vec{V}$.

3. Scalar multiplication: If c is a scalar and $\vec{V}$ is a vector then $c\vec{V}$ is also a vector.

4. Inverse: For every $\vec{V}$ there exists a $-\vec{V}$ such that $\vec{V} + (-\vec{V}) = \vec{0}$.

5. Associative: The addition operation in a linear vector space is associative, i.e., $(\vec{V} + \vec{W}) + \vec{X} = \vec{V} + (\vec{W} + \vec{X})$.

Example 3.1: Show that a set of vectors $(\vec{\alpha}, \vec{\beta}, \vec{\gamma}, \cdots)$, together with a set of positive scalars $(a, b, c, \cdots)$ cannot form a vector space.
Solution: If $(a, b, c, \cdots)$ are only positive then the inverse of a vector cannot exist and consequently a necessary property of vector space cannot be satisfied.

3.1.1.1   The C^n vector space

At this stage, we would like to draw the attention of the reader toward the conventional Cartesian coordinate system, where the unit vectors, $\hat{i}$, $\hat{j}$ and $\hat{k}$ form a complete set of orthonormal basis vectors. Under this basis we can describe an arbitrary vector $\vec{P} = x\hat{i} + y\hat{j} + z\hat{k}$ as (x, y, z) or as $\begin{pmatrix} x \\ y \\ z \end{pmatrix}$.

Thus the elements of the column matrix are essentially the coefficients of the basis vectors. Analogously to describe an arbitrary n-dimensional state/vector we need a complete set of n orthonormal basis vectors, and the state can be described by a set of n complex numbers, which are the coefficients of the n basis vectors (basis states). In general, the coefficients of the basis vectors can be complex numbers and the space spanned by the set of n orthonormal basis vectors is called C^n space. A special case of C^n space is one in which all the coefficients are real, and that space is known as R^n space.

The C^n vector space is a vector space where each vector is n dimensional. So we need n complex numbers to describe any vector of this space. In other words any object of this set is $n-tuples$ of complex numbers. For example, if z_i is a complex number then

$$\vec{a} = \begin{pmatrix} z_1 \\ z_2 \\ z_3 \\ \vdots \\ z_n \end{pmatrix}$$

is a vector in C^n space. From the definition of vector space we learn that the vectors in C^n space should satisfy the following relations:

$$\begin{pmatrix} z_1 \\ z_2 \\ z_3 \\ \vdots \\ z_n \end{pmatrix} + \begin{pmatrix} z_1' \\ z_2' \\ z_3' \\ \vdots \\ z_n' \end{pmatrix} = \begin{pmatrix} z_1 + z_1' \\ z_2 + z_2' \\ z_3 + z_3' \\ \vdots \\ z_n + z_n' \end{pmatrix},$$

and

$$c \begin{pmatrix} z_1 \\ z_2 \\ z_3 \\ \vdots \\ z_n \end{pmatrix} = \begin{pmatrix} cz_1 \\ cz_2 \\ cz_3 \\ \vdots \\ cz_n \end{pmatrix}.$$

For a discrete quantum system with n possible states, we will be interested in C^n space. Before we proceed further let us strengthen our idea of inner product and basis set.

3.1.1.2 Inner product space and Hilbert space

Before we introduce the inner product we wish to introduce a notation that will be used throughout this text. The notation is known as Dirac notation or bra-ket notation. An elementary idea of this notation can be provided as follows: $\langle \; \rangle$ is a "bracket"; now we may divide it into two parts as $\langle | \rangle$. Thus the "bra-ket" is divided in two parts and the left part (i.e.,

$\langle |$ is known as "bra" and the right part (i.e., $| \rangle$) is known as "ket". It is easy to remember the notation if we remember that the bra and ket in Dirac notation originate from the division of a conventional bracket into two parts. In this notation we denote a vector as $|\psi\rangle$. The conjugate to a ket vector $|\psi\rangle$ is denoted as $\langle\psi|$ and is called a bra vector. The following examples will further clarify the meaning of the notation.

Example 3.2: (a) If $|\psi\rangle = \begin{pmatrix} 0.5 \\ 0.25 \\ 0.25 \end{pmatrix}$ then $\langle\psi| = \begin{pmatrix} 0.5 & 0.25 & 0.25 \end{pmatrix}$.

(b) In this notation, we can express an arbitrary vector in three dimensional space $\overrightarrow{P} = x\hat{i} + y\hat{j} + z\hat{k} = \begin{pmatrix} x \\ y \\ z \end{pmatrix}$ as $|P\rangle = x|i\rangle + y|j\rangle + z|k\rangle$ where

$|i\rangle = \begin{pmatrix} 1 \\ 0 \\ 0 \end{pmatrix}$, $|j\rangle = \begin{pmatrix} 0 \\ 1 \\ 0 \end{pmatrix}$ and $|k\rangle = \begin{pmatrix} 0 \\ 0 \\ 1 \end{pmatrix}$ are unit vectors like $\hat{i}$, $\hat{j}$

and $\hat{k}$.

(c) In general an arbitrary vector in C^n vector space can be written in this

notation as $|\psi\rangle = \begin{pmatrix} z_1 \\ z_2 \\ \vdots \\ z_n \end{pmatrix}$ and then $\langle\psi| = \begin{pmatrix} z_1^* & z_2^* & \cdots & z_n^* \end{pmatrix}$. Thus

the conjugate transpose (or Hermitian conjugate) of a bra vector is the corresponding ket vector and vice versa. In a compact form we can write $|\psi\rangle^\dagger = \langle\psi|$ and $|\psi\rangle = \langle\psi|^\dagger$.

With this background of Dirac notation, we are now ready to introduce the inner product. The inner product is a generalization of the dot product. In a vector space, it is a way to multiply two vectors to yield a complex number $(|\psi\rangle, |\phi\rangle) \equiv \langle\psi|\phi\rangle \mapsto C$ that obeys the following rules:

1. $(|v_k\rangle, \Sigma_k a_k |w_k\rangle) = \Sigma_k a_k (|v_k\rangle, |w_k\rangle)$,

2. $(|v\rangle, |w\rangle) = (|w\rangle, |v\rangle)^*$, and

3. $(|v\rangle, |v\rangle) \geq 0$.

Norm of a vector $|v\rangle$ is defined as

$$||v|| = (|v\rangle, |v\rangle)^{\frac{1}{2}}. \tag{3.1}$$

From the third property of the inner product the norm of a vector is always greater than zero unless the vector is identically zero.

An inner product space is defined as a linear vector space in which an inner product can be defined for all elements of the space. A complex inner product space is called the Hilbert space. Thus the *Hilbert space is a complex vector space, where the inner product is well defined for all the elements of the space.*

Example 3.3: If $\langle i|j \rangle = \delta_{i,j}$ where $i, j \in \{0, 1\}$ then find the inner product of $|\psi_1\rangle = a|0\rangle + b|1\rangle$ and $|\psi_2\rangle = c|0\rangle + d|1\rangle$.

Solution: $\langle \psi_1|\psi_2 \rangle = ac\langle 0|0 \rangle + ad\langle 0|1 \rangle + bc\langle 1|0 \rangle + bd\langle 1|1 \rangle = ac + bd$.

3.1.1.3 Bases and linear independence

If there exists a set of vectors $\{|v_1\rangle, |v_2\rangle, \cdots, |v_n\rangle\}$ such that any vector $|v\rangle$ in the space can be written as a linear combination of the vectors in the set (i.e., $|v\rangle = \sum_{j=1}^{n} a_j|v_j\rangle$), then the set is called a spanning set. A spanning set does not always form a basis, but a basis always forms a spanning set[1]. Actually, a linearly independent spanning set forms a basis. The condition of linear independence of a set of vectors $\{|v_1\rangle, |v_2\rangle, \cdots, |v_n\rangle\}$ is mathematically stated as

$$\sum_{j=1}^{n} a_j|v_j\rangle = 0 \text{ iff all } a_j = 0.$$

In other words, if there does not exist any linear combination of the set of vectors $\{|v_1\rangle, |v_2\rangle, ..., |v_n\rangle\}$, which adds to zero non-trivially, then it forms a set of linearly independent vectors.

Now since the elements of the basis set are linearly independent of each other so they satisfy the following condition of orthogonality[2]

$$\langle v_i|v_j \rangle = 0 \text{ for all } i \neq j.$$

Until now we have not said anything about the magnitude of the vectors, but from our previous knowledge of vectors in three dimensional Cartesian space we know that normally we use a set of vectors having unit magnitude as basis vectors. Usually we follow the same convention in higher dimensions too, and if we have a vector in a particular direction we can easily obtain a unit vector in that direction by dividing the vector by its norm. A unit vector in the direction of $|v\rangle$ is obtained as

$$\text{unit vector in the direction of } |v\rangle = \frac{|v\rangle}{\sqrt{\langle v|v \rangle}}, \tag{3.2}$$

where the denominator is the norm of $|v\rangle$. It is the usual practice to use a set of mutually orthogonal unit vectors as a basis set. Such a basis set is known as an orthonormal basis set and the elements of the set satisfy

$$\langle v_i|v_j \rangle = \delta_{ij},$$

---

[1] A set of vectors that spans the space is also called complete.

[2] In general any two vectors $|v\rangle$ and $|w\rangle$ are orthogonal to each other if they satisfy $\langle v|w \rangle = 0$.

where δ_{ij} is Kronecker delta function:

$$\delta_{ij} = \left\{ \begin{array}{l} 0 \text{ for } i \neq j \\ 1 \text{ for } i = j \end{array} \right. \tag{3.3}$$

Example 3.4: In two dimensional quantum (vector) space we use a basis set $\{|0\rangle, |1\rangle\}$ to describe an arbitrary quantum state (qubit) as $|\psi\rangle = \alpha|0\rangle + \beta|1\rangle$. Here $|0\rangle$ and $|1\rangle$ satisfy

$$\langle 0|0\rangle = \langle 1|1\rangle = 1 \text{ and } \langle 0|1\rangle = \langle 1|0\rangle = 0. \tag{3.4}$$

Thus $\{|0\rangle, |1\rangle\}$ forms an orthonormal basis in two dimensional quantum vector space.

Example 3.5: Show that $\left\{|+\rangle = \frac{|0\rangle+|1\rangle}{\sqrt{2}}, |-\rangle = \frac{|0\rangle-|1\rangle}{\sqrt{2}}\right\}$ forms a basis in C^2 and then express an arbitrary vector $|\psi\rangle = \alpha|0\rangle + \beta|1\rangle$ in $\{|+\rangle, |-\rangle\}$.

Solution: $\{|+\rangle, |-\rangle\}$ forms a basis in C^2 as we can easily check that $\langle +|+\rangle = \frac{\langle 0|0\rangle+\langle 0|1\rangle+\langle 1|0\rangle+\langle 1|1\rangle}{2} = 1$ and similarly $\langle -|-\rangle = 1$ and $\langle +|-\rangle = \langle -|+\rangle = 0$. Now $|0\rangle = \frac{|+\rangle+|-\rangle}{\sqrt{2}}$ and $|1\rangle = \frac{|+\rangle-|-\rangle}{\sqrt{2}}$. Therefore, $|\psi\rangle = \alpha\frac{|+\rangle+|-\rangle}{\sqrt{2}} + \beta\frac{|+\rangle-|-\rangle}{\sqrt{2}} = \frac{\alpha+\beta}{\sqrt{2}}|+\rangle + \frac{\alpha-\beta}{\sqrt{2}}|-\rangle$.

Problem 3.1: Show that the set $\left\{\frac{|00\rangle+|11\rangle}{\sqrt{2}}, \frac{|00\rangle-|11\rangle}{\sqrt{2}}, \frac{|01\rangle+|10\rangle}{\sqrt{2}}, \frac{|01\rangle-|10\rangle}{\sqrt{2}}\right\}$ forms a complete set of orthonormal basis in C^4 space.

3.1.1.4 C^2 space: The space spanned by a single qubit

To describe a two level quantum system or a qubit we need a two dimensional complex Hilbert space, which we realize as C^2 - the set of all column vectors

$$\vec{a} = \left(\begin{array}{c} a_1 \\ a_2 \end{array} \right),$$

where a_1 and a_2 are the complex numbers. In C^2 the inner product (a, b) is defined as

$$(\vec{a}, \vec{b}) = a_1^* b_1 + a_2^* b_2, \tag{3.5}$$

This particular Hilbert space is of immense importance because it is the space spanned by the set of all possible single qubit states. In the matrix notation we write a qubit $|\psi\rangle$ as

$$|\psi\rangle = \alpha|0\rangle + \beta|1\rangle = \left(\begin{array}{c} \alpha \\ \beta \end{array} \right), \tag{3.6}$$

where $|\alpha|^2 + |\beta|^2 = 1$.

Here we would like to note that in the matrix form $|0\rangle$ and $|1\rangle$ are described as $|0\rangle = \left(\begin{array}{c} 1 \\ 0 \end{array} \right)$ and $|1\rangle = \left(\begin{array}{c} 0 \\ 1 \end{array} \right)$. This particular convention is used throughout this textbook.

3.1.1.5 Outer product

Let $|v\rangle$ be a vector in the vector space V and $|w\rangle$ be a vector in the vector space W. Then the outer product of $|w\rangle$ and $|v\rangle$ (which is denoted as $|w\rangle\langle v|$ in Dirac notation) is a linear map from V into W defined by

$$|w\rangle\langle v|(|v'\rangle) = |w\rangle\langle v|v'\rangle. \tag{3.7}$$

For example,

$$|0\rangle\langle 1| = \begin{pmatrix} 1 \\ 0 \end{pmatrix} \begin{pmatrix} 0 & 1 \end{pmatrix} = \begin{pmatrix} 0 & 1 \\ 0 & 0 \end{pmatrix}.$$

Since $\{|0\rangle, |1\rangle\}$ forms an orthonormal basis so we expect that $|0\rangle\langle 1|1\rangle = |0\rangle$ and $|0\rangle\langle 1|0\rangle = 0$. We can easily verify that as $|0\rangle\langle 1|1\rangle = \begin{pmatrix} 0 & 1 \\ 0 & 0 \end{pmatrix} \begin{pmatrix} 0 \\ 1 \end{pmatrix} = \begin{pmatrix} 1 \\ 0 \end{pmatrix} = |0\rangle$ and $|0\rangle\langle 1|0\rangle = \begin{pmatrix} 0 & 1 \\ 0 & 0 \end{pmatrix} \begin{pmatrix} 1 \\ 0 \end{pmatrix} = \begin{pmatrix} 0 \\ 0 \end{pmatrix} = 0.$

3.1.2 Linear operators

An operator A is said to be linear if and only if it satisfies the following properties:

$$A[f(x) + g(x)] = Af(x) + Ag(x) \tag{3.8}$$

and

$$A[cf(x)] = cAf(x), \tag{3.9}$$

where f and g are arbitrary functions and c is an arbitrary constant.
Example 3.6: $\frac{d}{dx}$ is a linear operator since,

$$\frac{d}{dx}[f(x) + g(x)] = \frac{d}{dx}f(x) + \frac{d}{dx}g(x)$$

and

$$\frac{d}{dx}[cf(x)] = c\frac{d}{dx}f(x).$$

But $(\ )^2$ is not a linear operator since,

$$(f(x) + g(x))^2 \neq (f(x))^2 + (g(x))^2.$$

Similarly, we can show that x^2, $\frac{d^2}{dx^2}$, etc. are linear operators and $\sqrt{\ }$, $\log()$, etc. are nonlinear operators. It is interesting to note that all the quantum mechanical operators of our interest are essentially linear.

3.1.3   Pauli matrices

The following set of four useful matrices, which acts on a two dimensional vector space (or on single qubit) are known as Pauli matrices:

$$\sigma_0 = I = \begin{pmatrix} 1 & 0 \\ 0 & 1 \end{pmatrix}, \tag{3.10}$$

$$\sigma_1 = \sigma_x = X = \begin{pmatrix} 0 & 1 \\ 1 & 0 \end{pmatrix}, \tag{3.11}$$

$$\sigma_2 = \sigma_y = Y = \begin{pmatrix} 0 & -i \\ i & 0 \end{pmatrix}, \tag{3.12}$$

and

$$\sigma_3 = \sigma_z = Z = \begin{pmatrix} 1 & 0 \\ 0 & -1 \end{pmatrix}. \tag{3.13}$$

The identity matrix is not always included in the set of Pauli matrices, but we have included it to make a complete set. The inclusion of an identity matrix in the set enables us to express any 2×2 complex matrix in terms of the Pauli matrices. These matrices also lead to the Bloch sphere[3] representation of 2×2 mixed states. It is interesting to note that all single-qubit quantum gates can be represented by 2×2 unitary matrices. The Pauli matrices are some of the most important single-qubit operations. For example, $\sigma_1 = \sigma_x$ represents the NOT gate. As these matrices are very useful in quantum computing, here we will note some of their properties.

For $i, j, k \in \{1, 2, 3\}$, Pauli matrices satisfy

$$\sigma_i^2 = I, \tag{3.14}$$

$$\sigma_i \sigma_j = -\sigma_j \sigma_i \text{ for } i \neq j, \tag{3.15}$$

$$Det(\sigma_i) = -1, \tag{3.16}$$

$$Tr(\sigma_i) = 0 \tag{3.17}$$

and

$$[\sigma_i, \sigma_j] = i\epsilon_{ijk}\sigma_k, \tag{3.18}$$

where ϵ_{ijk} is Levi-Civita symbol and $i \neq j \neq k$[4]. The value of Levi-Civita symbol is 1 if (i, j, k) is an even permutation of $(1, 2, 3)$, -1 for odd permutations and 0 for all other cases, to be more precise

$$\epsilon_{ijk} = \begin{cases} 1 & \text{if } (i, j, k) \text{ is } (1, 2, 3) \text{ or } (2, 3, 1) \text{ or } (3, 1, 2) \\ -1 & \text{if } (i, j, k) \text{ or } (1, 3, 2) \text{ or } (3, 2, 1) \text{ or } (2, 1, 3) \\ 0 & \text{otherwise} \end{cases}. \tag{3.19}$$

---

[3] Bloch sphere will be described at Subsection 3.3.10

[4] [A,B]=AB-BA defines commutation of A and B. If [A,B]=0, then we say that A commutes with B.

3.1.4 Gram-Schmidt procedure

Suppose we start with a non-orthonormal set of linearly independent vectors $\{|e_1\rangle, |e_2\rangle, \cdots, |e_n\rangle\}$. The Gram-Schmidt procedure provides us a simple prescription to generate an orthonormal basis set $\{|e_1'\rangle, |e_2'\rangle, \cdots, |e_n'\rangle\}$ from the set of nonorthonormal state vectors $\{|e_1\rangle, |e_2\rangle, \cdots, |e_n\rangle\}$. The prescription is as follows:

1. Normalize the first basis vector:
$$|e_1'\rangle = \frac{|e_1\rangle}{\||e_1\rangle\|}.$$

2. Find the projection of the second vector along $|e_1'\rangle$ and subtract it off to obtain
$$|e_2\rangle - \langle e_1'|e_2\rangle |e_1'\rangle.$$
This is orthogonal to $|e_1'\rangle$. Normalize it to obtain $|e_2'\rangle$. Thus
$$|e_2'\rangle = \frac{|e_2\rangle - \langle e_1'|e_2\rangle |e_1'\rangle}{\||e_2\rangle - \langle e_1'|e_2\rangle |e_1'\rangle\|}.$$

3. Subtract from $|e_3\rangle$, its projection along $|e_1'\rangle$ and $|e_2'\rangle$:
$$|e_3\rangle - \langle e_1'|e_3\rangle |e_1'\rangle - \langle e_2'|e_3\rangle |e_2'\rangle$$
and normalize the resultant vector to get $|e_3'\rangle$. Continue the process until we obtain $|e_n'\rangle$. In general
$$|e_{j+1}'\rangle = \frac{|e_{j+1}\rangle - \sum_{i=1}^{j}\langle e_i'|e_{j+1}\rangle |e_i'\rangle}{\left\||e_{j+1}\rangle - \sum_{i=1}^{j}\langle e_i'|e_{j+1}\rangle |e_i'\rangle\right\|}.$$

Example 3.7: Construct an orthonormal basis from the non-orthonormal set $\left\{|e_1\rangle, |e_2\rangle : |e_1\rangle = \begin{pmatrix} 2 \\ 1 \end{pmatrix}, |e_2\rangle = \begin{pmatrix} 1 \\ 3 \end{pmatrix}\right\}$.

Step 1: Normalize the first vector
$$|e_1'\rangle = \frac{\begin{pmatrix} 2 \\ 1 \end{pmatrix}}{\sqrt{\begin{pmatrix} 2 & 1 \end{pmatrix}\begin{pmatrix} 2 \\ 1 \end{pmatrix}}} = \frac{1}{\sqrt{5}}\begin{pmatrix} 2 \\ 1 \end{pmatrix}.$$

Step 2.1: Construct the second vector orthogonal to the normalized first vector
$$
\begin{aligned}
|e_2\rangle - \langle e_1'|e_2\rangle |e_1'\rangle &= \begin{pmatrix} 1 \\ 3 \end{pmatrix} - \begin{pmatrix} 0.4 & 0.2 \end{pmatrix}\begin{pmatrix} 1 \\ 3 \end{pmatrix}\begin{pmatrix} 2 \\ 1 \end{pmatrix} \\
&= \begin{pmatrix} 1 \\ 3 \end{pmatrix} - \begin{pmatrix} 2 \\ 1 \end{pmatrix} \\
&= \begin{pmatrix} -1 \\ 2 \end{pmatrix}.
\end{aligned}
$$

Step 2.2: Normalize the vector constructed in **Step 2.1** to obtain

$$|e_2'\rangle = \frac{\begin{pmatrix} -1 \\ 2 \end{pmatrix}}{\sqrt{\begin{pmatrix} -1 & 2 \end{pmatrix}\begin{pmatrix} -1 \\ 2 \end{pmatrix}}} = \frac{1}{\sqrt{5}}\begin{pmatrix} -1 \\ 2 \end{pmatrix}.$$

Thus the required orthonormal basis set is $\left\{ \frac{1}{\sqrt{5}}\begin{pmatrix} 2 \\ 1 \end{pmatrix}, \frac{1}{\sqrt{5}}\begin{pmatrix} -1 \\ 2 \end{pmatrix} \right\}.$

Example 3.8: Show that if $\{|e_1\rangle, |e_2\rangle\}$ are not linearly independent then the Gram-Schmidt procedure fails to produce an orthonormal basis set $\{|e_1'\rangle, |e_2'\rangle\}$.

Solution: As $|e_1\rangle$ and $|e_2\rangle$ are not linearly independent, assume that $|e_1\rangle = a|e_2\rangle$ where a is a constant. Also assume that $\||e_1\rangle\| = b$. Therefore, $|e_1'\rangle = \frac{1}{b}|e_1\rangle$ and $|e_2\rangle = \frac{b}{a}|e_1'\rangle$. Consequently $|e_2\rangle - \langle e_1'|e_2\rangle|e_1'\rangle = \frac{b}{a}|e_1'\rangle - \frac{b}{a}|e_1'\rangle = 0$. Thus the Gram-Schmidt procedure fails to produce an orthonormal basis set $\{|e_1'\rangle, |e_2'\rangle\}$ when $\{|e_1\rangle, |e_2\rangle\}$ are not linearly independent.

3.1.5   Eigenvalues and eigenvectors

If an operator is operated on a vector (state) and yields a scalar multiplied by the vector (state), then the equation is called the eigenvalue equation. The operator is called the eigen operator, the vector (state) is called the eigenvector (eigenstate) and the scalar is called the eigenvalue. Therefore, if we have

$$A|v\rangle = \lambda|v\rangle, \tag{3.20}$$

then A is the eigen operator, $|v\rangle$ is the eigenvector (eigenstate) and λ is the eigenvalue.

Example 3.9: If we consider $\frac{d}{dx}$ as the eigen operator then $\exp(nx)$ is a valid eigenstate having eigenvalue n, but $\sin(wx)$ is not an eigen function of the operator $\frac{d}{dx}$ as $\frac{d}{dx}(\sin(x)) = \cos(x) \neq \lambda \sin(x)$. Further, we can easily see that $\sin(wx)$ and $\cos(wx)$ are eigenstates of $\frac{d^2}{dx^2}$ with eigenvalues $-w^2$.

3.1.6   Hermitian operators

To define the Hermitian operator first we have to define adjoint operators. $A^\dagger$ is the adjoint operator of the operator A if A and $A^\dagger$ satisfies

$$(A^\dagger|v\rangle, |w\rangle) = (|v\rangle, A|w\rangle) \tag{3.21}$$

for all vectors $|v\rangle$ and $|w\rangle$ in the vector space V. The adjoint operators satisfy the following properties:

$$\left(A^\dagger\right)^\dagger = A, \ A^\dagger = (A^*)^T, \ (AB)^\dagger = B^\dagger A^\dagger. \tag{3.22}$$

From $A^\dagger = (A^*)^T$ we obtain a working definition of the adjoint operator $A^\dagger$ as the transpose of conjugate of operator A. We note this as a working definition because this is what we usually use when we work with the matrix form of operators. Now we can loosely define a Hermitian operator as an operator which satisfies

$$A = A^\dagger. \tag{3.23}$$

This is essentially the definition of a self adjoint operator. In a strict sense all the self adjoint operators defined by (3.23) are Hermitian, but all Hermitian operators are not necessarily self adjoint. But the Hermitian operators which are not self adjoint are not very important in the context of the present book so we exclude those from the present discussion and consider the definition (3.23) as the definition of a Hermitian operator.

Example 3.10: $(|1\rangle\langle0| + |0\rangle\langle1|)$ is a Hermitian operator as

$$(|1\rangle\langle0| + |0\rangle\langle1|)^\dagger = (|0\rangle\langle1| + |1\rangle\langle0|) = (|1\rangle\langle0| + |0\rangle\langle1|).$$

But $|1\rangle\langle0|$ is not Hermitian as $(|1\rangle\langle0|)^\dagger = |0\rangle\langle1| \neq |1\rangle\langle0|$.

3.1.7 Normal, unitary and positive operators

The following definitions are also important for the understanding of the quantum information theory:

1. **Normal operator:** Any operator that satisfies $A^\dagger A = AA^\dagger$ is known as a normal operator.

2. **Unitary operator:** If the operator A satisfies $A^\dagger A = AA^\dagger = I$, then A is called unitary. So it is easy to note that the unitary operators are essentially normal (but the converse is not true) and a unitary operator A must satisfy $A^{-1} = A^\dagger$. Thus the set of all unitary operators is a subset of the set of all normal operators.

3. **Positive operators:** An operator A is called a positive operator if it satisfies $(|v\rangle, A|v\rangle) \geq 0$ for all $|v\rangle$ in the vector space V. That means a positive operator does not have any negative eigenvalue. But a positive operator can have eigenvalue 0. Now if we exclude this possibility and demand that all the eigenvalues of A are positive and nonzero then the operator A is called a **positive definite operator**. Thus a positive definite operator A satisfies $(|v\rangle, A|v\rangle) > 0$ for all $|v\rangle$ in the vector space V.

3.1.8 Diagonalizable operator and spectral decomposition

An operator A is called diagonalizable if it can be expressed as

$$A = \sum_i \lambda_i |u_i\rangle\langle u_i|, \tag{3.24}$$

where the vectors $|u_i\rangle$ form a set of orthonormal eigenvectors of A with eigenvalues λ_i. Now it is straightforward to observe that (3.24) implies $\langle u_j|A|u_i\rangle = 0$ and $\langle u_i|A|u_i\rangle = \lambda_i$. Consequently, in the matrix form A can be written as

$$
A = \begin{pmatrix}
\lambda_1 & 0 & \cdots & 0 & 0 \\
0 & \lambda_2 & \cdots & 0 & 0 \\
\vdots & \vdots & \ddots & \vdots & \vdots \\
0 & 0 & \cdots & \lambda_{n-1} & 0 \\
0 & 0 & \cdots & 0 & \lambda_n
\end{pmatrix},
$$

which is a diagonal matrix as expected.

Example 3.11: Consider X operator. Its normalized eigenvectors are $|+\rangle = \frac{1}{\sqrt{2}}(|0\rangle + |1\rangle)$ and $|-\rangle = \frac{1}{\sqrt{2}}(|0\rangle - |1\rangle)$ with eigenvalues $+1$ and -1 respectively. We can easily crosscheck this as $X|+\rangle = \frac{1}{\sqrt{2}}(X|0\rangle + X|1\rangle) = \frac{1}{\sqrt{2}}(|1\rangle + |0\rangle) = |+\rangle$ and similarly, $X|-\rangle = \frac{1}{\sqrt{2}}(|1\rangle - |0\rangle) = -|-\rangle$. Thus in the diagonalized form X can be written as

$$
\begin{aligned}
X &= \tfrac{1}{2}(|0\rangle + |1\rangle)(\langle 0| + \langle 1|) - \tfrac{1}{2}(|0\rangle - |1\rangle)(\langle 0| - \langle 1|) \\
&= |+\rangle\langle +| - |-\rangle\langle -|.
\end{aligned}
$$

We can expand the above expression as follows to check that diagonal representation of X is equivalent to its conventional form:

$$
\begin{aligned}
X &= \tfrac{1}{2}(|0\rangle\langle 0| + |0\rangle\langle 1| + |1\rangle\langle 0| + |1\rangle\langle 1|) \\
&\quad - \tfrac{1}{2}(|0\rangle\langle 0| - |0\rangle\langle 1| - |1\rangle\langle 0| + |1\rangle\langle 1|) \\
&= |0\rangle\langle 1| + |1\rangle\langle 0| \\
&= \begin{pmatrix} 0 & 1 \\ 1 & 0 \end{pmatrix}.
\end{aligned}
$$

Now we see that the matrix of X is not diagonal. This observation often creates confusion. To circumvent such confusion among the readers we wish to note that the matrix form of an operator depends on the choice of basis. To be precise, in $\{|0\rangle, |1\rangle\}$ basis matrix form of X is

$$
X = \begin{pmatrix} \langle 0|X|0\rangle & \langle 0|X|1\rangle \\ \langle 1|X|0\rangle & \langle 1|X|1\rangle \end{pmatrix} = \begin{pmatrix} 0 & 1 \\ 1 & 0 \end{pmatrix}
$$

and similarly, in $\{|+\rangle, |-\rangle\}$ basis it is

$$
X = \begin{pmatrix} \langle +|X|+\rangle & \langle +|X|-\rangle \\ \langle -|X|+\rangle & \langle -|X|-\rangle \end{pmatrix} = \begin{pmatrix} 1 & 0 \\ 0 & -1 \end{pmatrix}, \tag{3.25}
$$

which is diagonal.

After the introduction of the normal operator in the previous subsection and the diagonalizable operator in the present subsection, we are now ready to introduce a connected but more popular and useful theorem, which is known as the spectral decomposition theorem. The theorem can be lucidly stated as: Any diagonalizable operator is normal and conversely, we can always diagonalize normal operators (in finite dimensional Hilbert space). In the context of the diagonalizable operator we have seen that in the diagonalized form, an operator A is diagonal in its own eigenbasis. It is almost a convention to refer to $A = \sum_i \lambda_i |u_i\rangle\langle u_i|$ written in its own eigenbasis as the spectral decomposition of A. The set of eigenvalues $\{\lambda_i\}$ of A is called the spectrum of A. As we know that the quantum mechanical operators are unitary and the unitary operators are normal, we may conclude that every finite dimensional quantum mechanical operator is diagonalizable in appropriate basis.

3.1.9 Tensor products

A tensor product space is a larger vector space formed from two smaller ones simply by combining elements from each in all possible ways that preserve both linearity and scalar multiplication. If V is a vector space of dimension n and W is a vector space of dimension m, then $V \otimes W$ is a vector space of dimension nm. Thus $C^2 \otimes C^4 = C^8$. The idea of a tensor product can be clarified with the following examples:

Example 3.12: Obtain the tensor product of Pauli matrices $X = \begin{pmatrix} 0 & 1 \\ 1 & 0 \end{pmatrix}$ and $Y = \begin{pmatrix} 0 & -i \\ i & 0 \end{pmatrix}$.

Solution: Tensor product of X and Y is

$$X \otimes Y = \begin{pmatrix} 0.Y & 1.Y \\ 1.Y & 0.Y \end{pmatrix} = \begin{pmatrix} 0 & 0 & 0 & -i \\ 0 & 0 & i & 0 \\ 0 & -i & 0 & 0 \\ i & 0 & 0 & 0 \end{pmatrix}.$$

Example 3.13: Express $|0\rangle \otimes |1\rangle$ as a column matrix.
Solution:

$$|0\rangle \otimes |1\rangle = |01\rangle = \begin{pmatrix} 1 \\ 0 \end{pmatrix} \otimes \begin{pmatrix} 0 \\ 1 \end{pmatrix} = \begin{pmatrix} 0 \\ 1 \\ 0 \\ 0 \end{pmatrix}.$$

The fact that a tensor product preserves linearity and scalar multiplication can be mathematically stated as

$$\left. \begin{aligned} |v\rangle \otimes (|w_1\rangle + |w_2\rangle) &= |v\rangle \otimes |w_1\rangle + |v\rangle \otimes |w_2\rangle) \\ (|w_1\rangle + |w_2\rangle) \otimes |v\rangle &= |w_1\rangle \otimes |v\rangle + |w_2\rangle \otimes |v\rangle \end{aligned} \right\} \Rightarrow \text{linearity},$$

and

$$z(|v\rangle \otimes |w\rangle) = (|v\rangle \otimes z|w\rangle) = (z|v\rangle \otimes |w\rangle) \Rightarrow \text{scalar multiplication}.$$

In our context a tensor product is often called a Kronecker product. For example, a very helpful Mathematica command is KroneckerProduct[A,B] which constructs the tensor product of matrix A and B.

Example 3.14: If $A = \begin{pmatrix} a & b \\ c & d \end{pmatrix}$ and $B = \begin{pmatrix} e & f \\ g & h \end{pmatrix}$ then find $A \otimes B$.

Solution:

$$A \otimes B = \begin{pmatrix} a\begin{pmatrix} e & f \\ g & h \end{pmatrix} & b\begin{pmatrix} e & f \\ g & h \end{pmatrix} \\ c\begin{pmatrix} e & f \\ g & h \end{pmatrix} & d\begin{pmatrix} e & f \\ g & h \end{pmatrix} \end{pmatrix} = \begin{pmatrix} ae & af & be & bf \\ ag & ah & bg & bh \\ ce & cf & de & df \\ cg & ch & dg & dh \end{pmatrix}.$$

Problem 3.2: Express $\frac{|01\rangle + |10\rangle}{\sqrt{2}}$ as a column matrix.

3.1.10   Trace

The trace[5] of an $n \times n$ square matrix A is defined as the sum of the elements on the main diagonal. Thus

$$Tr(A) = \sum_{j}^{n} A_{jj}. \tag{3.26}$$

Example 3.15: $Tr(X) = 0$, $Tr(Y) = 0$, $Tr(I_2) = 2$.

The trace operation has some interesting properties; let us list a few of them here:

1. Trace of transpose of a square matrix is the same as the trace of the matrix. Thus
$$Tr(A^T) = Tr(A).$$

   This is so because the diagonal elements of a square matrix do not change on transposition.

2. If A is an $m \times n$ matrix and B is an $n \times m$ matrix, then
$$Tr(AB) = Tr(BA).$$

3. The trace is a linear map, i.e.,
$$Tr(A + B) = Tr(A) + Tr(B)$$

   and
$$Tr(rA) = r\,Tr(A).$$

Problem 3.3: Prove that the trace has the cyclic property $Tr(ABC) = Tr(CAB) = Tr(BCA) \neq Tr(ACB)$.

---

[5]The use of the term trace arises from the German word "spur" which is synonymous with the English word "spoor".

3.2   A little bit of quantum mechanics

The goal of this section is to provide an introduction to the general mathematical structure of quantum mechanics. In this section we will briefly introduce the basic ideas of quantum mechanics and relate it to the mathematical tools developed in the previous section. This is not a textbook on quantum mechanics so our discussion of quantum mechanics will be very very short and we will only discuss those points that are absolutely essential for the understanding of quantum computation and quantum communication. Readers will find that most of the ideas of quantum computation and quantum communication can be visualized and appreciated without a very deep knowledge of quantum mechanics.

In classical mechanics we can measure dynamical observables like position, momentum, velocity and energy with arbitrary precision. Measurement of velocity is independent of measurement of position and one can simultaneously measure them with arbitrary precision. Now if we exactly know the position, acceleration and velocity of a particle at a particular time, we can predict its position and momentum at a later time. Therefore, the future is predictable in classical physics. This predictability or certainty is missing in quantum mechanics. In other words, the quantum mechanical world is probabilistic.

3.2.1   Basic ideas of quantum mechanics

In classical mechanics there exist several physical observables that can be measured. For example, position and momentum can be measured; consequently they are examples of physical observables. We can provide more examples to clarify the meaning of observables, but the idea gets clearer when we understand what is not an observable. For example, you can see me, but you cannot measure me so I am not an observable. You can measure my mass, length, position, etc. so mass, length, position, etc. are physical observables. Now think of a movie, you see the movie, but the movie itself is not an observable. Its duration is measurable, so time is an observable. We hope these examples clarify two points: (a) What an observable is and (b) In physics it is risky to interpret any phrase/word by its literal meaning. Keeping this idea in mind, let us list some simple rules of quantum mechanics.

1. In quantum mechanics there exists an operator corresponding to each dynamical observable of classical physics. For example, position, momentum and energy are classical observables. In quantum mechanics x represents a position operator along X direction, $-i\hbar\frac{d}{dx}$ represents momentum operator p_x along X direction and $i\hbar\frac{d}{dt}$ represents energy operator.

2. If two arbitrary quantum operators A and B do not commute (i.e., if

$AB \neq BA$) then they cannot be measured simultaneously with arbitrary accuracy. For example, position in X direction and momentum in Y direction commute so they can be measured simultaneously with arbitrary accuracy, but position in X direction and momentum in X direction do not commute so they cannot be measured simultaneously with arbitrary accuracy. Further, the noncommuting nature of the corresponding measurement operators does not allow one to simultaneously measure the polarization of a photon in the vertical-horizontal basis and in the diagonal basis. This provides us the basic building block of BB84 and a few other quantum cryptographic protocols.

3. These operators may be represented by matrices, and these operators satisfy eigenvalue equations of the form

$$A_{op}|\psi\rangle = \lambda|\psi\rangle \tag{3.27}$$

where A_{op} is the eigen operator, $|\psi\rangle$ is the eigen-state and λ is the eigenvalue. A measurement of the observable can only yield one of the eigenvalues of the operator that represents the particular observable and a measurement cannot yield an imaginary number. Consequently, the eigenvalues of the meaningful physical operators must be real. This requirement makes it essential that a quantum mechanical operator that represents a physical observable have to be Hermitian. Therefore, most of the time we will deal with Hermitian operators[6, 7]. It is expected that the eigenvalues are discrete and we have already noted that a measurement can only yield one of the eigenvalues. Consequently, all values of the observable are not possible. In other words, only discrete values of an observable are allowed. The meaning of the word quantum is discrete and this discrete nature of quantum mechanics intrinsically lies in the inherent operator algebra.

4. The eigenstate $|\psi\rangle$ introduced above is called wave function of the system when it is an eigenstate of the Hamiltonian (H) of the system. The wave function satisfies the following eigenvalue equation:

$$H|\psi\rangle = E|\psi\rangle. \tag{3.28}$$

---

[6]Sometimes we break the Hermitian operator into non-Hermitian components for example position operator x can be written as $x = \frac{1}{\sqrt{2}}[a + a^\dagger]$, which is Hermitian, but the operators a and $a^\dagger$ are not Hermitian. This does not violate the basic requirement since the annihilation operator a does not represent any physical observable. Further, there are quantum evolution operators which do not represent any physical observable and consequently they are not required to be Hermitian. For example, most of the quantum gates are not Hermitian.

[7]In the recent past it has been seen that the requirement of Hermiticity is sufficient, but not essential, and some non-Hermitian PT symmetric Hamiltonian can also give real spectrum.

The energy eigenvalues E provide the energy spectrum of the system. For a conservative system Hamiltonian H is the sum of kinetic energy $\frac{p^2}{2m}$ and potential energy $V(x)$ (here we have considered a one dimensional Hamiltonian). Now if we replace the classical observables by the corresponding operators then we have

$$
\begin{aligned}
H|\psi\rangle &= \left(\frac{\left(-i\hbar \frac{d}{dx}\right)\left(-i\hbar \frac{d}{dx}\right)}{2m} + V(x) \right) |\psi\rangle \\
&= \left(-\frac{\hbar^2}{2m} \frac{d^2}{dx^2} + V(x) \right) |\psi\rangle = E|\psi\rangle.
\end{aligned}
\tag{3.29}
$$

This is the well-known Schrodinger equation, which can be written in time independent form as

$$
\left(-\frac{\hbar^2}{2m} \frac{d^2}{dx^2} + V(x) \right) |\psi\rangle = E|\psi\rangle
\tag{3.30}
$$

and in time dependent form as

$$
H|\psi\rangle = i\hbar \frac{d}{dt}|\psi\rangle.
\tag{3.31}
$$

Now we can say that the wave function is the solution of the Schrodinger equation. A physical operator operates on the wave function and yields possible values of corresponding dynamical variable. For example, a momentum operator operating on the wave function of a system will give momentum eigenvalues and those eigenvalues are the only possible values of momentum in that system. Thus the wave function $|\psi\rangle$ contains all the information about the system. However, it does not have any physical meaning. Rather the meaning is associated with $|\psi(x)|^2 = \langle \psi(x)|\psi(x)\rangle$, which is the probability of getting the system in position x. Since the physical meaning is associated only with $|\psi(x)|$ so a global phase in wave function is meaningless and we can say that $|\psi(x)\rangle \equiv \exp(i\phi)|\psi(x)\rangle$.

3.2.2 A little more of quantum mechanics

The above stated basic ideas of quantum mechanics can be described in a more compact mathematical formalism with the help of the following three postulates:

1. **State space postulate:** The state of any isolated physical system is completely described by a unit vector $|\psi\rangle$ in a Hilbert space H (known as state space of the system).
   As far as this text is concerned the most important quantum system is the qubit. This is also the simplest quantum mechanical system as its state space is only two dimensional (see Subsection 3.1.1.4). We know that we need two orthonormal basis vectors to describe any

arbitrary state in this state space. As we consider $|0\rangle = \begin{pmatrix} 1 \\ 0 \end{pmatrix}$ and $|1\rangle = \begin{pmatrix} 0 \\ 1 \end{pmatrix}$ as the orthonormal basis so an arbitrary qubit can be described as

$$|\psi\rangle = \alpha|0\rangle + \beta|1\rangle = \alpha \begin{pmatrix} 1 \\ 0 \end{pmatrix} + \beta \begin{pmatrix} 0 \\ 1 \end{pmatrix} = \begin{pmatrix} \alpha \\ \beta \end{pmatrix}, \qquad (3.32)$$

where α and β are complex numbers. Since $|\psi\rangle$ is a unit vector (i.e., $\langle\psi|\psi\rangle = 1$), therefore $|\alpha|^2 + |\beta|^2 = 1$.

2. **Evolution postulate:** The time evolution of the state of a closed quantum system is described by a unitary transformation. To be precise, if the states of a closed quantum system at two different time t_1 and t_2 are $|\psi_1\rangle$ and $|\psi_2\rangle$ respectively then $|\psi_1\rangle$ and $|\psi_2\rangle$ are related by a unitary operator U such that

$$|\psi_2\rangle = U|\psi_1\rangle, \qquad (3.33)$$

where U depends on times t_1 and t_2 only.

In the previous subsection we have learned that the continuous time evolution of a closed quantum system is given by the time dependent Schrodinger equation (3.31). The solutions of (3.31) for a time-independent Hamiltonian at time t_1 and t_2 are related by

$$|\psi_2\rangle = e^{-\frac{iH}{\hbar}(t_2-t_1)}|\psi_1\rangle. \qquad (3.34)$$

Now it is easy to observe that

$$U = e^{-\frac{iH}{\hbar}(t_2-t_1)} \qquad (3.35)$$

is a unitary operator since H is a Hermitian operator. Thus the evolution postulate follows from the Schrodinger equation. In the context of quantum computing a unitary operator U represents a quantum gate. To be precise, if U operates on an m-qubit state and maps it into another m-qubit state, then U is an m-qubit quantum gate. For example, consider the Pauli operator $\sigma_x = X = \begin{pmatrix} 0 & 1 \\ 1 & 0 \end{pmatrix}$. This is a unitary operator which can operate on a single qubit. This operation transforms $|0\rangle = \begin{pmatrix} 1 \\ 0 \end{pmatrix}$ to $\begin{pmatrix} 0 & 1 \\ 1 & 0 \end{pmatrix}\begin{pmatrix} 1 \\ 0 \end{pmatrix} = \begin{pmatrix} 0 \\ 1 \end{pmatrix} = |1\rangle$ and $|1\rangle = \begin{pmatrix} 0 \\ 1 \end{pmatrix}$ to $\begin{pmatrix} 0 & 1 \\ 1 & 0 \end{pmatrix}\begin{pmatrix} 0 \\ 1 \end{pmatrix} = \begin{pmatrix} 1 \\ 0 \end{pmatrix} = |0\rangle$. Thus the unitary evolution operator X functions as a NOT gate. In short, an m-qubit quantum gate transforms an m-qubit quantum state to another m-qubit quantum state through an evolution operator U which can be described in matrix form by a $2^m \times 2^m$ matrix. Several examples of such unitary operators are provided in Chapter 4, where we describe the quantum gates and quantum circuits in detail.

So far we have considered the quantum system as closed. But in a realistic situation the system may interact with the environment. Unwanted interaction with the environment may lead to several problems, but even in absence of interaction with the environment, we need to obtain the result of computation and to do so we have to measure the quantum state. In the measurement process the quantum state has to interact with the measurement apparatus, and consequently it's not a closed system, and the evolution of the system at the time of measurement is not unitary. Therefore, we need an additional postulate to describe measurement.

3. **Measurement postulate:** A generalized quantum measurement is described by a set of measurement operators $\{M_m\}$, which operates on the state of the system to be measured. The label m refers to a particular outcome of the measurement. If the state of the system just before the experiment is described by the state vector $|\psi\rangle$ then the probability that a particular outcome indexed by m appears is

$$p(m) = \langle \psi | M_m^\dagger M_m | \psi \rangle \qquad (3.36)$$

and the state of the system after the measurement is

$$\frac{M_m |\psi\rangle}{\sqrt{p(m)}} = \frac{M_m |\psi\rangle}{\sqrt{\langle \psi | M_m^\dagger M_m | \psi \rangle}}. \qquad (3.37)$$

The measurement operators satisfy a completeness relation of the form $\sum_m M_m^\dagger M_m = I$.

This postulate describes a generalized notion of measurement. There exist a few special cases of this postulate. Before we describe them we would like to provide a simple example to clarify the meaning of measurement operator. Consider that the quantum measurement in two dimensions is described by the set of operators $\{M_0, M_1 : M_0 = |0\rangle\langle 0|, M_1 = |1\rangle\langle 1|\}$. Suppose these operators are used to measure the state of the qubit described by (3.32), then the probability of getting $|0\rangle$ as the outcome is

$$p(0) = ((\langle 0 | \alpha^* + \langle 1 | \beta^*)(|0\rangle\langle 0|0\rangle\langle 0|)(\alpha|0\rangle + \beta|1\rangle) = |\alpha|^2$$

and the probability of getting $|1\rangle$ as the output is $p(1) = |\beta|^2$. Since global phase is unimportant, the state of the system after the measurement is either $\frac{M_0(\alpha|0\rangle + \beta|1\rangle)}{\sqrt{p(0)}} = \frac{|0\rangle\langle 0|(\alpha|0\rangle + \beta|1\rangle)}{\sqrt{p(0)}} = \frac{\alpha}{|\alpha|}|0\rangle \approx |0\rangle$ or $\frac{M_1|\psi\rangle}{\sqrt{p(1)}} = |1\rangle$.

Projective measurement or Von Neumann measurement: In the generalized measurement postulate the measurement operators were restricted by the completeness relation only. Now if we impose another restriction on them by demanding that the measurement operators are orthogonal projectors, then we obtain projective measurement, which is often called Von Neumann measurement. This additional restriction implies that the measurement operators M_m are Hermitian (i.e., $M_m^\dagger = M_m$) and $M_m M_{m'} = \delta_{mm'} M_m$. Therefore, $M_m^\dagger M_m = M_m M_m = M_m$ and consequently it is easy to observe from (3.36) and (3.37) that the probability that a particular outcome indexed by m appears in a projective measurement is

$$p(m) = \langle \psi | M_m | \psi \rangle \tag{3.38}$$

and the state of the system after the measurement is

$$\frac{M_m | \psi \rangle}{\sqrt{p(m)}}. \tag{3.39}$$

We can describe this in a simplified manner if we consider that the input state of the measuring device is $\sum_m c_m | m \rangle$, where $\{ | m \rangle \}$ forms an orthonormal basis set and the orthogonal projection operators are $M_m = | m \rangle \langle m |$; then the probability of obtaining the output state $| m \rangle$ is $|c_m|^2$. Since the basis set chosen for the measurement describes the measurement process it is commonly mentioned via the phrase: "measured in basis $\{ | m \rangle \}$". Let us give an example to show that the measurement statistics depend on the choice of basis.

Example 3.16: Consider the state $| \psi \rangle = \frac{1}{\sqrt{2}} | 0 \rangle + \frac{1}{\sqrt{2}} | 1 \rangle$. If the state is measured in $\{ | 0 \rangle, | 1 \rangle \}$ basis then we obtain $p(0) = p(1) = \frac{1}{2}$. Now if we choose $\left\{ | + \rangle = \frac{|0\rangle + |1\rangle}{\sqrt{2}}, | - \rangle = \frac{|0\rangle - |1\rangle}{\sqrt{2}} \right\}$ as the basis set then we obtain $p(+) = 1$ and $p(-) = 0$. Thus the measurement statistics depend on the choice of basis.

Usually when we measure a single qubit system in $\{ | 0 \rangle, | 1 \rangle \}$ basis, or more generally an n-qubit system in

$$\{ | 0_1 0_2 \cdots 0_{n-1} 0_n \rangle, | 0_1 0_2 \cdots 0_{n-1} 1_n \rangle, \cdots, | 1_1 1_2 \cdots 1_{n-1} 1_n \rangle \}$$

basis, then we say that the system is measured in computational basis.

Example: 3.17: Let us measure the two qubit state

$$| \psi \rangle = \sqrt{\frac{1}{7}} | 00 \rangle + \sqrt{\frac{2}{7}} | 01 \rangle + \sqrt{\frac{3}{7}} | 10 \rangle + \sqrt{\frac{1}{7}} | 11 \rangle$$

in computational basis. The probability of different measurement outcomes are: $p(00) = p(11) = \frac{1}{7}$, $p(01) = \frac{2}{7}$ and $p(10) = \frac{3}{7}$. Thus the probability of getting $|0\rangle$ in the first qubit is $\frac{1}{7} + \frac{2}{7} = \frac{3}{7}$. Now we can rewrite the state as

$$\sqrt{\frac{3}{7}}|0\rangle \left(\sqrt{\frac{1}{3}}|0\rangle + \sqrt{\frac{2}{3}}|1\rangle\right) + \sqrt{\frac{4}{7}}|1\rangle \left(\sqrt{\frac{3}{4}}|0\rangle + \sqrt{\frac{1}{4}}|1\rangle\right)$$

and if we just measure the first qubit in the computational basis and obtain $|0\rangle$ then the second qubit collapses to the superposition state $\left(\sqrt{\frac{1}{3}}|0\rangle + \sqrt{\frac{2}{3}}|1\rangle\right)$. Similarly, if the measurement on the first qubit yields $|1\rangle$ then the second qubit collapses to $\left(\sqrt{\frac{3}{4}}|0\rangle + \sqrt{\frac{1}{4}}|1\rangle\right)$.

POVM: POVM is another special case of generalized measurement described in the measurement postulate. POVM stands for positive operator valued measure. This is a complete set of positive operators $E_m = M_m^\dagger M_m$ which satisfy $\sum_m E_m = I$ and where the probability of outcome m, given by $p(m) = \langle\psi|M_m^\dagger M_m|\psi\rangle = \langle\psi|E_m|\psi\rangle$. The operators E_m are known as POVM elements and $\{E_m\}$ is called a POVM.

Example 3.18: It is easy to see that $\{|0\rangle\langle0|, |1\rangle\langle1|\}$ is a POVM. If we consider a state $|\psi\rangle = \frac{1}{\sqrt{2}}(|0\rangle + |1\rangle)$ then $p(0) = p(1) = \frac{1}{2}$.

There is an important difference between projective measurement and POVM. The orthogonal projection operators commute in projective measurement, but POVM elements may or may not commute. Further, if we work in a d dimensional Hilbert space then the number of projection operators in projective measurement is d, but the number of elements in POVM can be greater than d. Let us provide an explicit example to elaborate this point.

Example 3.19: Consider a POVM containing three elements

$$\begin{aligned} E_1 &= \frac{\sqrt{2}}{1+\sqrt{2}}|1\rangle\langle1| \\ E_2 &= \frac{\sqrt{2}}{1+\sqrt{2}}\frac{(|0\rangle-|1\rangle)(\langle0|-\langle1|)}{2} \\ E_3 &= I - E_1 - E_2. \end{aligned} \tag{3.40}$$

As this POVM describes single qubit measurements so it works on two dimensional Hilbert space, but it contains three elements. Further if we expand E_2 as

$$E_2 = \frac{\sqrt{2}}{1+\sqrt{2}}\frac{(|0\rangle\langle0| - |1\rangle\langle0| - |0\rangle\langle1| + |1\rangle\langle1|)}{2}$$

then we can easily observe that

$$\begin{aligned} E_1 E_2 &= \frac{1}{(1+\sqrt{2})^2}(-|1\rangle\langle0| + |1\rangle\langle1|) \\ &\neq E_2 E_1 \\ &= \frac{1}{(1+\sqrt{2})^2}(-|0\rangle\langle1| + |1\rangle\langle1|). \end{aligned} \tag{3.41}$$

Thus $[E_1, E_2] \neq 0$, consequently, $[E_3, E_1] = -[E_2, E_1] = [E_1, E_2] \neq 0$ and $[E_3, E_2] = -[E_1, E_2] \neq 0$. Therefore, these POVM elements do not commute with each other.

This is an interesting POVM. This particular POVM may be used to distinguish between $|\psi_1\rangle = |0\rangle$ and $|\psi_2\rangle = \frac{1}{\sqrt{2}}(|0\rangle + |1\rangle)$. It is easy to observe that if we have been given $|\psi_1\rangle$ or $|\psi_2\rangle$ and we apply this POVM then the measurement outcome E_1 will indicate that the state is $|\psi_2\rangle$, the measurement outcome E_2 will indicate that the state is $|\psi_1\rangle$ and the measurement outcome E_3 will not conclude anything. In brief, we shall never make a mistake in identifying the state. To the contrary, if we try to distinguish $|\psi_1\rangle$ and $|\psi_2\rangle$ by a projective measurement in computational basis then we will never be able to identify input state $|0\rangle$ with certainty. This shows why POVMs play a very important role in state discrimination protocols. Further, we would like to note that POVM is beneficial when the post measurement state is not of much interest.

Cartoon 3.1: What is wrong in the argument of the student?

3.2.3 Density operator and density matrix

So far we have worked with the state vector representation of quantum mechanics. In such a representation the state vector $|\psi\rangle$ contains all the information about the physical system. However, it may not always be possible to ascribe a state vector to a quantum system that is not isolated. For rigorous treatment of such interacting quantum systems we need a density operator and its matrix representation, which is known as density matrix. The density matrix is very useful in quantum information processing, but before we describe specific uses of a density matrix in quantum information processing it is important to introduce the density matrix formalism.

3.2.3.1 Density operator of pure states

When the state of a quantum system can be represented as a linear superposition of the basis vectors $|n\rangle$ as

$$|\psi\rangle = \sum_n c_n |n\rangle \qquad (3.42)$$

then the state is said to be pure. The density operator ρ of a pure state is defined as

$$\rho = |\psi\rangle\langle\psi|. \qquad (3.43)$$

Using (3.42) in (3.43) we obtain

$$\rho = \sum_n \sum_m c_n c_m^* |n\rangle\langle m| = \sum_{n,m} \rho_{nm} |n\rangle\langle m|, \qquad (3.44)$$

where $\rho_{nm} = \langle n|\rho|m\rangle$ are the matrix elements of the density operator for the pure state. We know that if a measurement is done on (3.42) then the probability of getting the state $|n\rangle$ is $|c_n|^2$. This provides a physical meaning to the diagonal elements of density matrix as $\rho_{nn} = |c_n|^2$. Here the diagonal terms are necessarily nonnegative and hence the density operator is a positive operator.

If we write the amplitude of states in polar form as $c_j = r_j e^{i\phi_j}$ then the off-diagonal terms of the density matrix would be $\rho_{nm} = c_n c_m^* = r_n r_m e^{i(\phi_n - \phi_m)}$. These terms are called coherence as they depend on the relative phase between the amplitude of states (or coherence). These terms are very important for quantum computing because the largest problem of construction of a scalable quantum computer is our inability to maintain the coherence. The loss of coherence (i.e., the change in relative phase $\Delta\phi_{nm} = \phi_n - \phi_m$ between the amplitude of states) is called decoherence. In general, decoherence describes all such processes which lead to the loss of quantum information in a system.

The density operator of a pure state has the following properties:

$$Tr(\rho) = \sum_n \rho_{nn} = \sum_n |c_n|^2 = 1. \qquad (3.45)$$

Since $\rho^2 = |\psi\rangle\langle\psi|\psi\rangle\langle\psi| = |\psi\rangle\langle\psi| = \rho$ therefore

$$Tr(\rho^2) = 1. \qquad (3.46)$$

Example 3.20: Consider a pure state $|\psi\rangle = \sin(\theta)|0\rangle + \cos(\theta)|1\rangle$. The density operator for this state is

$$\rho = |\psi\rangle\langle\psi| = \sin^2\theta|0\rangle\langle 0| + \sin\theta\cos\theta|0\rangle\langle 1| + \sin\theta\cos\theta|1\rangle\langle 0| + \cos^2\theta|1\rangle\langle 1|,$$

which can be written in the matrix form as

$$\rho = \begin{pmatrix} \sin^2\theta & \sin\theta\cos\theta \\ \sin\theta\cos\theta & \cos^2\theta \end{pmatrix}. \qquad (3.47)$$

From (3.47) it is clear that the diagonal terms of ρ are nonnegative. Therefore, ρ is a positive operator and it has unit trace as $Tr(\rho) = \sin^2\theta + \cos^2\theta = 1$. Further, since

$$
\begin{aligned}
\rho^2 &= \begin{pmatrix} \sin^2\theta & \sin\theta\cos\theta \\ \sin\theta\cos\theta & \cos^2\theta \end{pmatrix} \begin{pmatrix} \sin^2\theta & \sin\theta\cos\theta \\ \sin\theta\cos\theta & \cos^2\theta \end{pmatrix} \\
&= \begin{pmatrix} \sin^4\theta + \sin^2\theta\cos^2\theta & \sin^3\theta\cos\theta + \sin\theta\cos^3\theta \\ \sin^3\theta\cos\theta + \sin\theta\cos^3\theta & \sin^2\theta\cos^2\theta + \cos^4\theta \end{pmatrix} \\
&= \begin{pmatrix} \sin^2\theta(\sin^2\theta + \cos^2\theta) & \sin\theta\cos\theta(\sin^2\theta + \cos^2\theta) \\ \sin\theta\cos\theta(\sin^2\theta + \cos^2\theta) & (\sin^2\theta + \cos^2\theta)\cos^2\theta \end{pmatrix} \\
&= \begin{pmatrix} \sin^2\theta & \sin\theta\cos\theta \\ \sin\theta\cos\theta & \cos^2\theta \end{pmatrix} = \rho
\end{aligned}
$$

therefore, $Tr(\rho^2) = 1$.

Summary

Quantum state $\Leftrightarrow$ Unit vector in a Hilbert space $\Leftrightarrow$ A matrix (density matrix) that acts on Hilbert space,

Quantum evolution $\Leftrightarrow$ Unitary operators,

Quantum measurement $\Leftrightarrow$ Projection.

3.2.3.2   Density operator of mixed states

A quantum system may not be in a pure state (i.e., it may not be possible to express it as (3.42)). In that case the state may be viewed as a mixture of states $|\psi_i\rangle$, which are not necessarily orthogonal to each other. Each $|\psi_i\rangle$ has a different expansion in the basis of eigenvectors $|n\rangle$ and $p_i \geq 0$ is the probability that the quantum system is found in a particular pure state $|\psi_i\rangle$. The density operator of such a mixed state is defined as

$$
\rho = \sum_i p_i |\psi_i\rangle\langle\psi_i|. \tag{3.48}
$$

The density operator for a mixed state (3.48) satisfies the condition of positivity and unit trace. We can prove this as follows.
    Proof:

$$
Tr(\rho) = \sum_i p_i Tr\left(|\psi_i\rangle\langle\psi_i|\right) = \sum_i p_i = 1,
$$

and

$$
\langle\phi|\rho|\phi\rangle = \sum_i p_i \langle\phi|\psi_i\rangle\langle\psi_i|\phi\rangle = \sum_i p_i |\langle\phi|\psi_i\rangle|^2 \geq 0,
$$

where $|\phi\rangle$ is an arbitrary state in the state space.

The set $\{p_i, |\psi_i\rangle\}$ describes an ensemble of pure states and there exists a unique density operator for an ensemble of states. But the converse is not true. Thus the same density operator may describe different ensembles of quantum states. Consequently, one should not predict the ensemble from the density operator. This point will be clarified in the following example.

Example 3.21: Let us consider an ensemble $\{p_1 = 0.64, |\psi_1\rangle = |0\rangle, p_2 = 0.36, |\psi_2\rangle = |1\rangle\}$. The density operator for this ensemble is

$$\rho_1 = 0.64|0\rangle\langle 0| + 0.36|1\rangle\langle 1|. \tag{3.49}$$

Now consider a second ensemble $\{p_1 = 0.5, |\psi_1\rangle = 0.8|0\rangle + 0.6|1\rangle, p_2 = 0.5, |\psi_2\rangle = 0.8|0\rangle - 0.6|1\rangle\}$. For this ensemble the density operator would be

$$
\begin{aligned}
\rho_2 &= 0.5|\psi_1\rangle\langle\psi_1| + 0.5|\psi_2\rangle\langle\psi_2| \\
&= 0.5 \left(0.64|0\rangle\langle 0| + 0.48|0\rangle\langle 1| + 0.48|1\rangle\langle 0| + 0.36|1\rangle\langle 1|\right) \\
&+ 0.5 \left(0.64|0\rangle\langle 0| - 0.48|0\rangle\langle 1| - 0.48|1\rangle\langle 0| + 0.36|1\rangle\langle 1|\right) \\
&= 0.64|0\rangle\langle 0| + 0.36|1\rangle\langle 1|.
\end{aligned}
\tag{3.50}
$$

Here we can easily see that $\rho_1 = \rho_2$ and consequently, two different ensembles may yield the same density operator. It is straightforward to see that $Tr(\rho_1) = Tr(\rho_2) = 1$ and the density operators are positive. We can also easily check that these states are mixed states as

$$
\begin{aligned}
Tr\left(\rho_1^2\right) = Tr\left(\rho_2^2\right) &= Tr\left(\begin{pmatrix} 0.64 & 0 \\ 0 & 0.36 \end{pmatrix}\begin{pmatrix} 0.64 & 0 \\ 0 & 0.36 \end{pmatrix}\right) \\
&= \begin{pmatrix} 0.4096 & 0 \\ 0 & 0.1296 \end{pmatrix} \\
&= 0.5392 < 1.
\end{aligned}
\tag{3.51}
$$

In summary, the density operator (of both pure state and mixed state) is a positive operator ($\langle\phi|\rho|\phi\rangle \geq 0$) on a Hilbert space with unit trace. A quantum state, which is characterized by the criterion $Tr(\rho^2) = 1$ is a pure state. For a mixed state $Tr(\rho^2) < 1$. Further, the state vector representation and the density operator representation are equivalent and the choice of a particular representation depends on convenience. As you proceed through the book it will become clear to you that there exist certain kinds of problems in quantum information processing where it is convenient to use density operators rather than state vectors.

3.2.4 The meaning of entanglement

If we have a quantum state $|\psi\rangle_A \in H_A$ and another state $|\psi\rangle_B \in H_B$ where H_A and H_B are the Hilbert spaces then the combined state $|\psi\rangle_{AB} \in H_A \otimes H_B$. The combined state may or may not be expressed as a tensor product of the subsystems. If we can write the combined state as

$$|\psi\rangle_{AB} = |\psi\rangle_A \otimes |\psi\rangle_B \tag{3.52}$$

then the combined state is separable. Otherwise, the combined state is entangled. Thus the condition for entanglement is

$$|\psi\rangle_{AB} \neq |\psi\rangle_A \otimes |\psi\rangle_B. \tag{3.53}$$

$|00\rangle = |0\rangle \otimes |0\rangle$, $\frac{1}{\sqrt{2}}(|01\rangle \pm |11\rangle) = \frac{1}{\sqrt{2}}(|0\rangle \pm |1\rangle) \otimes |1\rangle$ are examples of separable states and $\frac{|00\rangle \pm |11\rangle}{\sqrt{2}}$, $\frac{|01\rangle \pm |10\rangle}{\sqrt{2}}$ are examples of entangled states. Similarly, the condition of entanglement for mixed states can be expressed in terms of density matrix as

$$\rho \neq \sum_{j=1}^{n} p_j \sigma_j \otimes \xi_j, \tag{3.54}$$

where $\sigma_1 \cdots \sigma_n$ and $\xi_1 \cdots \xi_n$ are the states of first and second systems respectively and $\sum_{j=1}^{n} p_j = 1$. It is difficult to test whether a given density matrix is separable or not. Under some reductions it is an NP-hard problem [55]. Entangled states, which are primary resource for quantum computing, are more common in nature than the separable states. This can be visualized from the following simple example.

Example 3.22: A two qubit state may be expressed in general as

$$|\psi\rangle_{AB} = a|00\rangle + b|01\rangle + c|10\rangle + d|11\rangle \tag{3.55}$$

where a, b, c, d are constants which satisfy $|a|^2 + |b|^2 + |c|^2 + |d|^2 = 1$. Now if this state is separable then we must be able to write it as product of two qubits as

$$
\begin{aligned}
|\psi\rangle_{AB} &= (\alpha|0\rangle + \beta|1\rangle) \otimes (\alpha'|0\rangle + \beta'|1\rangle) \\
&= \alpha\alpha'|00\rangle + \alpha\beta'|01\rangle + \beta\alpha'|10\rangle + \beta\beta'|11\rangle \\
&= a|00\rangle + b|01\rangle + c|10\rangle + d|11\rangle.
\end{aligned} \tag{3.56}
$$

From (3.56) it is clear that the condition of separability is

$$ad = bc, \tag{3.57}$$

which is a very restricted case. Thus we can conclude that the entangled states are more common in nature than the separable states.

In the remaining part of the book we will see that entanglement is one of the most crucial resources for quantum computing and quantum communication. Keeping this in mind, it would be apt to mention one technique of generation of an entangled state that is frequently used in laboratories. To be precise, we will briefly mention how quantum states entangled in polarization states are produced in the laboratory. Usually it is done through a nonlinear optical process known as parametric down conversion. In this process a single photon of angular frequency ω_p from a pump laser is incident on a nonlinear crystal and simultaneously generates a pair of signal (s) and idler (i) photons of angular frequencies ω_s and ω_i respectively. Conservation of energy requires that the following condition is satisfied

$$\omega_p = \omega_s + \omega_i \tag{3.58}$$

and similarly conservation of momentum requires

$$\vec{k}_p = \vec{k}_s + \vec{k}_i, \tag{3.59}$$

where $\vec{k}_p$, $\vec{k}_s$ and $\vec{k}_i$ are the wave vectors of pump, signal and idler photons. Equation (3.59) implies that pump, signal and idler waves remain in phase throughout the nonlinear crystal. Consequently, if (3.58) and (3.59) are simultaneously satisfied then we say that the phase matching condition is satisfied. Now recall that the optical materials are dispersive, and dispersion means that the refractive index is a function of frequency. Therefore, in a dispersive medium waves of different frequencies move with different velocities and make it impossible to satisfy the phase matching condition. Interestingly, the nonlinear crystals are birefringent, too. As birefringence implies that the refractive index is a function of polarization of the light with respect to the crystal axes, we can use birefringence to balance dispersion and thus to satisfy phase matching condition. There are two ways in which the phase matching condition can be satisfied. In type-I phase matching the polarizations of the down converted photons are parallel to each other and orthogonal to the polarization of the pump field. We are not interested in this process. To the contrary, in type-II phase matching the down converted photons have orthogonal polarizations. Simply put, if one of them is horizontally polarized then the other has to be vertically polarized.

In most common experimental setups a BBO (β-Barium-borate) crystal is used as the nonlinear crystal. Wide phase matching range and large nonlinear coefficients make it very suitable for the present purpose. The pump laser emits a photon in the ultra-violet range. The photon is incident on the BBO crystal and as a result two down converted photons emerge. Here the type-II phase matching condition is satisfied which implies that the down converted photons emerge in cones of opposite polarization. Now if we place a virtual plane in front of these two cones then we will see two circles which intercept at two points. One circle represents horizontally polarized photons and the other represents vertically polarized photons. Now what happens in the intersection points? The photons at intersection points can have originated from each of the circles with equal probability. So they may be in a horizontally (H) polarized state or a vertically (V) polarized state with equal probability. If a measurement is made on the photon at one intersection point and the photon is found to be horizontally (H) polarized then the photon at the other intersection point must be vertically (V) polarized. In general the composite state of the photons present in intersection points is $|\psi\rangle = \frac{1}{\sqrt{2}}(|HV\rangle + |VH\rangle)$, which is an entangled state.

3.2.5   Bell's inequality and nonlocality

There is no true random number generator in the classical world. One may argue that this is not true because when we toss a coin we obtain a head or a tail at random. But is it really random? In other words, is the outcome of a toss probabilistic? A closer look into the event would tell us no. This is so because if we know the force applied on the coin, density of air, air pressure, gravity, direction and speed of the air flow, etc. then we can solve the equation of motion of the coin and predict the outcome of the toss. Thus a toss appears probabilistic because our description of the event is incomplete as we don't specify many variables which determine the result of the outcome of the event. Such variables, which affect the outcome of an experiment but are not considered to describe the experiment because of limitations of our knowledge about the finer description of the event, may be called hidden variables. Now a couple of questions arise: Is the nature really probabilistic as described by quantum mechanics or quantum mechanics is incomplete? Are there hidden variables which lead to the apparent probabilistic nature of quantum mechanics? These questions have been in existence since the beginning of quantum mechanics. However, it has not been clear how to formulate these questions in a precise mathematical way that would lead to experimental testing.

The issue at hand is: Is quantum randomness due to crypto-determinism (apparent probabilistic phenomena which are determinate when hidden variables are specified) or due to a strange state of affairs where a definite value did not exist, and was created "on the fly" when the system was interrogated. Quantum formalism suggests the latter, the property which we may call *nonrealism*. Einstein and his collaborators Podolsky and Rosen were famously uncomfortable with this kind of indeterminism. It turns out that locally it is not straightforward to formulate this question. Therefore, Einstein, Podolsky and Rosen gave an ingenious argument involving a nonlocal system.

In 1964, John Stewart Bell [56] proposed an interesting experiment, which provided us a scope to prove that quantum mechanics (and thus the nature) cannot be described by a local hidden variable theory. The experiment proposed by him can be understood clearly if we consider the following scenario:

1. A quantum experiment has created two particles, namely A and B.

2. The particle A is with Alice and the particle B is with Bob.

3. Alice and Bob each possess two sets of measuring devices. Alice's measuring devices are referred to as A_1 and A_2 and those of Bob are

referred to as B_1 and B_2.

4. Each measurement can have only two possible outcomes, say, $+1$ and -1 (the values are arbitrary and we can change it but ± 1 are convenient).

5. They agree that each of them will simultaneously and independently choose one of the measurement devices and will make a measurement with that.

In the above experiment, any local hidden variable theory would satisfy

$$E(A_1 B_1) + E(A_1 B_2) + E(A_2 B_1) - E(A_2 B_2) \leq 2, \qquad (3.60)$$

where $E(A_i B_j)$ is the expectation value that Alice has measured her particle with the measurement device A_i and Bob has measured his particle with the measurement device B_j. Now if we can show that quantum mechanics violates this inequality, which is known as Bell's inequality, then that would imply that it is not a local hidden variable theory. Further, it would imply existence of nonlocality and that will provide us applications of entanglement in teleportation, dense coding, quantum cryptography, etc. Thus this inequality is important from both the foundational perspective as well as the application perspective.

To prove the inequality Bell made the following two assumptions:

Realism: A measurement always reveals a pre-existing, definite value of a physical observable. The value may be unknown to us but the value certainly exists prior to the measurement. This is the notion of realism. Absence of realism implies nonrealism. This assumption contradicts the state vector collapse postulate of quantum mechanics. To be precise, a quantum measurement does not reveal a pre-existing value of a physical observable. While classical mechanics assumes realism, quantum mechanics does not. To understand the assumption clearly, consider a two level atomic system with 1 electron. Classically the electron is either in the ground state or in the excited state. We may not know whether it is in the ground state or not, but it exists in a certain state and measurement only reveals that. This is what happens in all realistic theories.

Localism: Alice's outcome is independent of Bob's setting, and vice versa, if Alice and Bob are making simultaneous measurements at spatially separated places. This is the notion of a local theory.
By local we mean that the effect of measurement on a particle cannot travel at a velocity larger than the velocity of light in vacuum. As Alice and Bob are making simultaneous measurements at spatially separated places, so in a local theory their measurements have to be independent, otherwise, the measurement outcome of Alice (Bob) has

to reach Bob (Alice) with a velocity greater than that of light, which is not consistent with the local theory. However, in a nonlocal theory measurement of one can affect the measurement outcome of the other without violating the postulate of special theory of relativity. Thus classical mechanics which is a local theory follows this assumption but quantum mechanics does not follow it.

We may now provide a simple proof of the inequality by expressing the sum of the expectation values as follows:

$$
\begin{aligned}
&E(A_1B_1) + E(A_1B_2) + E(A_2B_1) - E(A_2B_2) \\
={}& E(A_1B_1 + A_1B_2 + A_2B_1 - A_2B_2) \\
={}& E\left(A_1(B_1 + B_2) + A_2(B_1 - B_2)\right).
\end{aligned}
$$

Now as each measurement can have only two outcomes (± 1) so we have the following two possibilities:

1. $B_1 = B_2$, then $(B_1 - B_2) = 0$ and $(B_1 + B_2) = \pm 2$ (as either $B_1 = B_2 = 1$ or $B_1 = B_2 = -1$). Therefore,

$$
(A_1(B_1 + B_2) + A_2(B_1 - B_2)) = A_1(B_1 + B_2) = \pm 2A_1 = \pm 2.
$$

2. $B_1 = -B_2$, then $(B_1 + B_2) = 0$ and $(B_1 - B_2) = \pm 2$ (as either $B_1 = -B_2 = 1$ or $B_1 = -B_2 = -1$). Therefore,

$$
(A_1(B_1 + B_2) + A_2(B_1 - B_2)) = A_2(B_1 - B_2) = \pm 2A_2 = \pm 2.
$$

Now we may use the fact that both the possibilities yield $A_1(B_1 + B_2)$ $+A_2(B_1 - B_2) = \pm 2$, to write

$$
\begin{aligned}
&E(A_1B_1) + E(A_1B_2) + E(A_2B_1) - E(A_2B_2) \\
={}& E(A_1B_1 + A_1B_2 + A_2B_1 - A_2B_2) \\
={}& \textstyle\sum_{A_1,B_1,A_2,B_2} p(A_1B_1A_2B_2)(A_1B_1 + A_1B_2 + A_2B_1 - A_2B_2) \\
\leq{}& 2.
\end{aligned}
$$

In 1982, Aspect, Dalibard and Roger [57] experimentally showed the violation of Bell's inequality. Since then the violation of Bell's inequality has been observed experimentally in many different quantum systems. It is not surprising that quantum mechanics violates Bell's inequality, as both the assumptions (localism and realism) that lead to Bell's inequality are violated by quantum mechanics. Thus the observations of Aspect's experiment and other similar experiments cannot be explained by any local realistic theory (local hidden variable theory). Thus the nature is not local. This allows us to exploit nonlocal resources like entanglement to implement quantum teleportation, dense coding, quantum algorithms, quantum cryptography, etc.

In quantum mechanics maximum possible value of $E(A_1B_1)+E(A_1B_2)+E(A_2B_1)-E(A_2B_2)$ is $2\sqrt{2}$. Here it would be interesting to note that there may exist physical theories which are more nonlocal than quantum mechanics. For example, in the Popescu and Rohrlich box (PR box) we may achieve maximum violation of Bell's inequality by reaching $E(A_1B_1)+E(A_1B_2)+E(A_2B_1)-E(A_2B_2)=4$. Thus the PR box is more nonlocal than quantum mechanics. It is beyond the scope of the present book to discuss general nonlocal theories, but interested readers may find it tempting to explore the recent studies on PR (nonlocal) boxes. For a review one may see [58].

Finally, we would like to note that experimental verification of violation of Bell's inequality discards the possibilities of all local hidden variable theories of nature but it does not exclude the possibilities of nonlocal hidden variable theories. To date there does not exist any single inequality, the violation of which would discard all possible nonlocal hidden variable theories. However, Legget's inequality may be used to discard a subset of all possible nonlocal hidden variable theories [59]. The issues related to this topic are extremely important and interesting. For a quick review one may see [59] and references therein.

We have already seen that the entangled states, which are nonlocal in nature, are more common in nature than the separable states. But all entangled states are not of much use in quantum computing. Most of the applications of entangled states that are known in quantum computing use either maximally entangled states or nearly maximally entangled states. From the last sentence it appears that there exists at least one quantitative measure of entanglement. Yes, there exists a set of quantitative measures of entanglement. But before we introduce them we need to introduce the notions of Bell measurement, partial trace, Schmidt decomposition, partial transpose, fidelity, etc. To be precise, we need to learn a few more algebraic techniques to proceed further.

3.3 A little more algebra for quantum computing

3.3.1 Bell measurement and entanglement

In C^4 space an important basis set is

$$\left\{ |\psi^+\rangle = \frac{|00\rangle+|11\rangle}{\sqrt{2}}, \; |\psi^-\rangle = \frac{|00\rangle-|11\rangle}{\sqrt{2}}, \right.$$
$$\left. |\phi^+\rangle = \frac{|01\rangle+|10\rangle}{\sqrt{2}}, \; |\phi^-\rangle = \frac{|01\rangle-|10\rangle}{\sqrt{2}} \right\}. \tag{3.61}$$

This complete set of orthonormal basis states is known as Bell basis and the elements of this basis set are called Bell states. When a projective measurement is done in this basis it is usually referred to as Bell measurement.

Bell measurement is an important resource in quantum teleportation and secure quantum communication. Readers will obtain a better idea about that later in this book. Here one can easily note that these Bell states are entangled in computational basis. Now we would like to show a very interesting application of Bell measurement. Consider a state $|\psi\rangle = |00\rangle$. This state is separable and a measurement in computational basis will keep the state unchanged. But since $|\psi\rangle = |00\rangle = \frac{1}{\sqrt{2}} \left(\frac{|00\rangle + |11\rangle}{\sqrt{2}} \right) + \frac{1}{\sqrt{2}} \left(\frac{|00\rangle - |11\rangle}{\sqrt{2}} \right) = \frac{1}{\sqrt{2}}|\psi^+\rangle + \frac{1}{\sqrt{2}}|\psi^-\rangle$, therefore a Bell measurement would yield $|\psi^\pm\rangle$ with probability $p(\psi^\pm) = \frac{1}{2}$. It is clear from this example that a Bell measurement can transform a separable state into an entangled state.

Now we will show that a Bell measurement can be used to swap entanglement. For example, consider a tripartite state

$$|\psi\rangle_{ABC} = |\psi^+\rangle_{AB}|0\rangle_C = \frac{|000\rangle_{ABC} + |110\rangle_{ABC}}{\sqrt{2}},$$

where the first two qubits are entangled and the third qubit is separable. We can rewrite this state as

$$|\psi\rangle_{ABC} = \frac{|0\rangle_A \left(|\psi^+\rangle_{BC} + |\psi^-\rangle_{BC} \right) + |1\rangle_A \left(|\phi^+\rangle_{BC} - |\phi^-\rangle_{BC} \right)}{2}.$$

Now a Bell measurement on the last two qubits will yield one of the Bell states, and consequently the last two qubits will get entangled and the first qubit will become separable. For example, if the Bell measurement on the last two qubits yields $|\psi^+\rangle_{BC}$ then the total state after the measurement is $|0\rangle_A|\psi^+\rangle_{BC}$. Thus the initial entanglement present between the first two qubits is now transferred to the last two qubits. Two systems that have never interacted with each other can become entangled using this procedure. Specifically, we may think that Alice and Bob share an entangled state (say $|\psi^+\rangle_{AB}$). Particle A is with Alice and B is with Bob and they are spatially separated. Alice prepares another entangled state (say $|\psi^+\rangle_{CD}$). The particles C and D have never interacted with Bob's particle B. But now if Alice measures particles A and C in an entangled basis (say Bell basis) that would bring B and D into an entangled state. To visualize this more clearly note that

$$|\psi^+\rangle_{AB}|\psi^+\rangle_{CD} = \tfrac{1}{2} \left\{ |\psi^+\psi^+\rangle + |\phi^+\phi^+\rangle + |\phi^-\phi^-\rangle + |\psi^-\psi^-\rangle \right\}_{ACBD} .$$

Thus whatever be the outcome of Bell measurement of Alice (on particle A and C), the measurement would bring the other two particles into a Bell state. This simple idea of entanglement swapping through Bell measurement is very useful in quantum information processing. Because of this process, two systems that have never interacted can become entangled. Further, it is easy to observe that the repeated application of entanglement swapping can help us to transfer entanglement between distant sites.

3.3.2 Partial trace

Partial trace is a generalized version of trace. It is easy to understand if we consider state $|\psi\rangle_{AB} \in H_A \otimes H_B$. Even if the states are entangled, the state of the first qubit can in general be described by a density operator ρ^A on H_A. Popularly, ρ^A is called the reduced density operator. The mathematical operation that calculates the reduced density operator is the partial trace. The reduced density operator ρ^A can be defined in terms of the density operator of the composite system as

$$\rho^A \equiv Tr_B(\rho^{AB}), \tag{3.62}$$

where Tr_B is the partial trace over system B, which is defined as

$$Tr_B\left(|a_1\rangle\langle a_2| \otimes |b_1\rangle\langle b_2|\right) \equiv |a_1\rangle\langle a_2|Tr\left(|b_1\rangle\langle b_2|\right). \tag{3.63}$$

Using the cyclic property of the trace

$$Tr\left(|b_1\rangle\langle b_2|\right) = Tr\left(\langle b_2|b_1\rangle\right) = \langle b_2|b_1\rangle. \tag{3.64}$$

We can simplify (3.63) as

$$Tr_B\left(|a_1\rangle\langle a_2| \otimes |b_1\rangle\langle b_2|\right) \equiv |a_1\rangle\langle a_2|\langle b_2|b_1\rangle. \tag{3.65}$$

This operation (Tr_B) is often called tracing out system B.

Example 3.23: Trace out the second qubit of the two qubit entangled state $|\psi\rangle = \frac{1}{\sqrt{2}}\left(|00\rangle + |11\rangle\right)$.

Solution: The density matrix for this state is

$$\rho = |\psi\rangle\langle\psi| = \frac{1}{2}\left(|00\rangle\langle00| + |00\rangle\langle11| + |11\rangle\langle00| + |11\rangle\langle11|\right). \tag{3.66}$$

Now we can compute the partial trace as

$$
\begin{aligned}
\rho^A &= Tr_B(\rho) \\
&= \tfrac{1}{2}Tr_B\left(|00\rangle\langle00| + |00\rangle\langle11| + |11\rangle\langle00| + |11\rangle\langle11|\right) \\
&= \tfrac{1}{2}\left(|0\rangle\langle0|Tr(|0\rangle\langle0|) + |0\rangle\langle1|Tr(|0\rangle\langle1|)\right) \\
&\quad + |1\rangle\langle0|Tr(|1\rangle\langle0|) + |1\rangle\langle1|Tr(|1\rangle\langle1|)) \\
&= \tfrac{1}{2}\left(|0\rangle\langle0|\langle0|0\rangle + |0\rangle\langle1|\langle1|0\rangle + |1\rangle\langle0|\langle0|1\rangle + |1\rangle\langle1|\langle1|1\rangle\right) \\
&= \tfrac{1}{2}\left(|0\rangle\langle0| + |1\rangle\langle1|\right).
\end{aligned}
\tag{3.67}
$$

Problem 3.4: Trace out the first qubit of all four Bell states.

Partial trace can be computed directly in the matrix form. For example, if we consider the density matrix of an arbitrary bipartite system

$$\rho = \begin{pmatrix} a_{11} & a_{12} & a_{13} & a_{14} \\ a_{21} & a_{22} & a_{23} & a_{24} \\ a_{31} & a_{32} & a_{33} & a_{34} \\ a_{41} & a_{42} & a_{43} & a_{44} \end{pmatrix} \tag{3.68}$$

then the partial trace

$$
\begin{aligned}
\rho^A &= Tr_B\left(\rho\right) \\
&= \begin{pmatrix}
Tr\begin{pmatrix} a_{11} & a_{12} \\ a_{21} & a_{22} \end{pmatrix} & Tr\begin{pmatrix} a_{13} & a_{14} \\ a_{23} & a_{24} \end{pmatrix} \\
Tr\begin{pmatrix} a_{31} & a_{32} \\ a_{41} & a_{42} \end{pmatrix} & Tr\begin{pmatrix} a_{33} & a_{34} \\ a_{43} & a_{44} \end{pmatrix}
\end{pmatrix} \\
&= \begin{pmatrix}
a_{11} + a_{22} & a_{13} + a_{24} \\
a_{31} + a_{42} & a_{33} + a_{44}
\end{pmatrix}.
\end{aligned}
\tag{3.69}
$$

This is simple and direct. Consequently, in the rest of the book we refer to this method as direct method. For example, the density matrix of the previous example (i.e., (3.66)) can be written in matrix form as

$$
\rho = \frac{1}{2}\begin{pmatrix}
1 & 0 & 0 & 1 \\
0 & 0 & 0 & 0 \\
0 & 0 & 0 & 0 \\
1 & 0 & 0 & 1
\end{pmatrix}.
\tag{3.70}
$$

Therefore, using (3.69) we can quickly write

$$
\rho^A = \frac{1}{2}\begin{pmatrix} 1 & 0 \\ 0 & 1 \end{pmatrix},
\tag{3.71}
$$

which is same as the result obtained via multiple steps in (3.67).

3.3.2.1   Quantum bit commitment and quantum coin tossing

Quantum bit commitment (QBC) and quantum coin tossing are two very interesting and connected problems. It is very easy to understand these problems, but extremely difficult (and perhaps impossible) to solve. Let us first describe the bit commitment problem. Consider that Alice has a bit $b \in \{0,1\}$, which she wishes to commit to Bob, but she wants to disclose the bit value at a later time. For example, imagine that Alice has two balls. One is white and the other one is black and she has to give one of the balls to Bob. In the morning she has decided and committed which ball she wants to give to Bob, but she wants to announce it only in the evening. Then we need a process to keep her commitment secret until evening and also ensure that she cannot change her mind in between (say in the afternoon). Thus we need concealing (i.e., keeping the committed bit secret) and binding (i.e., ensuring that Alice cannot change her mind after making the commitment) of the bit value. Classically the solution of the problem can be thought of as follows: Alice writes her bit value on a paper and puts the paper (or puts the ball of a specific color) in a box. She locks the box and sends it to Bob, but keeps the key of the lock with herself. At a later time when Alice wishes to disclose her bit value then she sends the key to Bob.

Since the box is with Bob, Alice cannot change the bit value. Consequently, the binding condition is satisfied. However, the concealing condition is not satisfied unconditionally. Classically there is always a probability that before Alice sends the key, Bob breaks the lock and observes the committed bit. Thus classically secure bit commitment is not possible. We have already mentioned that unconditionally secure quantum communication is possible. So if Alice uses a quantum key (say a key generated by BB84 protocol) and uses that to lock the committed bit, then from the simple minded logic it appears that quantum bit commitment is possible. But in reality it is not so as the protocol has to be two party protocol involving Alice and Bob only. If they implement BB84 to generate the key to be used, then a copy of that key is already with Bob. Thus the simple minded idea of using a quantum key will not succeed to provide two party protocol of QBC. In fact, in 1984, Bennett and Brassard proposed a protocol of QBC along with the famous proposal of BB84 protocol [18], but almost simultaneously it was realized that the proposed QBC was insecure. An improved protocol was introduced in 1993. The protocol is known as the Brassard, Crepeau, Jozsa and Langlois (BCJL) 93 protocol [60]. In 1996 a cheating strategy was introduced and BCJL 93 failed. Subsequently several proposals appeared that claimed that QBC protocols cannot be unconditionally secure. Now it is widely believed that secure QBC is not possible. However, people still have doubt about the proof of the impossibility of secure QBC and every year several new proposals of QBC appear. It is an active and interesting field of study and the problem is still open.

A similar problem is remote coin tossing. In this problem Alice and Bob are at two different places and they wish to decide over something by tossing a coin. But they cannot meet and neither trusts each other or any third party. Thus we need to design a two-party remote coin tossing protocol such that neither Alice nor Bob can cheat the other. The problem of implementing remote coin tossing is that if Alice announces her outcome first then Bob will demand that his bet was the same as that of Alice's outcome. On the other hand if Bob announces the bet first then Alice will demand that the outcome of the toss is the opposite. Now a simple classical strategy works as follows: Alice tosses the coin and puts either the coin or the result in a box and locks it. She keeps the key and sends the box to Bob. After receiving the box, Bob sends his bet to Alice. Alice cannot change the outcome any more as the result is with Bob, and Bob cannot change his bet by observing the outcome of Alice as the box is locked and the key is with Alice. Now Alice sends the key to Bob after receiving the bet of Bob. This strategy would work if we could ensure that it is impossible for Bob to break the lock of the box within a stipulated time. But classically no such unconditionally secure lock can be prepared. The remote coin tossing problem is interesting for several reasons. Most importantly it has applications in many extremely important practical situations. Here we would like to provide a few interesting examples of such situations:

Consider that the United States and Russia have decided to destroy some nuclear weapons but nobody wants to disclose the number first. In this situation we need a version of coin tossing protocol. Or consider that Nokia and Samsung want to develop a quantum mobile phone in collaboration but nobody wants to announce first the amount of initial investment they want to make. In such collaborative business we need a secure coin tossing protocol.

3.3.3   Schmidt decomposition

Schmidt decomposition theorem: Suppose $|\psi\rangle$ is a pure state of tensor product space $H_A \otimes H_B$. Then there exists an orthonormal basis $\{|i_A\rangle\}$ for H_A, and an orthonormal basis $\{|i_B\rangle\}$ for H_B and nonnegative real numbers $\{p_i\}$ such that

$$|\psi\rangle = \sum_i p_i |i_A\rangle |i_B\rangle. \tag{3.72}$$

The coefficients p_i are called Schmidt coefficients and for a normalized state $|\psi\rangle$ we must have $\sum_i p_i^2 = 1$. The expansion (3.72) is known as the Schmidt decomposition. The Schmidt number (also called Schmidt rank) is the number of nonzero Schmidt coefficients p_i. The Schmidt number can be used for detection of entanglement in bipartite systems, as the Schmidt number is 1 for all separable states and for entangled states it is always greater than 1.

This is a very important theorem as it can be used to detect entanglement. Consequently, we need to understand it clearly. To do so, let us assume that $\{|i_A\rangle\}$ and $\{|j_B\rangle\}$ are any arbitrary orthonormal bases in H_A and H_B respectively. In that case $\{|i_A\rangle \otimes |j_B\rangle\}$ is a basis set in $H_A \otimes H_B$ and an arbitrary state in $H_A \otimes H_B$ can be written as

$$|\psi\rangle = \sum_{i,j} p_{ij} |i_A\rangle |j_B\rangle, \tag{3.73}$$

where p_{ij} is in general complex. If H_A is m dimensional and H_B is n dimensional then the basis set $\{|i_A\rangle \otimes |j_B\rangle\}$ is mn dimensional and $|\psi\rangle$ in (3.73) is superposition of mn basis vectors. Now we can illustrate the meaning of Schmidt decomposition by comparing (3.72) with (3.73). The comparison shows that in (3.73) different indices (i and j) are used on the two sets of basis vectors to incorporate all the cross-terms, and the expansion coefficients are complex. But in (3.72) the expansion coefficients are real and all the cross-terms have vanished, and consequently the summation is over a single index (i). Further, the maximum number of terms in (3.72) is the minimum of m and n. Thus Schmidt decomposition expands an arbitrary vector $|\psi\rangle$ in $H_A \otimes H_B$ in a special basis such that all the coefficients of expansion are real and no cross-term appears. The Schmidt decomposition theorem tells us that it is always possible to construct such a special basis set. Let us elaborate on this idea with a few simple examples.

Example 3.24: Consider $|\psi\rangle = |00\rangle$. As all the expansion coefficients are real and there is no cross-term so the state is already in Schmidt decomposed form. For both H_A and H_B the Schmidt basis is the computational basis. The same will be true for the states $|11\rangle$ and $|\psi^+\rangle = \frac{|00\rangle+|11\rangle}{\sqrt{2}}$. Further note that for $|00\rangle$ and $|11\rangle$ there is only one nonzero Schmidt coefficient and thus the Schmidt number is 1. Consequently, the states are separable. But in the case of $|\psi^+\rangle = \frac{|00\rangle+|11\rangle}{\sqrt{2}}$, the Schmidt number is 2 and thus the state is entangled.

Example 3.25: Consider $|\psi\rangle = \frac{1}{2}(|00\rangle - |01\rangle - |10\rangle + |11\rangle)$. Since it contains cross-terms in computational basis, the Schmidt bases used in the previous example cannot be valid Schmidt bases here. But it is straightforward to see that $|\psi\rangle = \frac{1}{2}(|00\rangle - |01\rangle - |10\rangle + |11\rangle) = |-\rangle|-\rangle$. For both H_A and H_B the Schmidt basis is diagonal (Hadamard) basis. The Schmidt number is 1 as there is only one nonzero Schmidt coefficient. Clearly the state is separable.

Example 3.26: Consider $|\psi\rangle = \frac{1}{2}(|00\rangle + |01\rangle + |10\rangle - |11\rangle) = \frac{|+\rangle|0\rangle+|-\rangle|1\rangle}{\sqrt{2}}$. Here the Schmidt basis in H_A is diagonal basis but the Schmidt basis in H_B is computational basis. Here the Schmidt number is 2 and thus the state is entangled. You can also reach the same conclusion by using a simple separability criterion (3.57) as here $ad = -\frac{1}{4} \neq bc = \frac{1}{4}$. However, it must be noted that applicability of (3.57) is limited to two qubit systems whereas Schmidt number characterizes entanglement of bipartite systems in general (H_A and H_B can be of any arbitrary dimension).

So far we have given simple examples where you can identify the Schmidt basis by observation (guess) but it may not be always that simple, so we need to provide a clear prescription of how to obtain Schmidt bases and Schmidt coefficients for an arbitrary state. To do so, first we note that the density matrix for the arbitrary state $|\psi\rangle$ in the Schmidt basis is

$$
\begin{aligned}
|\psi\rangle\langle\psi| &= \left(\sum_i p_i |i_A\rangle|i_B\rangle\right) \otimes \left(\sum_j p_j \langle j_A|\langle j_B|\right) \\
&= \sum_{i,j} p_i p_j |i_A\rangle|i_B\rangle\langle j_A|\langle j_B|.
\end{aligned}
\tag{3.74}
$$

Now we may use (3.65) to trace out the system B and obtain reduced density matrix for system A as

$$
\begin{aligned}
Tr_B\left(|\psi\rangle\langle\psi|\right) &= \sum_{i,j} p_i p_j |i_A\rangle\langle j_A|\langle j_B|i_B\rangle \\
&= \sum_{i,j} p_i p_j |i_A\rangle\langle j_A|\delta_{ij} \\
&= \sum_i p_i^2 |i_A\rangle\langle i_A|.
\end{aligned}
\tag{3.75}
$$

Similarly, $Tr_A\left(|\psi\rangle\langle\psi|\right) = \sum_i p_i^2 |i_B\rangle\langle i_B|$. Thus the reduced density operators are diagonal in the Schmidt bases. Earlier we have seen that the density operator is diagonalizable since it is a normal operator and the spectral decomposition theorem tells us that a diagonalizable operator is diagonal in its own eigenbasis. Consequently, the reduced density operator will also be diagonal in its own eigenbasis and we may use that eigenbasis as the Schmidt basis of the subsystem and inserting that in the given state we may find out the Schmidt basis of the other subsystem. Thus, if a composite system is given then we may find out Schmidt decomposition by the following simple steps:

Step 1: Trace out the second subsystem and obtain the reduced density operator for the first subsystem.

Step 2: Compute the eigenvectors and the eigenvalues of the reduced density operator obtained in the previous step. The square roots of the eigenvalues are the Schmidt coefficients and the eigenvectors are the Schmidt basis of the first subsystem.

Step 3: Rewrite the given state such that the first subsystem is in the Schmidt basis states obtained in Step 2. This will automatically reveal the Schmidt basis of the other subsystem.

Let us provide an example to elaborate the protocol described above.
Example 3.27: Consider the state $|\psi\rangle = \frac{1}{2}(|00\rangle - |01\rangle - |10\rangle + |11\rangle)$ discussed above and follow the steps.

Step 1: We compute

$$\rho^A = Tr_B(|\psi\rangle\langle\psi|) = {}_B\langle 0|\psi\rangle\langle\psi|0\rangle_B + {}_B\langle 1|\psi\rangle\langle\psi|1\rangle_B$$
$$= \frac{1}{2}(|0\rangle\langle 0| - |0\rangle\langle 1| - |1\rangle\langle 0| + |1\rangle\langle 1|)$$

and express ρ_A in matrix form as

$$\begin{aligned}
\rho^A &= \frac{1}{2}(|0\rangle\langle 0| - |0\rangle\langle 1| - |1\rangle\langle 0| + |1\rangle\langle 1|) \\
&= \frac{1}{2}\begin{pmatrix} 1 \\ 0 \end{pmatrix}\begin{pmatrix} 1 & 0 \end{pmatrix} - \frac{1}{2}\begin{pmatrix} 1 \\ 0 \end{pmatrix}\begin{pmatrix} 0 & 1 \end{pmatrix} \\
&\quad - \frac{1}{2}\begin{pmatrix} 0 \\ 1 \end{pmatrix}\begin{pmatrix} 1 & 0 \end{pmatrix} + \frac{1}{2}\begin{pmatrix} 0 \\ 1 \end{pmatrix}\begin{pmatrix} 0 & 1 \end{pmatrix} \\
&= \frac{1}{2}\begin{pmatrix} 1 & 0 \\ 0 & 0 \end{pmatrix} - \frac{1}{2}\begin{pmatrix} 0 & 1 \\ 0 & 0 \end{pmatrix} - \frac{1}{2}\begin{pmatrix} 0 & 0 \\ 1 & 0 \end{pmatrix} + \frac{1}{2}\begin{pmatrix} 0 & 0 \\ 0 & 1 \end{pmatrix} \\
&= \frac{1}{2}\begin{pmatrix} 1 & -1 \\ -1 & 1 \end{pmatrix}.
\end{aligned}$$

Step 2: Find eigenvalues and eigenvectors of ρ^A. Eigenvalues are the roots of the characteristic equation

$$Det\begin{pmatrix} \frac{1}{2} - \lambda & -\frac{1}{2} \\ -\frac{1}{2} & \frac{1}{2} - \lambda \end{pmatrix} = 0$$

where Det stands for determinant and we need the roots of $\left(\frac{1}{2} - \lambda\right)^2 - \frac{1}{4} = 0$ which implies $\lambda^2 - \lambda + \frac{1}{4} - \frac{1}{4} = \lambda(\lambda - 1) = 0$. Clearly the eigenvalues are 1 and 0. Thus the Schmidt coefficients are 1 and 0 and the

Schmidt number is 1 indicating that the state is separable. Now we need to find out eigenvectors. The eigenvalue equation corresponding to eigenvalue 1 is

$$\frac{1}{2}\begin{pmatrix} 1 & -1 \\ -1 & 1 \end{pmatrix}\begin{pmatrix} x_1 \\ y_1 \end{pmatrix} = 1\begin{pmatrix} x_1 \\ y_1 \end{pmatrix}$$

$$\Rightarrow \frac{1}{2}\begin{pmatrix} x_1 - y_1 \\ -x_1 + y_1 \end{pmatrix} = \begin{pmatrix} x_1 \\ y_1 \end{pmatrix}.$$

Thus we need simultaneous solution of

$$x_1 - y_1 = 2x_1 \Rightarrow x_1 = -y_1$$

and

$$-x_1 + y_1 = 2y_1 \Rightarrow x_1 = -y_1.$$

Apparently the solution is not unique, but we need a normalized eigenvector so we need to solve $x_1 = -y_1$ together with $x_1^2 + y_1^2 = 1$, which yield $x_1 = \frac{1}{\sqrt{2}}$ and $y_1 = -\frac{1}{\sqrt{2}}$. Similarly, for eigenvalue 0 we need to solve $x_2 = y_2$ together with $x_2^2 + y_2^2 = 1$, which yield $x_2 = \frac{1}{\sqrt{2}}$ and $y_2 = \frac{1}{\sqrt{2}}$. Thus the Schmidt basis in H_A is

$$\left\{ \begin{pmatrix} \frac{1}{\sqrt{2}} \\ -\frac{1}{\sqrt{2}} \end{pmatrix}, \begin{pmatrix} \frac{1}{\sqrt{2}} \\ \frac{1}{\sqrt{2}} \end{pmatrix} \right\} = \{|-\rangle, |+\rangle\}.$$

Step 3: Now that we know the Schmidt coefficients and Schmidt basis of the first system we can write the original state in terms of them as

$$\begin{aligned} |\psi\rangle &= \frac{1}{2}(|00\rangle - |01\rangle - |10\rangle + |11\rangle) \\ &= 1|-\rangle_A \otimes \begin{pmatrix} x_3 \\ y_3 \end{pmatrix}_B + 0|+\rangle_A \otimes \begin{pmatrix} x_4 \\ y_4 \end{pmatrix}_B \\ &= \frac{1}{\sqrt{2}}\begin{pmatrix} x_3 \\ -x_3 \\ y_3 \\ -y_3 \end{pmatrix}_{AB} \\ &= \frac{1}{\sqrt{2}}(x_3|00\rangle - x_3|01\rangle + y_3|10\rangle - y_3|11\rangle)_{AB}. \end{aligned}$$

It is straightforward to see that $x_3 = \frac{1}{\sqrt{2}}$ and $y_3 = -\frac{1}{\sqrt{2}}$. Therefore,

$$\begin{pmatrix} x_3 \\ y_3 \end{pmatrix}_B = \frac{1}{\sqrt{2}}\begin{pmatrix} 1 \\ -1 \end{pmatrix}_B = |-\rangle_B \text{ and } |\psi\rangle = |-\rangle_A|-\rangle_B \text{ as expected.}$$

Now we can recognize that the Schmidt basis in H_B is diagonal basis but we formally obtain the same by considering that $\begin{pmatrix} x_4 \\ y_4 \end{pmatrix}_B$ is normalized, i.e., $x_4^2 + y_4^2 = 1$ and it is orthogonal to $\begin{pmatrix} x_3 \\ y_3 \end{pmatrix}_B$, i.e.,

$$\begin{pmatrix} x_4 & y_4 \end{pmatrix}_B \begin{pmatrix} x_3 \\ y_3 \end{pmatrix}_B = \frac{1}{\sqrt{2}}\begin{pmatrix} x_4 & y_4 \end{pmatrix}\begin{pmatrix} 1 \\ -1 \end{pmatrix} = \frac{1}{\sqrt{2}}(x_4 - y_4)_B =$$

$0, \Rightarrow x_4 = y_4$. Solving $x_4 = y_4$ together with $x_4^2 + y_4^2 = 1$ we obtain

$$\begin{pmatrix} x_4 \\ y_4 \end{pmatrix}_B = \frac{1}{\sqrt{2}} \begin{pmatrix} 1 \\ 1 \end{pmatrix}_B = |+\rangle_B.$$

The above example explicitly elaborates how to obtain Schmidt decomposition of a given state. Given any composite state we just need to follow the above mentioned three steps to obtain Schmidt decomposition of the state. We can in fact obtain a compact formula for a bipartite system. For an arbitrary density matrix (3.68) we have already provided the general form of ρ^A in (3.69) which can be further simplified as follows:

$$\rho^A = \begin{pmatrix} a_{11} + a_{22} & a_{13} + a_{24} \\ a_{31} + a_{42} & a_{33} + a_{44} \end{pmatrix} = \begin{pmatrix} a & b \\ b^* & 1 - a \end{pmatrix}$$

where for simplification we have used $a_{11} + a_{22} = a$, $a_{13} + a_{24} = b$, $a_{31} + a_{42} = c = b^*$ and $a_{33} + a_{44} = 1 - a$ as ρ^A is a Hermitian operator having unit trace. Now **Step 1** of Schmidt decomposition is done. For **Step 2** we need eigenvalues and eigenvectors of $\begin{pmatrix} a & b \\ b^* & 1 - a \end{pmatrix}$. To obtain the eigenvalues we need to solve the characteristic equation $(a - \lambda)(1 - a - \lambda) = |b|^2$ or $\lambda^2 - \lambda + (a - a^2 - |b|^2) = 0$ which is a simple quadratic equation whose roots are

$$\lambda = \frac{1}{2} \left(1 \pm \sqrt{1 + 4|b|^2 + 4a^2 - 4a} \right) = \frac{1}{2} \left(1 \pm \sqrt{4|b|^2 + (1 - 2a)^2} \right).$$

(3.76)

Thus we have a general form of square of Schmidt coefficients and for the special case $b = 0$, the eigenvalues are a and $1 - a$. In that case the corresponding Schmidt coefficients will be $\sqrt{a}$, $\sqrt{1 - a}$. Now using these two eigenvalues (3.76) we can easily find the corresponding two eigenvectors under different conditions as follows:

(i) $b \neq 0$:

$$\left\{ \begin{pmatrix} -\frac{1 - 2a + \sqrt{1 + 4|b|^2 + 4a^2 - 4a}}{2b^*} \\ 1 \end{pmatrix}, \begin{pmatrix} -\frac{1 - 2a - \sqrt{1 + 4|b| + 4a^2 - 4a}}{2b^*} \\ 1 \end{pmatrix} \right\}, \quad (3.77)$$

(ii) $b = 0$:

$$\left\{ \begin{pmatrix} 0 \\ 1 \end{pmatrix}, \begin{pmatrix} 1 \\ 0 \end{pmatrix} \right\}, \quad (3.78)$$

which are the Schmidt basis for the first system. For a quick check consider the density matrix (3.70) of the Bell state $|\psi^+\rangle$. Here $b = b^* = 0$ and $a = \frac{1}{2}$ which yields the eigenvalues as $(a, 1 - a) = \left(\frac{1}{2}, \frac{1}{2} \right)$. Thus Schmidt coefficients are $\left(\frac{1}{\sqrt{2}}, \frac{1}{\sqrt{2}} \right)$. The eigenvectors (3.77) are orthogonal to each other, but they are not normalized. Normalization is a trivial step. See the example below and visualize the symmetry; the above results can often help us to quickly obtain the Schmidt decomposition of a given state.

Example 3.28: Express $|\psi\rangle = \frac{1}{\sqrt{3}}(|00\rangle + |01\rangle + |10\rangle)$ in Schmidt decomposed form.

Solution:

$$
\begin{aligned}
\rho &= |\psi\rangle\langle\psi| \\
&= \tfrac{1}{3}(|00\rangle\langle00| + |00\rangle\langle01| + |00\rangle\langle10| + 0 \times |00\rangle\langle11| \\
&+ \quad |01\rangle\langle00| + |01\rangle\langle01| + |01\rangle\langle10| + 0 \times |01\rangle\langle11| \\
&+ \quad |10\rangle\langle00| + |10\rangle\langle01| + |10\rangle\langle10| + 0 \times |10\rangle\langle11| \\
&+ \quad 0 \times |11\rangle\langle00| + 0 \times |11\rangle\langle01| + 0 \times |11\rangle\langle10| + 0 \times |11\rangle\langle11|).
\end{aligned}
$$
(3.79)

Here we have intentionally written terms with coefficients "0" to show you how to expand $\rho = |\psi\rangle\langle\psi|$ in a systematic way (please follow the sequence of the appearance of the outer products) so that you can obtain the matrix form of ρ without any effort. In (3.79) there are 4 rows and each row has 4 elements; we just need to take the coefficients in sequence and we will obtain a 4×4 matrix which is the matrix form of ρ. So in our case

$$
\rho = \frac{1}{3}\begin{pmatrix} 1 & 1 & 1 & 0 \\ 1 & 1 & 1 & 0 \\ 1 & 1 & 1 & 0 \\ 0 & 0 & 0 & 0 \end{pmatrix}.
$$
(3.80)

Now using (3.69) we directly obtain

$$
\rho^A = \frac{1}{3}\begin{pmatrix} 2 & 1 \\ 1 & 1 \end{pmatrix}.
$$

Thus in this case $a = \frac{2}{3}$ and $b = \frac{1}{3}$. Now using (3.76) we obtain the eigenvalues of ρ^A as $\frac{1}{2} + \frac{\sqrt{5}}{6} = 0.873$ and $\frac{1}{2} - \frac{\sqrt{5}}{6} = 0.127$. Therefore, Schmidt rank is 2 and Schmidt coefficients are $\sqrt{0.873} = 0.934$ and $\sqrt{0.127} = 0.356$. Similarly, using (3.77) we obtain corresponding eigenvectors as $\begin{pmatrix} \frac{1+\sqrt{5}}{2} \\ 1 \end{pmatrix} = \begin{pmatrix} 1.618 \\ 1 \end{pmatrix}$ and $\begin{pmatrix} \frac{1-\sqrt{5}}{2} \\ 1 \end{pmatrix} = \begin{pmatrix} -0.618 \\ 1 \end{pmatrix}$. Normalizing these two eigenvectors we obtain the Schmidt basis for the subsystem A as $\left\{ \begin{pmatrix} 0.851 \\ 0.526 \end{pmatrix}, \begin{pmatrix} -0.526 \\ 0.851 \end{pmatrix} \right\}$. To obtain the Schmidt basis of the subsystem B we expand the given state in terms of the obtained Schmidt coefficients and the Schmidt basis of subsystem A and assume that Alternatively, basis of subsystem B is $\left\{ \begin{pmatrix} x_1 \\ y_1 \end{pmatrix}, \begin{pmatrix} x_2 \\ y_2 \end{pmatrix} \right\}$. Now

$$
\begin{aligned}
|\psi\rangle &= \frac{1}{\sqrt{3}} \begin{pmatrix} 1 \\ 1 \\ 1 \\ 0 \end{pmatrix} = \begin{pmatrix} 0.577 \\ 0.577 \\ 0.577 \\ 0 \end{pmatrix} \\
&= 0.934 \begin{pmatrix} 0.851 \\ 0.526 \end{pmatrix} \otimes \begin{pmatrix} x_1 \\ y_1 \end{pmatrix} + 0.356 \begin{pmatrix} -0.526 \\ 0.851 \end{pmatrix} \begin{pmatrix} x_2 \\ y_2 \end{pmatrix} \\
&= \begin{pmatrix} 0.795x_1 - 0.187x_2 \\ 0.795y_1 - 0.187y_2 \\ 0.491x_1 + 0.303x_2 \\ 0.491y_1 + 0.303y_2 \end{pmatrix}.
\end{aligned}
$$

Thus we need simultaneous solution of

$$
\begin{aligned}
0.795x_1 - 0.187x_2 &= 0.577 \\
0.795y_1 - 0.187y_2 &= 0.577 \\
0.491x_1 + 0.303x_2 &= 0.577 \\
0.491y_1 + 0.303y_2 &= 0
\end{aligned}
$$

which can be obtained as $x_1 = 0.851$, $y_1 = 0.526$, $x_2 = 0.526$, $y_2 = -0.851$. Now we can present our final expression for Schmidt decomposition of $|\psi\rangle = \frac{1}{\sqrt{3}}(|00\rangle + |01\rangle + |10\rangle)$ as

$$
|\psi\rangle = 0.934 \begin{pmatrix} 0.851 \\ 0.526 \end{pmatrix} \otimes \begin{pmatrix} 0.851 \\ 0.526 \end{pmatrix} + 0.357 \begin{pmatrix} -0.526 \\ 0.851 \end{pmatrix} \otimes \begin{pmatrix} 0.526 \\ -0.851 \end{pmatrix}.
$$

3.3.4   Partial transpose and test of entanglement

As it appears from the name, partial transpose is the transpose taken with respect to one party of a composite bipartite system. In other words, only part of the composite system is transposed. For example, if we have a general density matrix $\rho = \rho^{AB}$ in $H_A \otimes H_B$ and we wish to take transpose with respect to system A only then partial transpose can be viewed as a map $T(\rho) \otimes I$ that transposes the first system and keeps the second system unchanged. It can be defined simply as

$$
\text{PT}^A : \begin{array}{l} |1x\rangle\langle 0y| \Rightarrow |0x\rangle\langle 1y| \\ |0x\rangle\langle 1y| \Rightarrow |1x\rangle\langle 0y|, \end{array} \tag{3.81}
$$

where $x, y \in \{0, 1\}$. Similarly,

$$
\text{PT}^B : \begin{array}{l} |x1\rangle\langle y0| \Rightarrow |x0\rangle\langle y1| \\ |x0\rangle\langle y1| \Rightarrow |x1\rangle\langle y0|. \end{array} \tag{3.82}
$$

It is easy to observe that PT^A maps $|00\rangle\langle 11|$ to $|10\rangle\langle 01|$ and $|11\rangle\langle 00|$ to $|01\rangle\langle 10|$. Now if we consider the Bell state $|\psi^+\rangle = \frac{|00\rangle + |11\rangle}{\sqrt{2}}$ then the corresponding density matrix is $\rho = \frac{1}{2}(|00\rangle\langle 00| + |00\rangle\langle 11| + |11\rangle\langle 00| + |11\rangle\langle 11|)$

and partial transpose with respect to the system A will map ρ to $\rho^{T_A} = \frac{1}{2}(|00\rangle\langle00| + |10\rangle\langle01| + |01\rangle\langle10| + |11\rangle\langle11|)$. Thus in matrix form

$$\rho = \frac{1}{2}\begin{pmatrix} 1 & 0 & 0 & 1 \\ 0 & 0 & 0 & 0 \\ 0 & 0 & 0 & 0 \\ 1 & 0 & 0 & 1 \end{pmatrix}$$

and

$$\rho^{T_A} = \frac{1}{2}\begin{pmatrix} 1 & 0 & 0 & 0 \\ 0 & 0 & 1 & 0 \\ 0 & 1 & 0 & 0 \\ 0 & 0 & 0 & 1 \end{pmatrix}.$$

Similarly from (3.82) we obtain

$$\rho^{T_B} = \frac{1}{2}(|00\rangle\langle00| + |01\rangle\langle10| + |10\rangle\langle01| + |11\rangle\langle11|) = \rho^{T_A}.$$

Now we can become more comfortable with the idea of partial transpose through a few more examples.

Example 3.29: Consider a general density matrix

$$\begin{aligned} \rho &= (a_{11}|00\rangle\langle00| + a_{12}|00\rangle\langle01| + a_{13}|00\rangle\langle10| + a_{14}|00\rangle\langle11| \\ &+ a_{21}|01\rangle\langle00| + a_{22}|01\rangle\langle01| + a_{23}|01\rangle\langle10| + a_{24}|01\rangle\langle11| \\ &+ a_{31}|10\rangle\langle00| + a_{32}|10\rangle\langle01| + a_{33}|10\rangle\langle10| + a_{34}|10\rangle\langle11| \\ &+ a_{41}|11\rangle\langle00| + a_{42}|11\rangle\langle01| + a_{43}|11\rangle\langle10| + a_{44}|11\rangle\langle11|) \\ &= \begin{pmatrix} a_{11} & a_{12} & a_{13} & a_{14} \\ a_{21} & a_{22} & a_{23} & a_{24} \\ a_{31} & a_{32} & a_{33} & a_{34} \\ a_{41} & a_{42} & a_{43} & a_{44} \end{pmatrix} \end{aligned}$$

and take partial transpose with respect to the first system (A) to yield

$$\begin{aligned} \rho^{T_A} &= (a_{11}|00\rangle\langle00| + a_{12}|00\rangle\langle01| + a_{13}|10\rangle\langle00| + a_{14}|10\rangle\langle01| \\ &+ a_{21}|01\rangle\langle00| + a_{22}|01\rangle\langle01| + a_{23}|11\rangle\langle00| + a_{24}|11\rangle\langle01| \\ &+ a_{31}|00\rangle\langle10| + a_{32}|00\rangle\langle11| + a_{33}|10\rangle\langle10| + a_{34}|10\rangle\langle11| \\ &+ a_{41}|01\rangle\langle10| + a_{42}|01\rangle\langle11| + a_{43}|11\rangle\langle10| + a_{44}|11\rangle\langle11|) \\ &= (a_{11}|00\rangle\langle00| + a_{12}|00\rangle\langle01| + a_{31}|00\rangle\langle10| + a_{32}|00\rangle\langle11| \\ &+ a_{21}|01\rangle\langle00| + a_{22}|01\rangle\langle01| + a_{41}|01\rangle\langle10| + a_{42}|01\rangle\langle11| \\ &+ a_{13}|10\rangle\langle00| + a_{14}|10\rangle\langle01| + a_{33}|10\rangle\langle10| + a_{34}|10\rangle\langle11| \\ &+ a_{23}|11\rangle\langle00| + a_{24}|11\rangle\langle01| + a_{43}|11\rangle\langle10| + a_{44}|11\rangle\langle11|) \\ &= \begin{pmatrix} a_{11} & a_{12} & a_{31} & a_{32} \\ a_{21} & a_{22} & a_{41} & a_{42} \\ a_{13} & a_{14} & a_{33} & a_{34} \\ a_{23} & a_{24} & a_{43} & a_{44} \end{pmatrix}. \end{aligned} \tag{3.83}$$

Example 3.30: Density matrix for the state $|\psi\rangle = \frac{1}{\sqrt{3}}(|00\rangle + |01\rangle + |10\rangle)$

is obtained in (3.80) as $\rho = \frac{1}{3}\begin{pmatrix} 1 & 1 & 1 & 0 \\ 1 & 1 & 1 & 0 \\ 1 & 1 & 1 & 0 \\ 0 & 0 & 0 & 0 \end{pmatrix}$ obtain ρ^{T_A}.

Solution: Using (3.83) we can directly obtain $\rho^{T_A} = \frac{1}{3}\begin{pmatrix} 1 & 1 & 1 & 1 \\ 1 & 1 & 0 & 0 \\ 1 & 0 & 1 & 0 \\ 1 & 0 & 0 & 0 \end{pmatrix}$.

Example 3.31: Use (3.83) to obtain partial transpose of all the Bell states.

Solution: As

$$|\psi^\pm\rangle\langle\psi^\pm| = \frac{1}{2}(|00\rangle\langle00| \pm |00\rangle\langle11| \pm |11\rangle\langle00| + |11\rangle\langle11|)$$

$$= \frac{1}{2}\begin{pmatrix} 1 & 0 & 0 & \pm1 \\ 0 & 0 & 0 & 0 \\ 0 & 0 & 0 & 0 \\ \pm1 & 0 & 0 & 1 \end{pmatrix}$$

therefore, using (3.83) we obtain

$$|\psi^\pm\rangle\langle\psi^\pm|^{T_A} = \frac{1}{2}\begin{pmatrix} 1 & 0 & 0 & 0 \\ 0 & 0 & \pm1 & 0 \\ 0 & \pm1 & 0 & 0 \\ 0 & 0 & 0 & 1 \end{pmatrix}. \qquad (3.84)$$

Similarly,

$$|\phi^\pm\rangle\langle\phi^\pm| = \frac{1}{2}(|01\rangle\langle01| \pm |01\rangle\langle10| \pm |10\rangle\langle01| + |10\rangle\langle10|)$$

$$= \frac{1}{2}\begin{pmatrix} 0 & 0 & 0 & 0 \\ 0 & 1 & \pm1 & 0 \\ 0 & \pm1 & 1 & 0 \\ 0 & 0 & 0 & 0 \end{pmatrix}$$

and using (3.83) we obtain

$$|\phi^\pm\rangle\langle\phi^\pm|^{T_A} = \frac{1}{2}\begin{pmatrix} 0 & 0 & 0 & \pm1 \\ 0 & 1 & 0 & 0 \\ 0 & 0 & 1 & 0 \\ \pm1 & 0 & 0 & 0 \end{pmatrix}. \qquad (3.85)$$

Example 3.32: Use (3.84) and (3.85) to obtain partial transpose of the Werner state[8] which is defined as

$$\rho_W = F|\psi^+\rangle\langle\psi^+| + \frac{1-F}{3}(|\psi^-\rangle\langle\psi^-| + |\phi^+\rangle\langle\phi^+| + |\phi^-\rangle\langle\phi^-|), \qquad (3.86)$$

---

[8]This is a very important state as it can be obtained from a maximally entangled state in presence of some kind of isotropic noise. Such a scenario is not uncommon in practice. We may think that Alice communicates a state $|\psi^+\rangle$ to Bob. But due to noise Bob receives a mixed state which is an ensemble of Bell states. The noise is isotropic in the sense that $p(|\psi^-\rangle\langle\psi^-|) = p(|\phi^\pm\rangle\langle\phi^\pm|)$.

where $0 \leq F \leq 1$ is known as singlet fidelity.

Solution:

$$\rho_W^{T_A} = F|\psi^+\rangle\langle\psi^+|^{T_A} + \frac{1-F}{3}\left(|\psi^-\rangle\langle\psi^-|^{T_A} + |\phi^+\rangle\langle\phi^+|^{T_A} + |\phi^-\rangle\langle\phi^-|^{T_A}\right)$$

$$= \frac{F}{2}\begin{pmatrix} 1 & 0 & 0 & 0 \\ 0 & 0 & 1 & 0 \\ 0 & 1 & 0 & 0 \\ 0 & 0 & 0 & 1 \end{pmatrix} + \frac{1-F}{6}\begin{pmatrix} 1 & 0 & 0 & 0 \\ 0 & 0 & -1 & 0 \\ 0 & -1 & 0 & 0 \\ 0 & 0 & 0 & 1 \end{pmatrix}$$

$$+ \frac{1-F}{6}\begin{pmatrix} 0 & 0 & 0 & 1 \\ 0 & 1 & 0 & 0 \\ 0 & 0 & 1 & 0 \\ 1 & 0 & 0 & 0 \end{pmatrix} + \frac{1-F}{6}\begin{pmatrix} 0 & 0 & 0 & -1 \\ 0 & 1 & 0 & 0 \\ 0 & 0 & 1 & 0 \\ -1 & 0 & 0 & 0 \end{pmatrix}$$

$$= \frac{1}{6}\begin{pmatrix} 2F+1 & 0 & 0 & 0 \\ 0 & 2-2F & 4F-1 & 0 \\ 0 & 4F-1 & 2-2F & 0 \\ 0 & 0 & 0 & 2F+1 \end{pmatrix}.$$

$$(3.87)$$

The importance of partial transpose lies in the fact that it is useful to detect entanglement. Corresponding criterion is known as positive partial transpose (PPT) criterion or Peres-Horodecki criterion. The criterion states that if $\rho = \rho^{AB}$ is separable, then ρ^{T_A} has nonnegative eigenvalues (i.e., partial transpose is positive). Alternatively, if ρ^{T_A} has a negative eigenvalue (i.e., if partial transpose is not positive), then ρ must be entangled. In general, this is a necessary criterion. However, in 2×2, 2×3 and 3×2 dimensional cases PPT criterion is sufficient, too. This implies that in a higher dimension (i.e., dimension $> 2 \times 3$) if at least one eigenvalue is negative then the state is entangled but if all the eigenvalues are nonnegative then we cannot conclude whether the state is separable or entangled. In other words, PPT is a necessary but not sufficient condition for separability for larger Hilbert spaces and in larger Hilbert spaces there exist entangled states whose density matrices are positive under the partial transpose operation.

Example 3.33: We have already obtained ρ^{T_A} for the states (a) $|\psi^+\rangle = \frac{|00\rangle + |11\rangle}{\sqrt{2}}$ and (b) $|\psi\rangle = \frac{1}{\sqrt{3}}(|00\rangle + |01\rangle + |10\rangle)$. Now use PPT criterion to show that these two states are entangled.

Solution: (a) See ρ^{T_A} for $|\psi^+\rangle = \frac{|00\rangle + |11\rangle}{\sqrt{2}}$ is $\frac{1}{2}\begin{pmatrix} 1 & 0 & 0 & 0 \\ 0 & 0 & 1 & 0 \\ 0 & 1 & 0 & 0 \\ 0 & 0 & 0 & 1 \end{pmatrix}$, whose

eigenvalues are $\{-\frac{1}{2}, \frac{1}{2}, \frac{1}{2}, \frac{1}{2}\}$. Existence of a negative eigenvalue is the signature of the entanglement.

(b) Similarly, ρ^{T_A} for $|\psi\rangle = \frac{1}{\sqrt{3}}(|00\rangle + |01\rangle + |10\rangle)$ is $\frac{1}{3}\begin{pmatrix} 1 & 1 & 1 & 1 \\ 1 & 1 & 0 & 0 \\ 1 & 0 & 1 & 0 \\ 1 & 0 & 0 & 0 \end{pmatrix}$, whose eigenvalues are $\left\{ \frac{1}{6} \left(3 + \sqrt{5}\right), -\frac{1}{3}, \frac{1}{3}, \frac{1}{6} \left(3 - \sqrt{5}\right) \right\}$. Existence of a negative eigenvalue is the signature of the entanglement.

Example 3.34: Is the state $\rho = \frac{1}{2} \left(|00\rangle\langle00| + |11\rangle\langle11| \right)$ separable?

As the state has only diagonal terms $\rho^{T_A} = \rho = \frac{1}{2} \begin{pmatrix} 1 & 0 & 0 & 0 \\ 0 & 0 & 0 & 0 \\ 0 & 0 & 0 & 0 \\ 0 & 0 & 0 & 1 \end{pmatrix}$ whose

eigenvalues are $\left\{ \frac{1}{2}, \frac{1}{2}, 0, 0 \right\}$. Therefore, the partial transpose is positive and consequently the state is separable.

Example 3.35: Use PPT criterion to show that the Werner state defined in (3.86) is entangled if $F > \frac{1}{2}$.

Solution: Partial transpose of the Werner state $(\rho_W^{T_A})$ is already obtained in (3.87). The eigenvalues of $\rho_W^{T_A}$ are $\left\{ \frac{1}{2}(1 - 2F), \frac{1}{6}(1 + 2F), \frac{1}{6}(1 + 2F), \frac{1}{6}\{(1 + 2F)\} \right\}$. As $0 < F < 1$ so we can obtain negative eigenvalue of $\rho_W^{T_A}$ iff $1 - 2F < 0$ or, $F > \frac{1}{2}$. Thus the Warner state is entangled for $F > \frac{1}{2}$.

3.3.5 Entanglement witness

We have seen that PPT criterion, separability criterion (3.57) and Schmidt number may be used to decide whether a system is entangled. The task to decide whether a given bipartite state ρ_{AB} is entangled or not is essentially equivalent to witnessing the correlation. Classically the task is trivial. If we observe that the mutual information $I(X : Y)$ is positive then there exists correlation between X and Y. For pure state classical and quantum correlations are not separable and consequently it is not difficult to test whether a given bipartite state ρ_{AB} is entangled or not. For example, the same can be done by using the Schmidt number. However, the task is relatively difficult for mixed states as we need to subtract classical correlation from the total correlation. To be precise, we need to detect quantum correlations alone. Several techniques are proposed to obtain quantum correlation. Each has some advantages and some disadvantages. It is beyond the scope of the present textbook to discuss them in detail. However, we wish to note that the entanglement witness is the key concept in relation to the detection of separability of mixed states. Keeping this in mind, we will briefly describe the meaning of entanglement witness.

If there exists a Hermitian operator W whose expectation values are different for separable and entangled states then W is called entanglement witness. Here we will not construct witness operators but we will note some simple ideas that can help interested readers to understand the concept in further detail. Let us consider that T, E and S are set of all allowed density matrices, set of density matrices of all entangled states and set of density matrices of all separable states respectively. Obviously $E \subset T$ and $S \subset T$. Now we may recall that the states are points in the state space (vector space) and so geometrically S (E) is a set of points. Now a set of points is called convex if for every pair of points within the set, every point on the straight line segment that joins them is also within the set. In other words a set is convex if linear superposition of any two elements of the set is also an element of the same set. S is convex. Thus if ρ_{AB} and σ_{AB} are separable states then so is $p\rho_{AB} + (1-p)\sigma_{AB}$, where $0 < p < 1$. If a state is not in the convex set S then it must be entangled. The question is how do we know whether a given state is in S or not. A corollary of Hahn–Banach theorem[9] is that given a convex set and a point outside it, there exists a plane such that the point is on one side of it and the set is on the other side [8].

To visualize the idea geometrically we may think of a plane P and a unit vector $|w\rangle$ which is perpendicular (orthogonal) to P in a vector space V. Now any vector $|\psi\rangle$ lying in plane P must be orthogonal to $|w\rangle$. Thus P is the set of all vectors $|\psi\rangle$ such that $\langle\psi|w\rangle = 0$. Thus for a given state ρ we may define a plane orthogonal to it as $\langle\psi|\rho\rangle = 0$. Now we follow [8] and consider Hermitian operators as vectors to obtain $\langle A_1|\rho\rangle = Tr\left(A_1^\dagger\rho\right) = Tr\left(A_1\rho\right)$. The Hermitian operator A_1 defines a plane in T if $Tr\left(A_1\rho\right) = 0$. If we can find a specific operator A_1 such that all the entangled states are on one side of this plane and all the separable states are on the other side of the plane then $A_1 = W$ is the required witness operator. Now the question is which side of the plane contains the separable (entangled) states? This is a matter of convention. The usual convention is that for all separable states $Tr(W\rho) \geq 0$ and for all entangled states $Tr(W\rho) < 0$. It is interesting to note that it can be proved that there exists a witness operator for every entangled state. However, it may be very difficult to find the operator.

Example 3.36: Given that an equal mixture of any two maximally entangled states is a separable state, use this fact to show that the Werner state defined in (3.86) is separable for $F = \frac{1}{2}$.

Solution: When $F = \frac{1}{2}$ then the Werner state is

$$
\begin{aligned}
\rho_W &= F|\psi^+\rangle\langle\psi^+| + \frac{1-F}{3}\left(|\psi^-\rangle\langle\psi^-| + |\phi^+\rangle\langle\phi^+| + |\phi^-\rangle\langle\phi^-|\right), \\
&= 3 \times \frac{1}{6}|\psi^+\rangle\langle\psi^+| + \frac{1}{6}\left(|\psi^-\rangle\langle\psi^-| + |\phi^+\rangle\langle\phi^+| + |\phi^-\rangle\langle\phi^-|\right), \\
&= \frac{1}{6}\left(|\psi^+\rangle\langle\psi^+| + |\psi^-\rangle\langle\psi^-|\right) + \frac{1}{6}\left(|\psi^+\rangle\langle\psi^+| + |\phi^+\rangle\langle\phi^+|\right) \\
&\quad + \frac{1}{6}\left(|\psi^+\rangle\langle\psi^+| + |\phi^-\rangle\langle\phi^-|\right) \\
&= \frac{1}{3}(\rho_1 + \rho_2 + \rho_3),
\end{aligned}
$$

where $\rho_1 = \frac{1}{2}\left(|\psi^+\rangle\langle\psi^+| + |\psi^-\rangle\langle\psi^-|\right)$, $\rho_2 = \frac{1}{2}\left(|\psi^+\rangle\langle\psi^+| + |\phi^+\rangle\langle\phi^+|\right)$ and $\rho_3 = \frac{1}{2}\left(|\psi^+\rangle\langle\psi^+| + |\phi^-\rangle\langle\phi^-|\right)$. Thus from the given fact that an equal mixture of any two maximally entangled states is a separable state we may conclude that ρ_1, ρ_2 and ρ_3 are separable. Further, since the set of separable states is convex so ρ_W is a separable state (as it is a linear superposition of ρ_1, ρ_2 and ρ_3).

---

[9]Detailed proof of the theorem is complex and is not required for our purpose.

3.3.6   State discrimination

How far is London from Paris and New York? If I ask you this question then you will tell me two distances, which will show that London is closer to Paris than to New York. It's common practice to measure proximity in this way. But what about two quantum states? If I ask you: How close are $|0\rangle$ and $|+\rangle$ states? Or, how close are $\rho_1 = 0.64|0\rangle\langle 0| + 0.36|1\rangle\langle 1|$ and $\rho_2 = 0.36|0\rangle\langle 0| + 0.64|1\rangle\langle 1|$? Can you provide me a quantitative answer? Yes, there exist a large number of quantitative measures of distance between two states. Among this large number of measures, fidelity and trace distance are the most popular measures. So we will mainly restrict ourselves to these two measures only. However, we will briefly mention Bures distance function, too. The classical analogue of trace distance and fidelity have been frequently used in classical information theory for a long time.

3.3.6.1   Trace distance and fidelity

Classically trace distance between two probability distributions $\{p_x\}$ and $\{q_x\}$ is defined as

$$D\left(p_x, q_x\right) = \frac{1}{2}\sum_x |p_x - q_x|. \tag{3.88}$$

Similarly, classical fidelity defines the distance between two probability distributions $\{p_x\}$ and $\{q_x\}$ as

$$F\left(p_x, q_x\right) = \sum_x \sqrt{p_x q_x}. \tag{3.89}$$

Before we generalize these measures to quantum information it would be apt to provide some simple examples to clear the notion of distance in the context of information theory. Here are a few simple examples.

Example 3.37: (a) Show that the classical trace distance between the probability distributions $(p, 1-p)$ and $(q, 1-q)$ is $|p-q|$. (b) Compute the trace distance between $(0.7, 0.3)$ and $(0.48, 0.52)$.

Solution: (a) $D\left(p_x, q_x\right) = \frac{1}{2}\left[|p-q| + |(1-p) - (1-q)|\right] = |p-q|$.

(b)$D\left(p_x, q_x\right) = \frac{1}{2}\left(|0.7 - 0.48| + |0.3 - 0.52|\right) = \frac{1}{2}(0.22 + 0.22) = 0.22$. Note that the answer to (b) can also be obtained directly from the answer to (a).

Example 3.38: Classical fidelity defines the distance between two probability distributions $\{p_x\}$ and $\{q_x\}$ as $F\left(p_x, q_x\right) = \sum_x \sqrt{p_x q_x}$. Now use this definition to find out the fidelity between $(0.7, 0.3)$ and $(0.48, 0.52)$.

Solution: Here $F(p_x, q_x) = \sqrt{(0.7 \times 0.48 + 0.3 \times 0.52)} = 0.575$.

Quantum trace distance

The notion of classical trace distance can be extended to quantum states, but to do so we need the basic idea of *trace norm*. The trace norm of an operator A is just the trace of the norm of A. Thus the trace norm of an operator A is

$$\|A\|_1 = Tr|A| = Tr\{\sqrt{A^\dagger A}\}.$$

We are interested in Hermitian operators (especially in density operators), which can be diagonalized by spectral decomposition technique as described in Section 3.1.8. Spectral decomposition allows us to write a Hermitian operator A as $A = \sum_i \alpha_i |i\rangle\langle i|$ and this directly leads to $\|A\|_1 = \sum_i |\alpha_i|$. Thus it is always convenient to diagonalize the operator (unless it is already diagonal) first to compute the trace norm. With this background we are ready to introduce trace distance. The trace distance between quantum states ρ_1 and ρ_2 is defined as

$$D(\rho_1, \rho_2) = \frac{1}{2}Tr|\rho_1 - \rho_2| = \frac{1}{2}\|(\rho_1 - \rho_2)\|_1. \qquad (3.90)$$

This is an excellent measure of closeness (proximity) between two quantum states. It has following the properties:

1. $D(\rho_1, \rho_2) \geq 0$ with equality iff $\rho_1 = \rho_2$.

2. $D(\rho_1, \rho_2) \leq 1$ with equality iff ρ_1 and ρ_2 are orthogonal.

3. $D(\rho_1, \rho_2) = D(\rho_2, \rho_1)$.

We can elaborate the idea with some more examples.

Example 3.39: If ρ_1 and ρ_2 commute then show that the (quantum) trace distance between ρ_1 and ρ_2 is equal to the classical trace distance between the eigenvalues of ρ_1 and ρ_2.

Solution: If two states commute then they are diagonal in the same basis. So we can write

$$\rho_1 = \sum_x p_x |x\rangle\langle x| \qquad (3.91)$$

and

$$\rho_2 = \sum_x q_x |x\rangle\langle x|. \qquad (3.92)$$

Therefore,

$$
\begin{aligned}
D(\rho_1, \rho_2) &= \tfrac{1}{2}Tr|\sum_x (p_x - q_x)|x\rangle\langle x|| \\
&= \tfrac{1}{2}\sum_x |p_x - q_x| \\
&= D(p_x, q_x).
\end{aligned}
$$

Example 3.40: How close is $\rho_1 = 0.64|0\rangle\langle 0| + 0.36|1\rangle\langle 1|$ and $\rho_2 = 0.36|0\rangle\langle 0| + 0.64|1\rangle\langle 1|$? Show that they are not orthogonal.

Solution: Here $\rho_1 = \begin{pmatrix} 0.64 & 0 \\ 0 & 0.36 \end{pmatrix}$ and $\rho_2 = \begin{pmatrix} 0.36 & 0 \\ 0 & 0.64 \end{pmatrix}$. There-

fore, $\rho_1 - \rho_2 = \begin{pmatrix} 0.28 & 0 \\ 0 & -0.28 \end{pmatrix} = (\rho_1 - \rho_2)^\dagger$. Therefore, $D(\rho_1, \rho_2) = 0.28$.

Since $D(\rho_1, \rho_2) \neq 1$, the states are not orthogonal.

Alternatively, we may note that the diagonal matrices commute, so the given ρ_1 and ρ_2 commute and eigenvalues of these density matrices are their diagonal terms. Consequently we can directly use the classical formula of trace distance as $D(\rho_1, \rho_2) = D(p_x, q_x) = \frac{1}{2}\sum_x |p_x - q_x| = \frac{1}{2}(|0.64 - 0.36| + |0.36 - 0.64|) = 0.28$.

Example 3.41: Consider two orthogonal pure states $|0\rangle$ and $|1\rangle$ and compute $D(\rho_1, \rho_2)$.

Solution: Here the density matrices are $\rho_1 = |0\rangle\langle 0| = \begin{pmatrix} 1 & 0 \\ 0 & 0 \end{pmatrix}$ and $\rho_2 = |1\rangle\langle 1| = \begin{pmatrix} 0 & 0 \\ 0 & 1 \end{pmatrix}$. Therefore, $\rho_1 - \rho_2 = \begin{pmatrix} 1 & 0 \\ 0 & -1 \end{pmatrix}$ and $D(\rho_1, \rho_2) = \frac{1}{2}Tr\left| \begin{pmatrix} 1 & 0 \\ 0 & -1 \end{pmatrix} \right| = \frac{1}{2}Tr\begin{pmatrix} 1 & 0 \\ 0 & 1 \end{pmatrix} = 1$ as expected from the property 2 of $D(\rho_1, \rho_2)$.

Quantum fidelity

We have discussed trace distance in detail; now it would be apt to introduce quantum fidelity. Quantum fidelity of the states ρ_1 and ρ_2 is defined as[10]

$$F(\rho_1, \rho_2) = Tr\sqrt{(\rho_1)^{\frac{1}{2}}\rho_2(\rho_1)^{\frac{1}{2}}}. \tag{3.93}$$

There exist a few special cases where we may obtain more useful and simpler expressions of fidelity. To begin with let us consider that one of the states is a pure state $|\psi\rangle$ (say $\rho_1 = |\psi\rangle\langle\psi|$) and the other state is an arbitrary state ρ_2. Now for a pure state $(\rho_1)^{\frac{1}{2}} = \rho_1 = |\psi\rangle\langle\psi|$. Therefore,

$$\begin{aligned} F(|\psi\rangle, \rho_2) &= Tr\sqrt{(\rho_1)^{\frac{1}{2}}\rho_2(\rho_1)^{\frac{1}{2}}} = Tr\sqrt{|\psi\rangle\langle\psi|\rho_2|\psi\rangle\langle\psi|} \\ &= \sqrt{\langle\psi|\rho_2|\psi\rangle}Tr\sqrt{\langle\psi|\psi\rangle} = \sqrt{\langle\psi|\rho_2|\psi\rangle}. \end{aligned} \tag{3.94}$$

Thus the fidelity is the square root of the overlap between $|\psi\rangle$ and ρ_2. Now we may assume that the second state is also a pure state. In that case we may consider $\rho_2 = |\phi\rangle\langle\phi|$ and (3.93) further simplifies to

$$F(|\psi\rangle, |\phi\rangle) = \sqrt{\langle\psi|\rho_2|\psi\rangle} = \sqrt{\langle\psi|\phi\rangle\langle\phi|\psi\rangle} = |\langle\psi|\phi\rangle|. \tag{3.95}$$

This simple result is very useful in the context of teleportation and quantum cryptography.

Now we consider another case, where ρ_1 and ρ_2 commute. In such a case they are diagonal in the same basis and we can express them as we

---

[10]In some literature $F(\rho_1, \rho_2) = \left(Tr\sqrt{(\rho_1)^{\frac{1}{2}}\rho_2(\rho_1)^{\frac{1}{2}}}\right)^2$ is used as the definition of fidelity. Both the definitions are correct and both the definitions yield the same physical conclusions.

have done in (3.91) and (3.92). In this special case

$$F(\rho_1,\rho_2) = Tr\left(\sum_x \sqrt{p_x q_x}|x\rangle\langle x|\right) = \sum_x \sqrt{p_x q_x} = F(p_x, q_x).$$

Thus if the states commute then the quantum fidelity $F(\rho_1,\rho_2)$ reduces to the classical fidelity $F(p_x, q_x)$ between the eigenvalues of ρ_1 and ρ_2. This is analogous to the trace distance as we have already observed that if the states commute then the (quantum) trace distance between ρ_1 and ρ_2 is equal to the classical trace distance between the eigenvalues of ρ_1 and ρ_2.

Bures distance

The Bures distance function is another measure of distance between quantum states. This makes use of fidelity and is defined as $d_B(\rho_1,\rho_2) = \sqrt{2(1 - F(\rho_1,\rho_2))}$. As $0 \le F(\rho_1,\rho_2) \le 1$ so $0 \le d_B(\rho_1,\rho_2) \le \sqrt{2}$.

Example 3.42: Obtain fidelity and Bures distance between $|+\rangle$ and $|0\rangle$.

Solution: Here both the states are pure. So $F(\rho_1,\rho_2) = |\langle+|0\rangle| = \frac{1}{\sqrt{2}}$ and consequently $d_B(\rho_1,\rho_2) = \sqrt{2(1 - F(\rho_1,\rho_2))} = \sqrt{2 - \sqrt{2}} = 0.765$.

Example 3.43: Compute the fidelity between $\frac{4}{5}|0\rangle\langle0|+\frac{1}{5}|1\rangle\langle1|$ and $\frac{3}{5}|0\rangle\langle0|+\frac{2}{5}|1\rangle\langle1|$.

Solution: Here $\rho_1 = \frac{4}{5}|0\rangle\langle0| + \frac{1}{5}|1\rangle\langle1| = \begin{pmatrix} \frac{4}{5} & 0 \\ 0 & \frac{1}{5} \end{pmatrix}$ and $\rho_2 = \frac{3}{5}|0\rangle\langle0| + \frac{2}{5}|1\rangle\langle1| = \begin{pmatrix} \frac{3}{5} & 0 \\ 0 & \frac{2}{5} \end{pmatrix}$. First we check that $\rho_1^2 = \begin{pmatrix} \frac{16}{25} & 0 \\ 0 & \frac{1}{25} \end{pmatrix}$ and $\rho_2^2 = \begin{pmatrix} \frac{9}{25} & 0 \\ 0 & \frac{4}{25} \end{pmatrix}$. Hence none of the states are pure state and we need to use general formula for the calculation of fidelity. Now[11]

$$(\rho_1)^{\frac{1}{2}} \rho_2 (\rho_1)^{\frac{1}{2}} = \begin{pmatrix} \frac{2}{\sqrt{5}} & 0 \\ 0 & \frac{1}{\sqrt{5}} \end{pmatrix} \begin{pmatrix} \frac{3}{5} & 0 \\ 0 & \frac{2}{5} \end{pmatrix} \begin{pmatrix} \frac{2}{\sqrt{5}} & 0 \\ 0 & \frac{1}{\sqrt{5}} \end{pmatrix}$$
$$= \begin{pmatrix} \frac{12}{25} & 0 \\ 0 & \frac{2}{25} \end{pmatrix}.$$

Therefore, $F(\rho_1,\rho_2) = Tr\sqrt{(\rho_1)^{\frac{1}{2}} \rho_2 (\rho_1)^{\frac{1}{2}}} = Tr\begin{pmatrix} \frac{2\sqrt{3}}{5} & 0 \\ 0 & \frac{\sqrt{2}}{5} \end{pmatrix} = 0.976$.

Since the diagonal matrices commute, we can also use the classical formula and directly obtain $F(\rho_1,\rho_2) = F(p_x,q_x) = \sum_x \sqrt{p_x q_x} = \sqrt{\frac{4}{5} \times \frac{3}{5}} + \sqrt{\frac{1}{5} \times \frac{2}{5}} = \frac{2\sqrt{3}+\sqrt{2}}{5} = 0.976$.

---

[11]It is easy to find out the square root of a diagonal matrix. However, it is not an easy task to find out $\rho^{\frac{1}{2}}$ for an arbitrary density matrix. We'll frequently need to compute $\rho^{\frac{1}{2}}$ and it would be easier for the readers either to write a small program to compute the square root of a matrix or to use Mathematica command MatrixPower$[\rho, \frac{1}{2}]$.

3.3.7   Measures of entanglement

There exist different quantitative measures of amount of entanglement. For example, Schimdt measure, entanglement of formation, concurrence, negativity, etc. are commonly used as measures of entanglement. Here we briefly introduce these measures with a set of examples. The examples will clarify the computational methods that are usually followed to quantify the amount of entanglement.

3.3.7.1   Schmidt measure

Schmidt number can be used as a coarse measure of amount of entanglement. Such a measure is referred to as Schmidt measure (also called Hartley strength) and is defined as

$$E_S\left(|\psi\rangle\right) = \log_2\left(\text{Schmidt number of } |\psi\rangle\right).$$

This provides the amount of entanglement in a unit called e-bits (entangled bits) in analogy to qubits. In the previous subsection we have already seen that the Schmidt number of Bell states is 2 and consequently, $E_S = 1$ e-bit for them. Thus the amount of entanglement present in a Bell state is the unit of entanglement in this measure.

3.3.7.2   Von Neumann entropy of the subsystem

We have already learned about Shannon entropy as a quantitative measure of classical information or of ignorance. Similarly, we can quantify our ignorance about a quantum system by the Von Neumann entropy, which is defined as

$$E = S(\rho) = -Tr(\rho \log_2 \rho). \tag{3.96}$$

Alternatively, it can be defined as

$$E = S(\rho) = -\sum_i \lambda_i \log_2 \lambda_i, \tag{3.97}$$

where λ_i are the eigenvalues of ρ. Von Neumann entropy $S(\rho)$ is a quantum analog of classical Shannon entropy and this analogy can be extended to Shannon's first (noiseless) coding theorem (see Subsection 1.1.3) and a bound on quantum data compression can be obtained as: The best data rate R achievable is $S(\rho)$. This bound on quantum data compression was obtained by Schumacher, and consequently, it is known as Schumacher's quantum noiseless coding theorem. $S(\rho)$ is actually the uncertainty of a quantum state before measurement. In the Hilbert space of dimension d, we have $0 \leq S(\rho) \leq \log_2 d$, where $S(\rho) = 0$ for pure state and $S(\rho) = \log_2 d$ for a completely mixed state. $S(\rho)$ also satisfies the following interesting property:

$$S\left(\rho^{AB}\right) \leq S\left(\rho^A\right) + S\left(\rho^B\right)$$

with
$$S\left(\rho^A \otimes \rho^B\right) = S\left(\rho^A\right) + S\left(\rho^B\right).$$

Further, $E = S(\rho^A) = S(\rho^B)$ provides a quantitative measure of entanglement. Let us clarify this point through a specific example.

John Von Neumann was one of the greatest scientists of the last century. He was born on 28th December 1903 in Budapest, Hungary. At the age of 22, he received his Ph.D. in mathematics from Pázmány Péter University in Budapest, but his father did not consider a career in mathematics financially lucrative, so he insisted that Von Neumann simultaneously obtain a diploma in Chemical engineering from ETH Zurich, Switzerland. In 1930, Von Neumann immigrated to the United States. Subsequently he joined the Institute for Advanced Study, Princeton University as a professor of mathematics in 1933. He continued working at Princeton until his death on February 8, 1957. He died under military security because the security agencies were afraid that he might disclose some military secrets while heavily medicated. His contribution to the foundation of quantum mechanics, construction of the atom bomb and hydrogen bomb, construction of the digital computer, design of different algorithms, and the development of the theory of cellular automata, the Monte Carlo method of simulation and game theory are enormous.

Once he heard a lecture of Heisenberg on matrix mechanics. He was not completely satisfied with the matrix treatment of quantum mechanics used by Heisenberg and he decided to develop his own version of quantum mechanics. This led to the discovery of operator theory. His work provided a strong mathematical foundation to quantum mechanics and also established the equivalence of the Schrodinger picture and Heisenberg picture of quantum mechanics. The mathematical tools for quantum mechanics (i.e., the algebra of linear Hermitian operators on Hilbert spaces) used in the present book in particular and in quantum computing in general were mostly developed by him.

Photo credit: Photograph by Alan W. Richards, courtesy AIP Emilio Segre Visual Archives. With permission.

Example 3.44: Consider an entangled state of the form

$$|\psi(\theta)\rangle_{AB} = \cos\theta|00\rangle_{AB} + \sin\theta|11\rangle_{AB}. \tag{3.98}$$

The subsystem A can be exclusively described by

$$
\begin{aligned}
\rho^A &= Tr_B(\rho^{AB}) \\
&= Tr_B\left(\cos^2\theta|00\rangle\langle00| + \cos\theta\sin\theta|00\rangle\langle11|\right. \\
&\quad + \left.\cos\theta\sin\theta|11\rangle\langle00| + \sin^2\theta|11\rangle\langle11|\right) \\
&= \left(\cos^2\theta|0\rangle\langle0|Tr(|0\rangle\langle0|) + \cos\theta\sin\theta|0\rangle\langle1|Tr(|0\rangle\langle1|)\right. \\
&\quad + \left.\cos\theta\sin\theta|1\rangle\langle0|Tr(|1\rangle\langle0|) + \sin^2\theta|1\rangle\langle1|Tr(|1\rangle\langle1|)\right) \\
&= \left(\cos^2\theta|0\rangle\langle0|\langle0|0\rangle + \cos\theta\sin\theta|0\rangle\langle1|\langle1|0\rangle\right. \\
&\quad + \left.\cos\theta\sin\theta|1\rangle\langle0|\langle0|1\rangle + \sin^2\theta|1\rangle\langle1|\langle1|1\rangle\right) \\
&= \left(\cos^2\theta|0\rangle\langle0| + \sin^2\theta|1\rangle\langle1|\right).
\end{aligned}
\tag{3.99}
$$

It would be easy to calculate the Von Neumann entropy $E(\theta)$ for this system if we work in matrix notation. In the matrix form ρ^A can be written as

$$
\rho^A = \begin{pmatrix} \cos^2\theta & 0 \\ 0 & \sin^2\theta \end{pmatrix}. \tag{3.100}
$$

Therefore,

$$
\begin{aligned}
E(\theta) &= -Tr(\rho^A \log_2 \rho^A) \\
&= -Tr\left(\begin{pmatrix} \cos^2\theta & 0 \\ 0 & \sin^2\theta \end{pmatrix} \log_2 \begin{pmatrix} \cos^2\theta & 0 \\ 0 & \sin^2\theta \end{pmatrix}\right) \\
&= -Tr\left(\begin{pmatrix} \cos^2\theta & 0 \\ 0 & \sin^2\theta \end{pmatrix} \begin{pmatrix} \log_2\cos^2\theta & 0 \\ 0 & \log_2\sin^2\theta \end{pmatrix}\right) \\
&= -Tr\left(\begin{pmatrix} \cos^2\theta & 0 \\ 0 & \sin^2\theta \end{pmatrix} \begin{pmatrix} 2\log_2\cos\theta & 0 \\ 0 & 2\log_2\sin\theta \end{pmatrix}\right) \\
&= -Tr\begin{pmatrix} 2\cos^2\theta\log_2\cos\theta & 0 \\ 0 & 2\sin^2\theta\log_2\sin\theta \end{pmatrix} \\
&= -2\left(\cos^2\theta\log_2\cos\theta + \sin^2\theta\log_2\sin\theta\right).
\end{aligned}
\tag{3.101}
$$

It is straightforward to check that $\rho^A = \rho^B$ in this case and the eigenvalues of $\rho^A = \rho^B = \begin{pmatrix} \cos^2\theta & 0 \\ 0 & \sin^2\theta \end{pmatrix}$ are $\cos^2\theta$ and $\sin^2\theta$. We can substitute these eigenvalues in (3.97) to directly obtain (3.101). We can easily find out the minima and maxima of $E(\theta)$ and see how it measures entanglement. See

$$
\begin{aligned}
\frac{dE}{d\theta} &= -2\frac{d}{d\theta}\left(\cos^2\theta\log_2\cos\theta + \sin^2\theta\log_2\sin\theta\right) \\
&= -2\left(-2\cos\theta\sin\theta\log_2\cos\theta - \cos\theta\sin\theta\right. \\
&\quad + \left.2\sin\theta\cos\theta\log_2\sin\theta + \sin\theta\cos\theta\right) \\
&= -2\left(-2\cos\theta\sin\theta\log_2\cos\theta + 2\sin\theta\cos\theta\log_2\sin\theta\right) \\
&= -2\left(2\sin\theta\cos\theta(\log_2\sin\theta - \log_2\cos\theta)\right) \\
&= 2\sin2\theta\log_2\cot\theta.
\end{aligned}
$$

Now $\frac{dE}{d\theta}$ has a set of zeroes at $\sin 2\theta = 0$ or $\theta = \frac{n\pi}{2}$. At these points the entropy is $E(\frac{n\pi}{2}) = 0$. These points are minima of the entropy. Our original state (3.98) reduces to separable states at these points. For example, $|\psi(\theta)\rangle_{AB} = |00\rangle$ for $\theta = 0$ and $|\psi(\theta)\rangle_{AB} = |11\rangle$ for $\theta = \frac{\pi}{2}$. The other set of zeroes of $\frac{dE}{d\theta}$ correspond to $\log_2 \cot \theta = 0$ or $\cot \theta = 1$ or $\theta = \frac{\pi}{4} \pm 2n\pi$. At these points $E(\theta) = 1$, which corresponds to the maxima of Von Neumann entropy. A bipartite quantum state (i.e., a two particle system) for which the Von Neumann entropy is unity, is called a maximally entangled state. For example, if we consider $n = 0$, i.e., $\theta = \frac{\pi}{4}$ then the state (3.98) reduces to $|\psi\rangle_{AB} = \frac{|00\rangle+|11\rangle}{\sqrt{2}}$ which is a Bell state. One can follow the same procedure and show that all the Bell states are maximally entangled (i.e., Von Neumann entropy of the subsystem is maximum for them). Thus the Von Neumann entropy of the subsystem shows that the Bell state $|\psi^+\rangle$ is more entangled than a state $a|00\rangle_{AB} + b|11\rangle_{AB}$, where $|a| \neq \frac{1}{\sqrt{2}}$ and $|a|^2 + |b|^2 = 1$.

Now you can easily see that (3.98) is already in Schmidt decomposed form and if the state is entangled (i.e., if $\theta \neq 0$ and $\theta \neq \frac{n\pi}{2}$) then for all values of θ its Schmidt rank is 2 and entanglement content is 1 e-bit. For example, according to Schmidt measure, amount of entanglement in $\frac{1}{\sqrt{2}}|00\rangle + \frac{1}{\sqrt{2}}|11\rangle$ and in $0.8|00\rangle + 0.6|11\rangle$ is the same (1 e-bit). Thus Schmidt measure could not distinguish between maximally entangled state and a non-maximally entangled state. This is why we mentioned that Schmidt measure is a crude measure of entanglement. The fact that the Von Neumann entropy is a better measure will be clarified in the next example.

Example 3.45: Use (3.101) to compute the amount of entanglement in (a) $\frac{1}{\sqrt{2}}|00\rangle + \frac{1}{\sqrt{2}}|11\rangle$ and (b) $0.8|00\rangle + 0.6|11\rangle$.

Solution: From (3.101) amount of entanglement E in a state $|\psi(\theta)\rangle = \cos\theta|00\rangle + \sin\theta|11\rangle$ is $-2\left(\cos^2\theta \log_2\cos\theta + \sin^2\theta\log_2\sin\theta\right)$. Now in (a) $\cos(\theta) = \sin(\theta) = \frac{1}{\sqrt{2}}$, therefore $E = -2\left(\frac{1}{2}\log_2\left(\frac{1}{2}\right)^{\frac{1}{2}} + \frac{1}{2}\log_2\left(\frac{1}{2}\right)^{\frac{1}{2}}\right) = -2\log_2\left(\frac{1}{2}\right)^{\frac{1}{2}} = \log_2 2 = 1$. Similarly, in (b) we have $\cos(\theta) = 0.8$, $\sin(\theta) = 0.6$, therefore $E = -2\left(0.64\log_2 0.8 + 0.36\log_2 0.6\right) = 0.9427$.

3.3.7.3 Negativity

There is another popularly used measure of entanglement known as negativity, which is defined as

$$N(\rho) = \frac{1}{2}\left(\|\rho^{T_A}\|_1 - 1\right) = |\sum_i \lambda_i| = \frac{1}{2}\sum_j \left(|\lambda_j| - 1\right),$$

where i runs over the subset of negative eigenvalues of the density operator ρ^{T_A} and j runs over all the eigenvalues of ρ^{T_A}. The negativity vanishes if and only if all the eigenvalues of ρ^{T_A} are nonnegative. Thus a nonvanishing value of negativity indicates entanglement. However, there exists a class of entangled states which cannot be characterized by negativity.

Example 3.46: Use negativity to quantify the amount of entanglement in the following states: (a) $|\psi^+\rangle = \frac{|00\rangle+|11\rangle}{\sqrt{2}}$ and (b) $|\psi\rangle = \frac{1}{\sqrt{3}}(|00\rangle + |01\rangle + |10\rangle)$.

Solution: (a) In Section 3.3.4 we have already shown that ρ^{T_A} for $|\psi^+\rangle =$

$\frac{|00\rangle+|11\rangle}{\sqrt{2}}$ is $\frac{1}{2}\begin{pmatrix} 1 & 0 & 0 & 0 \\ 0 & 0 & 1 & 0 \\ 0 & 1 & 0 & 0 \\ 0 & 0 & 0 & 1 \end{pmatrix}$, whose eigenvalues are $\{-\frac{1}{2}, \frac{1}{2}, \frac{1}{2}, \frac{1}{2}\}$. There

is only one negative eigenvalue which is $-\frac{1}{2}$. Therefore, $N(\rho) = |\sum_i \lambda_i| = |-\frac{1}{2}| = \frac{1}{2}$.

Alternatively, since $\rho^{T_A} = \frac{1}{2}\begin{pmatrix} 1 & 0 & 0 & 0 \\ 0 & 0 & 1 & 0 \\ 0 & 1 & 0 & 0 \\ 0 & 0 & 0 & 1 \end{pmatrix} = (\rho^{T_A})^\dagger$, so $(\rho^{T_A})^\dagger \rho^{T_A} =$

$\frac{1}{4}\begin{pmatrix} 1 & 0 & 0 & 0 \\ 0 & 0 & 1 & 0 \\ 0 & 1 & 0 & 0 \\ 0 & 0 & 0 & 1 \end{pmatrix}\begin{pmatrix} 1 & 0 & 0 & 0 \\ 0 & 0 & 1 & 0 \\ 0 & 1 & 0 & 0 \\ 0 & 0 & 0 & 1 \end{pmatrix} = \frac{1}{4}\begin{pmatrix} 1 & 0 & 0 & 0 \\ 0 & 1 & 0 & 0 \\ 0 & 0 & 1 & 0 \\ 0 & 0 & 0 & 1 \end{pmatrix} = \frac{1}{4}I.$ There-

fore, $\sqrt{(\rho^{T_A})^\dagger \rho^{T_A}} = \frac{1}{2}\begin{pmatrix} 1 & 0 & 0 & 0 \\ 0 & 1 & 0 & 0 \\ 0 & 0 & 1 & 0 \\ 0 & 0 & 0 & 1 \end{pmatrix}$ and $\|\rho^{T_A}\|_1 = Tr\left(\sqrt{(\rho^{T_A})^\dagger \rho^{T_A}}\right) =$

2 and consequently, negativity $N(\rho) = \frac{1}{2}$.

(b) In Section 3.3.4 we have shown that ρ^{T_A} for $|\psi\rangle = \frac{1}{\sqrt{3}}(|00\rangle + |01\rangle + |10\rangle)$ is

$$\frac{1}{3}\begin{pmatrix} 1 & 1 & 1 & 1 \\ 1 & 1 & 0 & 0 \\ 1 & 0 & 1 & 0 \\ 1 & 0 & 0 & 0 \end{pmatrix}$$

whose eigenvalues are $\{\frac{1}{6}(3+\sqrt{5}), -\frac{1}{3}, \frac{1}{3}, \frac{1}{6}(3-\sqrt{5})\}$. There is only one negative eigenvalue which is $-\frac{1}{3}$. Therefore, $N(\rho) = |\sum_i \lambda_i| = |-\frac{1}{3}| = \frac{1}{3}$. As expected the negativity of this state is less than the negativity of the Bell state which is maximally entangled.

Example 3.47: Compute the negativity of quantum state $0.8|00\rangle + 0.6|11\rangle$ and use the obtained value to compare the amount of entanglement of this state with that of $|\psi^+\rangle = \frac{|00\rangle+|11\rangle}{\sqrt{2}}$.

Solution: The density matrix for the state $0.8|00\rangle + 0.6|11\rangle$ is

$$\rho = \begin{pmatrix} 0.64 & 0 & 0 & 0.48 \\ 0 & 0 & 0 & 0 \\ 0 & 0 & 0 & 0 \\ 0.48 & 0 & 0 & 0.36 \end{pmatrix}.$$

Therefore,

$$
\rho^{T_A} = \begin{pmatrix} 0.64 & 0 & 0 & 0 \\ 0 & 0 & 0.48 & 0 \\ 0 & 0.48 & 0 & 0 \\ 0 & 0 & 0 & 0.36 \end{pmatrix}
$$

whose eigenvalues are $\{0.64, -0.48, 0.48, 0.36\}$. Therefore, $N(\rho) = |-0.48| = 0.48$. This is less compared to the negativity (0.5) of the maximally entangled state $|\psi^+\rangle$. In other words, negativity shows that the amount of entanglement in $0.8|00\rangle + 0.6|11\rangle$ is less than that in $|\psi^+\rangle$.

3.3.7.4 Concurrence and entanglement of formation

In a bipartite system concurrence provides a nice quantitative measure of amount of entanglement. It is defined as

$$
C(|\psi\rangle) = |\langle\psi|\tilde{\psi}\rangle|, \tag{3.102}
$$

where $|\tilde{\psi}\rangle = Y \otimes Y|\psi\rangle^*$ and $|\psi\rangle^*$ is the complex conjugate of the state vector $|\psi\rangle$. Alternatively, concurrence can be computed from the density operator of the state ρ as

$$
C(\rho) = max\{0, \lambda_1 - \lambda_2 - \lambda_3 - \lambda_4\} \tag{3.103}
$$

where $\lambda_1 \geq \lambda_2 \geq \lambda_3 \geq \lambda_4$ are the eigenvalues of the matrix $R = \sqrt{\rho^{\frac{1}{2}}\tilde{\rho}\rho^{\frac{1}{2}}}$ and $\tilde{\rho} = Y \otimes Y\rho^*Y \otimes Y$. A nonzero concurrence is a signature of entanglement. For a separable state concurrence is zero and for a maximally entangled state it is 1. Now we may define entanglement of formation in terms of concurrence as follows:

$$
E(\rho) = H_2 \left(\frac{1 + \sqrt{1 - (C(\rho))^2}}{2} \right) \tag{3.104}
$$

where H_2 is the binary entropy function $H_2(x) = -x \log_2 x - (1-x) \log_2(1-x)$. For an entangled state, entanglement of formation $E(\rho)$ mathematically characterizes the amount of resources required to create that particular entangled state (i.e., ρ). Further, we may note that for a pure state entanglement of formation is the same as the Von Neumann entropy of the subsystems. Thus for a pure state $E(\rho) = -Tr(\rho^A \log_2 \rho^A) = -Tr(\rho^B \log_2 \rho^B)$.

Example 3.48: Find concurrence and entanglement of formation for the state $|\psi^+\rangle = \frac{|00\rangle + |11\rangle}{\sqrt{2}}$.

Solution: First we note that

$$
Y \otimes Y = \begin{pmatrix} 0 & -i \\ i & 0 \end{pmatrix} \otimes \begin{pmatrix} 0 & -i \\ i & 0 \end{pmatrix} = \begin{pmatrix} 0 & 0 & 0 & -1 \\ 0 & 0 & 1 & 0 \\ 0 & 1 & 0 & 0 \\ -1 & 0 & 0 & 0 \end{pmatrix}.
$$

Now we can compute the concurrence in two different ways.

Method 1: Note that $|\psi^+\rangle = \frac{1}{\sqrt{2}}\begin{pmatrix} 1 \\ 0 \\ 0 \\ 1 \end{pmatrix} = |\psi\rangle^*$ and consequently

$$|\tilde{\psi}\rangle = Y \otimes Y |\psi\rangle^* = \frac{1}{\sqrt{2}}\begin{pmatrix} 0 & 0 & 0 & -1 \\ 0 & 0 & 1 & 0 \\ 0 & 1 & 0 & 0 \\ -1 & 0 & 0 & 0 \end{pmatrix}\begin{pmatrix} 1 \\ 0 \\ 0 \\ 1 \end{pmatrix} = \frac{-1}{\sqrt{2}}\begin{pmatrix} 1 \\ 0 \\ 0 \\ 1 \end{pmatrix} \equiv |\psi^+\rangle.$$

In the last step we have ignored the global phase. Now concurrence is simply $C = |\langle\psi|\tilde{\psi}\rangle| = |\langle\psi^+|\psi^+\rangle| = 1$.

Method 2: We already know that the density matrix for $|\psi^+\rangle$ is $\rho = \frac{1}{2}\begin{pmatrix} 1 & 0 & 0 & 1 \\ 0 & 0 & 0 & 0 \\ 0 & 0 & 0 & 0 \\ 1 & 0 & 0 & 1 \end{pmatrix}$. As it is a pure state $\rho^{\frac{1}{2}} = \rho = \frac{1}{2}\begin{pmatrix} 1 & 0 & 0 & 1 \\ 0 & 0 & 0 & 0 \\ 0 & 0 & 0 & 0 \\ 1 & 0 & 0 & 1 \end{pmatrix}$.

Therefore,

$$
\begin{aligned}
\tilde{\rho} &= Y \otimes Y \rho^* Y \otimes Y \\
&= \frac{1}{2}\begin{pmatrix} 0 & 0 & 0 & -1 \\ 0 & 0 & 1 & 0 \\ 0 & 1 & 0 & 0 \\ -1 & 0 & 0 & 0 \end{pmatrix}\begin{pmatrix} 1 & 0 & 0 & 1 \\ 0 & 0 & 0 & 0 \\ 0 & 0 & 0 & 0 \\ 1 & 0 & 0 & 1 \end{pmatrix}\begin{pmatrix} 0 & 0 & 0 & -1 \\ 0 & 0 & 1 & 0 \\ 0 & 1 & 0 & 0 \\ -1 & 0 & 0 & 0 \end{pmatrix} \\
&= \frac{1}{2}\begin{pmatrix} 1 & 0 & 0 & 1 \\ 0 & 0 & 0 & 0 \\ 0 & 0 & 0 & 0 \\ 1 & 0 & 0 & 1 \end{pmatrix} = \rho,
\end{aligned}
$$

$$
\begin{aligned}
R^2 &= \rho^{\frac{1}{2}}\tilde{\rho}\rho^{\frac{1}{2}} \\
&= \frac{1}{8}\begin{pmatrix} 1 & 0 & 0 & 1 \\ 0 & 0 & 0 & 0 \\ 0 & 0 & 0 & 0 \\ 1 & 0 & 0 & 1 \end{pmatrix}\begin{pmatrix} 1 & 0 & 0 & 1 \\ 0 & 0 & 0 & 0 \\ 0 & 0 & 0 & 0 \\ 1 & 0 & 0 & 1 \end{pmatrix}\begin{pmatrix} 1 & 0 & 0 & 1 \\ 0 & 0 & 0 & 0 \\ 0 & 0 & 0 & 0 \\ 1 & 0 & 0 & 1 \end{pmatrix} \\
&= \frac{1}{2}\begin{pmatrix} 1 & 0 & 0 & 1 \\ 0 & 0 & 0 & 0 \\ 0 & 0 & 0 & 0 \\ 1 & 0 & 0 & 1 \end{pmatrix}
\end{aligned}
$$

and consequently,

$$R = \sqrt{\rho^{\frac{1}{2}}\tilde{\rho}\rho^{\frac{1}{2}}} = \frac{1}{2}\begin{pmatrix} 1 & 0 & 0 & 1 \\ 0 & 0 & 0 & 0 \\ 0 & 0 & 0 & 0 \\ 1 & 0 & 0 & 1 \end{pmatrix},$$

whose eigenvalues are $\{1, 0, 0, 0\}$. Therefore, $C(\rho) = max\{0, \lambda_1 - \lambda_2 - \lambda_3 - \lambda_4\} = max\{0, 1 - 0 - 0 - 0\} = max\{0, 1\} = 1$. Thus from both the methods we have obtained $C = 1$. Now we can compute entanglement of formation. As $x = \frac{1 + \sqrt{1 - (C(\rho))^2}}{2} = \frac{1}{2}$. Therefore

$$E(\rho) = H_2(x) = -\frac{1}{2} \log_2 \frac{1}{2} - \frac{1}{2} \log_2 \frac{1}{2} = 1.$$

We would also like to note that in this example, we could have quickly obtained R as in this particular case $R = \sqrt{\rho^{\frac{1}{2}} \tilde{\rho} \rho^{\frac{1}{2}}} = \sqrt{\rho^{\frac{1}{2}} \rho \rho^{\frac{1}{2}}} = \rho$.

We have already mentioned that for a pure state entanglement of formation $E(\rho)$ is the same as the Von Neumann entropy of the subsystem and in Subsection 3.3.7.2 we have already seen that $E = 1$ for $|\psi^+\rangle$. Here we have obtained the same result using concurrence. Actually entanglement of formation is a monotone of concurrence and their extremal values are the same. To be precise, for a separable state $C(\rho) = E(\rho) = 0$ and for a maximally entangled state $C(\rho) = E(\rho) = 1$.

Example 3.49: Find concurrence and entanglement of formation for the state $0.8|00\rangle + 0.6|11\rangle$.

Solution: Method 1: Here $|\psi\rangle = \begin{pmatrix} 0.8 \\ 0 \\ 0 \\ 0.6 \end{pmatrix} = |\psi\rangle^*$ and consequently

$$|\tilde{\psi}\rangle = Y \otimes Y |\psi\rangle^* = \begin{pmatrix} 0 & 0 & 0 & -1 \\ 0 & 0 & 1 & 0 \\ 0 & 1 & 0 & 0 \\ -1 & 0 & 0 & 0 \end{pmatrix} \begin{pmatrix} 0.8 \\ 0 \\ 0 \\ 0.6 \end{pmatrix} = \begin{pmatrix} 0.6 \\ 0 \\ 0 \\ 0.8 \end{pmatrix}.$$

In the last step we have ignored the global phase. Now concurrence is simply $C(|\psi\rangle) = C(\rho) = |\langle \psi | \tilde{\psi} \rangle| = 0.96$. Therefore, $x = \frac{1 + \sqrt{1 - (C(\rho))^2}}{2} = 0.64$ and $E(\rho) = H_2(x) = -0.64 \log_2 0.64 - 0.36 \log_2 0.36 = 0.9427$. As expected, the value of entanglement of formation is the same as the Von Neumann entropy of the subsystem obtained in Subsection 3.3.7.2.

Method 2: The density matrix for $|\psi\rangle = 0.8|00\rangle + 0.6|11\rangle$ is

$$\rho = \begin{pmatrix} 0.64 & 0 & 0 & 0.48 \\ 0 & 0 & 0 & 0 \\ 0 & 0 & 0 & 0 \\ 0.48 & 0 & 0 & 0.36 \end{pmatrix}.$$

Clearly here $\rho = \rho^*$ and as the given state is pure consequently $\rho^{\frac{1}{2}} = \rho =$

$$\begin{pmatrix} 0.64 & 0 & 0 & 0.48 \\ 0 & 0 & 0 & 0 \\ 0 & 0 & 0 & 0 \\ 0.48 & 0 & 0 & 0.36 \end{pmatrix}. \text{ Therefore,}$$

$$
\begin{aligned}
\tilde{\rho} &= Y \otimes Y \rho^* Y \otimes Y \\
&= \begin{pmatrix} 0 & 0 & 0 & -1 \\ 0 & 0 & 1 & 0 \\ 0 & 1 & 0 & 0 \\ -1 & 0 & 0 & 0 \end{pmatrix}
\begin{pmatrix} 0.64 & 0 & 0 & 0.48 \\ 0 & 0 & 0 & 0 \\ 0 & 0 & 0 & 0 \\ 0.48 & 0 & 0 & 0.36 \end{pmatrix}
\begin{pmatrix} 0 & 0 & 0 & -1 \\ 0 & 0 & 1 & 0 \\ 0 & 1 & 0 & 0 \\ -1 & 0 & 0 & 0 \end{pmatrix} \\
&= \begin{pmatrix} 0.36 & 0 & 0 & 0.48 \\ 0 & 0 & 0 & 0 \\ 0 & 0 & 0 & 0 \\ 0.48 & 0 & 0 & 0.64 \end{pmatrix},
\end{aligned}
$$

$$
\begin{aligned}
R^2 &= \rho^{\frac{1}{2}} \tilde{\rho} \rho^{\frac{1}{2}} \\
&= \begin{pmatrix} 0.64 & 0 & 0 & 0.48 \\ 0 & 0 & 0 & 0 \\ 0 & 0 & 0 & 0 \\ 0.48 & 0 & 0 & 0.36 \\ 0.64 & 0 & 0 & 0.48 \\ 0 & 0 & 0 & 0 \\ 0 & 0 & 0 & 0 \\ 0.48 & 0 & 0 & 0.36 \end{pmatrix}
\begin{pmatrix} 0.36 & 0 & 0 & 0.48 \\ 0 & 0 & 0 & 0 \\ 0 & 0 & 0 & 0 \\ 0.48 & 0 & 0 & 0.64 \end{pmatrix} \\
&= \begin{pmatrix} 0.589824 & 0. & 0. & 0.442368 \\ 0. & 0. & 0. & 0. \\ 0. & 0. & 0. & 0. \\ 0.442368 & 0. & 0. & 0.331776 \end{pmatrix}
\end{aligned}
$$

and consequently,

$$
R = \sqrt{\rho^{\frac{1}{2}} \tilde{\rho} \rho^{\frac{1}{2}}} = \begin{pmatrix} 0.6144 & 0. & 0. & 0.4608 \\ 0. & 0. & 0. & 0. \\ 0. & 0. & 0. & 0. \\ 0.4608 & 0. & 0. & 0.3456 \end{pmatrix}
$$

whose eigenvalues are $\{0.96, 0, 0, 0\}$. Therefore $C(\rho) = max\{0, \lambda_1 - \lambda_2 - \lambda_3 - \lambda_4\} = max\{0, 0.96 - 0 - 0 - 0\} = max\{0, 0.96\} = 0.96$.

In the last two cases we have shown method 2 to make you familiar with the general procedure for computation of concurrence. Although we have chosen pure states in our examples, concurrence is a measure of 2-qubit mixed state entanglement, and is thus more general than other measures, proposed only for pure states, like the entropy of entanglement. Method 2 will be more useful in computing concurrence of mixed states. Actually, in the case of pure states, we can compute concurrence very easily if we note that $Y \otimes Y$ is transformation which maps $|00\rangle \rightarrow -|11\rangle, |01\rangle \rightarrow |10\rangle, |10\rangle \rightarrow$

$|01\rangle, |11\rangle \rightarrow -|00\rangle$, then given a $|\psi\rangle$, you can directly write $|\tilde{\psi}\rangle$ and obtain $C(|\psi\rangle)$ in a single step.

Example 3.50: Compute the concurrence and entanglement of formation for the state $|\psi\rangle = \frac{1}{\sqrt{3}} (|00\rangle + |01\rangle + |10\rangle)$.

Solution: Here $|\psi\rangle = |\psi\rangle^*$ and $|\tilde{\psi}\rangle = Y \otimes Y |\psi\rangle^* = \frac{1}{\sqrt{3}} (|01\rangle + |10\rangle - |11\rangle)$.

Now concurrence is simply $C(|\psi\rangle) = |\langle\psi|\tilde{\psi}\rangle| = \frac{2}{3}$.

Therefore, $x = \frac{1+\sqrt{1-(C(\rho))^2}}{2} = 0.8727$ and consequently,

$$E(\rho) = H_2(x) = -0.8727 \log_2 0.8727 - 0.1273 \log_2 0.1273 = 0.55.$$

Before we end this section we would like to note that the Von Neumann entropy is also called entanglement entropy as it measures the entanglement between the observed subsystem and the rest of the system. Further, we would like to note that so far we have given examples of bipartite entanglements only. However, entanglement is not limited to bipartite cases; there are several examples of multipartite entanglement and their applications. A few popular examples of multipartite maximally entangled quantum states are listed below (the names of the states are written in the subscripts):

$$
\begin{aligned}
|\psi\rangle_{GHZ} &= \frac{|000\rangle \pm |111\rangle}{\sqrt{2}}, \\
|\psi\rangle_{CAT} &= \frac{|00\cdots0\rangle \pm |11\cdots1\rangle}{\sqrt{2}}, \\
|\psi\rangle_{3\,\text{qubit}\,W} &= \frac{|100\rangle + |010\rangle + |001\rangle}{\sqrt{3}}, \\
|\psi\rangle_{4\,\text{qubit}\,W} &= \frac{|1000\rangle + |0100\rangle + |0010\rangle + |0001\rangle}{2}, \\
|\psi\rangle_{4\,\text{qubit Cluster}} &= \frac{|0000\rangle + |0101\rangle + |1010\rangle - |1111\rangle}{2}, \\
|\psi\rangle_{\text{Brown}} &= \frac{|000\rangle|\phi^-\rangle + |010\rangle|\psi^-\rangle + |100\rangle|\phi^+\rangle + |111\rangle|\psi^+\rangle}{2},
\end{aligned}
\tag{3.105}
$$

where $|\psi^\pm\rangle$ and $|\phi^\pm\rangle$ are the Bell states defined in Subsection 3.3.1.

3.3.8 State purification

Purification or state purification is a frequently used mathematical trick. This has no direct physical significance. This allows us to associate a pure state with a mixed state. Specifically, if an arbitrary mixed state ρ^A of a quantum system A is given then it is always possible to introduce a reference system R and define a pure state $|AR\rangle$ for the joint system AR such that $\rho^A = Tr_R(|AR\rangle\langle AR|)$. Thus the given arbitrary mixed state can be obtained by reduction from a suitably constructed fictitious larger composite quantum state. The procedure to construct $|AR\rangle$ is known as purification. Let us now explain how to construct R and $|AR\rangle$ for the given ρ^A. Since ρ^A is a valid density operator, it is Hermitian and using spectral decomposition technique (see 3.1.8) we can express it as $\rho^A = \sum_i p_i |i_A\rangle\langle i_A|$. Now to purify the given state ρ^A we introduce a system R which has the same state space as system A, with orthonormal basis states $|i_R\rangle$, and define a pure state for the combined system as

$$|AR\rangle = \sum_i \sqrt{p_i}|i_A\rangle|i_R\rangle. \tag{3.106}$$

Now we can easily see that

$$
\begin{aligned}
Tr_R\left(|AR\rangle\langle AR|\right) &= \sum_{ij} \sqrt{p_i p_j}|i_A\rangle\langle j_A|Tr\left(|i_R\rangle\langle j_R|\right)\\
&= \sum_{ij} \sqrt{p_i p_j}|i_A\rangle\langle j_A|\delta_{i.j}\\
&= \sum_i p_i|i_A\rangle\langle i_A|\\
&= \rho^A.
\end{aligned}
$$

Let us visualize the process with an example.

Example 3.51: Consider a mixed state $\rho^A = 0.64|0\rangle\langle 0| + 0.36|1\rangle\langle 1|$. By definition $|AR\rangle = \sqrt{0.64}|00\rangle + \sqrt{0.36}|11\rangle = 0.8|00\rangle + 0.6|11\rangle$. Therefore, $\rho_{AR} = 0.64|00\rangle\langle 00| + 0.48|00\rangle\langle 11| + 0.48|11\rangle\langle 00| + 0.36|11\rangle\langle 11|$. Consequently,

$$
\begin{aligned}
\rho^2_{AR} &= \left((0.64)^2 + (0.48)^2\right)|00\rangle\langle 00| + (0.48 \times 0.36 + 0.48 \times 0.64)|00\rangle\langle 11|\\
&+ (0.48 \times 0.64 + 0.48 \times 0.36)|11\rangle\langle 00| + \left((0.48)^2 + (0.36)^2\right)|11\rangle\langle 11|\\
&= 0.64|00\rangle\langle 00| + 0.48|00\rangle\langle 11| + 0.48|11\rangle\langle 00| + 0.36|11\rangle\langle 11|
\end{aligned}
$$

which implies $Tr(\rho^2_{AR}) = 0.64 + 0.36 = 1$. Thus ρ_{AR} is a pure state.

Now let us look at (3.72) and (3.106). We can easily recognize the symmetry between the Schmidt decomposition and purification. The procedure adopted here to purify a mixed state of system A is to define a pure state whose Schmidt basis for system A is just the basis in which the mixed state is diagonal, with the Schmidt coefficients being the square root of the eigenvalues of the density operator being purified.

3.3.9 Holevo bound

In Chapters 7 and 8 we will see that in dense coding and in all protocols of secure direct quantum communications Alice encodes her message into a set of mutually orthogonal states so that Bob can deterministically discriminate them. Further, in Chapter 6 (cf. Solved Example 7) we will learn that noise can transform a pure state into a mixed state. The same may be the effect of eavesdropping. Keeping these facts in mind we describe a special situation. Let us consider that Alice encodes her messages into a general mixed state[12] ρ_x with probabilities p_x, and then sends the state to Bob. Equivalently, we may think that Alice encodes her messages into pure states $|\psi_i\rangle$, but either Eve's attack or the noise present in the channel transforms them into mixed states. In both cases, Bob receives a mixed state ρ_x with probability p_x. Now Bob's task is to discriminate between mixed

---

[12]$\{\rho_x\}$ are arbitrary quantum states, they are not necessarily mutually orthogonal and pure.

states in general (a special case may be to discriminate between nonorthogonal pure states). To discriminate the state Bob has to do a measurement. We consider a general scenario where Bob performs a measurement on the state using the measurement operators $\{M_y\}$ such that $\sum_y M_y = 1$ and obtains an outcome Y. Now the question is how much information Bob can obtain from his measurement on the mixed state. Holevo bound provides an upper bound on the information accessible to Bob. To be precise, the Holevo bound states that the mutual information between the probability distribution of the states prepared by Alice (i.e., $p(x)$) and the probability distribution of the measurement outcomes of Bob (i.e., $p(y)$) is bounded by

$$I(X:Y) \leq S(\rho) - \sum_x p_x S(\rho_x), \qquad (3.107)$$

where $\rho = \sum_x p_x \rho_x$. Often Alice and Bob are referred to as originator and recipient, respectively. The right-hand side of the above equation is often called Holevo quantity and is usually denoted as χ. Thus $\chi = S(\rho) - \sum_x p_x S(\rho_x)$.

Now recall that in Subsection 1.1.2 we described a situation in which Alice prepares a random key X and sends it to Bob by some means. However, Bob receives Y, which is a noisy version of X. An unauthenticated user Eve, who tries to obtain the key, obtains Z, which is another noisy version of X. In this context we mentioned that in a realistic situation, Alice and Bob can establish a secret key if and only if $I(X:Y) > \min\{I(X:Z), I(Y:Z)\}$. Now if Alice sends the key using a mixed state then she can use Holevo quantity χ to compute the upper bounds on $I(X:Z)$ and $I(Y:Z)$. She may subsequently substitute them (i.e., the upper bounds on $I(X:Z)$ and $I(Y:Z)$) in the relation $I(X:Y) > \min\{I(X:Z), I(Y:Z)\}$ to obtain the tolerable noise limit of a protocol. Alice and Bob use part of the key to compute the noise and if they find the noise level is within the tolerable limit then in principle they can obtain a secure key. Further, we may add that Holevo quantity χ gives us an upper bound on the accessible classical information communicated via quantum states. We recover classicality when the states are orthogonal. Thus the computation of Holevo quantity χ plays a crucial role in the security and efficiency of the quantum communication protocols. The concept of Holevo bound will be clearer if we provide a simple example.

Example 3.52: Consider that Alice has a random number generator. If the random number generator yields 0 then she sends Bob $|0\rangle$ and if the random number generator yields 1 then she sends Bob $\cos\theta|0\rangle + \sin\theta|1\rangle$. The random number generator yields 0 and 1 with equal probability. Now compute χ.

Solution: Here

$$\rho_1 = |0\rangle\langle 0| = \begin{pmatrix} 1 & 0 \\ 0 & 0 \end{pmatrix}$$

which has eigenvalues $\{1, 0\}$. Therefore, $S(\rho_1) = -1\log_2 1 - 0\log_2 0 = 0$.

Similarly

$$\rho_2 = \begin{pmatrix} \cos\theta \\ \sin\theta \end{pmatrix} \begin{pmatrix} \cos\theta & \sin\theta \end{pmatrix} = \begin{pmatrix} \cos^2\theta & \sin\theta\cos\theta \\ \sin\theta\cos\theta & \sin^2\theta \end{pmatrix}$$

which also has eigenvalues $\{1,0\}$ and consequently $S(\rho_2) = 0$. Here $\chi = S(\rho)$ as in this case $S(\rho_1) = S(\rho_2) = 0$. Now

$$\rho = \sum_i p_i\rho_i = \frac{1}{2}(\rho_0 + \rho_1) = \begin{pmatrix} 1+\cos^2\theta & \sin\theta\cos\theta \\ \sin\theta\cos\theta & \sin^2\theta \end{pmatrix}$$

whose eigenvalues are $\lambda_1 = \frac{1+\cos\theta}{2}$ and $\lambda_2 = \frac{1-\cos\theta}{2} = 1 - \lambda_1$. Therefore,

$$\chi = -\frac{1+\cos\theta}{2}\log_2\frac{1+\cos\theta}{2} - \frac{1-\cos\theta}{2}\log_2\frac{1-\cos\theta}{2}.$$

Let us recall that the binary entropy $H(p) = -p\log_2 -(1-p)\log_2(1-p)$. Thus in this particular case the Holevo quantity $\chi = H\left(\frac{1+\cos\theta}{2}\right)$. Now we know that $H(p)$ is maximum when $p = \frac{1}{2}$ (see Fig. 1.1). Therefore χ is maximum when $\theta = \frac{\pi}{2}$. In that particular case Bob obtains 1 bit of information and he can discriminate the states prepared by Alice as the prepared states are orthogonal. In all other cases $\chi < 1$ bit and consequently Bob cannot discriminate the states prepared by Alice with certainty. Once again we have seen that nonorthogonal states cannot be discriminated with certainty.

Here it would be apt to note that how much information Bob obtains through his measurement depends on the choice of $\{M_y\}$. Consider that for a particular case Bob has constructed a POVM through which he obtains information equal to Holevo quantity. In that case there is no scope of further improvement, otherwise he may try to construct another POVM to obtain more information.

3.3.10   Bloch sphere

The Bloch sphere is a geometric representation of the pure state space of a two-level quantum mechanical system or in other words this unit sphere is a geometric representation of qubit states as points on its surface.

Let us start with a qubit in general. The state vector of the qubit is $|\psi\rangle = \alpha|0\rangle + \beta|1\rangle$ such that $|\alpha|^2 + |\beta|^2 = 1$. Since the coefficients α and β are complex in general, we can use polar decomposition technique to write $\alpha = r_1\exp(i\theta_1)$ and $\beta = r_2\exp(i\theta_2)$, where r_1 and r_2 are real and $|\alpha|^2 + |\beta|^2 = r_1^2 + r_2^2 = 1$. Thus the state vector of the qubit can be rewritten as

$$\begin{aligned} |\psi\rangle &= r_1\exp(i\theta_1)|0\rangle + r_2\exp(i\theta_2)|1\rangle \\ &= \exp(i\theta_1)\left(r_1|0\rangle + r_2\exp\left(i\left(\theta_2 - \theta_1\right)\right)|1\rangle\right) \\ &= r_1|0\rangle + r_2\exp\left(i\left(\theta_2 - \theta_1\right)\right)|1\rangle. \end{aligned}$$

In the last step we have neglected the global phase as it does not have any measurable effect. Now we may change the parameters and write $r_1 = \cos\left(\frac{\theta}{2}\right)$, $r_2 = \sin\left(\frac{\theta}{2}\right)$ and $\theta_2 - \theta_1 = \phi$. After this parametric transformation the state vector of the qubit becomes

$$|\psi\rangle = \cos\left(\frac{\theta}{2}\right)|0\rangle + \sin\left(\frac{\theta}{2}\right)\exp(i\phi)|1\rangle. \qquad (3.108)$$

The numbers θ and ϕ define a point on a unit three dimensional sphere (see Fig. 3.1a), known as the Bloch sphere. Thus the state vector of the qubit is depicted as a point on the surface of the Bloch sphere. Points on the surface of the Bloch sphere can also be expressed in Cartesian coordinates. From Fig. 3.1b we can easily see that the Cartesian coordinates of a point on the Bloch sphere can be expressed as

$$(v_1, v_2, v_3) = (\sin\theta\cos\phi, \sin\theta\sin\phi, \cos\theta). \qquad (3.109)$$

This means that the projections of the tip of the state vector on X, Y, Z axes are $v_1 = \sin\theta\cos\phi$, $v_2 = \sin\theta\sin\phi$, $v_3 = \cos\theta$ respectively. (v_1, v_2, v_3) represent a vector and the vector is known as the Bloch vector. In spherical polar coordinates we can describe an arbitrary qubit by using only two variables (θ and ϕ) because the state vector is constrained by the normalization condition (unit norm condition). We can see that for $\theta = 0, \phi = 0$, the state vector is $|0\rangle$. Thus the north pole of the sphere is the state $|0\rangle$ and the Bloch vector for this state is $(0, 0, 1)$. Now we can see that the point opposite to $|\psi\rangle = \cos\left(\frac{\theta}{2}\right)|0\rangle + \sin\left(\frac{\theta}{2}\right)\exp(i\phi)|1\rangle$ in the Bloch sphere is

$$\begin{aligned} |\chi\rangle &= \cos\left(\frac{\pi-\theta}{2}\right)|0\rangle + \sin\left(\frac{\pi-\theta}{2}\right)\exp(i(\phi+\pi))|1\rangle \\ &= \cos\left(\frac{\pi-\theta}{2}\right)|0\rangle - \sin\left(\frac{\pi-\theta}{2}\right)\exp(i\phi)|1\rangle. \end{aligned} \qquad (3.110)$$

Consequently,

$$\begin{aligned} \langle\psi|\chi\rangle &= \cos\left(\frac{\theta}{2}\right)\cos\left(\frac{\pi-\theta}{2}\right) - \sin\left(\frac{\theta}{2}\right)\sin\left(\frac{\pi-\theta}{2}\right) \\ &= \cos\left(\frac{\theta}{2} + \frac{\pi-\theta}{2}\right) \\ &= \cos\left(\frac{\pi}{2}\right) = 0. \end{aligned} \qquad (3.111)$$

Thus the opposite points in the Bloch sphere are orthogonal. This implies that the south pole of the Bloch sphere would be $|1\rangle$. Therefore, the Bloch vector of the state vector $|1\rangle$ is $(0, 0, -1)$.

Now we will show the relation between these Cartesian components and the density matrix. To do so, let us first write the density matrix for a pure one qubit state

$$|\psi\rangle = \alpha|0\rangle + \beta|1\rangle = \begin{pmatrix} \alpha \\ \beta \end{pmatrix}.$$

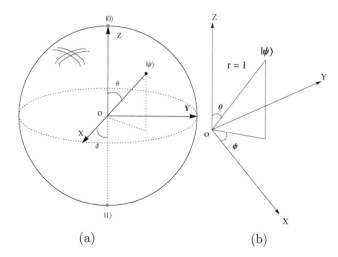

Figure 3.1: (a) Bloch sphere representation of a qubit and (b) spherical polar coordinates.

The density matrix for this state is

$$
\begin{aligned}
\rho &= |\psi\rangle\langle\psi| \\
&= \begin{pmatrix} \alpha \\ \beta \end{pmatrix} \begin{pmatrix} \alpha^* & \beta^* \end{pmatrix} \\
&= \begin{pmatrix} |\alpha|^2 & \alpha\beta^* \\ \beta\alpha^* & |\beta|^2 \end{pmatrix}.
\end{aligned}
\tag{3.112}
$$

Equation (3.108) allows us to consider $\alpha = \cos\left(\frac{\theta}{2}\right)$ and $\beta = \sin\left(\frac{\theta}{2}\right)\exp(i\phi)$. Under this consideration the single qubit density operator reduces to

$$
\begin{aligned}
\rho &= \begin{pmatrix} \cos^2\frac{\theta}{2} & \cos\frac{\theta}{2}\sin\frac{\theta}{2}\exp(-i\phi) \\ \cos\frac{\theta}{2}\sin\frac{\theta}{2}\exp(i\phi) & \sin^2\frac{\theta}{2} \end{pmatrix} \\
&= \frac{1}{2}\begin{pmatrix} 2\cos^2\frac{\theta}{2} & 2\cos\frac{\theta}{2}\sin\frac{\theta}{2}\exp(-i\phi) \\ 2\cos\frac{\theta}{2}\sin\frac{\theta}{2}\exp(i\phi) & 2\sin^2\frac{\theta}{2} \end{pmatrix} \\
&= \frac{1}{2}\begin{pmatrix} 1+\cos\theta & \sin\theta\exp(-i\phi) \\ \sin\theta\exp(i\phi) & 1-\cos\theta \end{pmatrix} \\
&= \frac{1}{2}\begin{pmatrix} 1+\cos\theta & \sin\theta\cos\phi - i\sin\theta\sin\phi \\ \sin\theta\cos\phi + i\sin\theta\sin\phi & 1-\cos\theta \end{pmatrix} \\
&= \frac{1}{2}\begin{pmatrix} 1+v_3 & v_1-iv_2 \\ v_1+iv_2 & 1-v_3 \end{pmatrix} \\
&= \frac{1}{2}\begin{pmatrix} 1 & 0 \\ 0 & 1 \end{pmatrix} + \frac{v_1}{2}\begin{pmatrix} 0 & 1 \\ 1 & 0 \end{pmatrix} + \frac{v_2}{2}\begin{pmatrix} 0 & -i \\ i & 0 \end{pmatrix} + \frac{v_3}{2}\begin{pmatrix} 1 & 0 \\ 0 & -1 \end{pmatrix}.
\end{aligned}
\tag{3.113}
$$

Now we can use (3.10-3.13) to obtain the density matrix of an arbitrary single qubit in terms of Pauli matrices as

$$\begin{aligned}
\rho &= \tfrac{1}{2}\left(I + v_1\sigma_1 + v_2\sigma_2 + v_3\sigma_3\right) \\
&= \frac{I + \vec{v}.\vec{\sigma}}{2}.
\end{aligned} \tag{3.114}$$

We can see that the $Tr(\rho) = 1$ as $Tr(\sigma_i) = 0$ and $Tr(I) = 2$. Let us also note that $Tr(\sigma_i\sigma_j) = 0$ for $i \neq j$ and $\sigma_i^2 = I$. Now we can use these simple relations to derive an analytic expression for $Tr(\rho^2)$ as

$$\begin{aligned}
Tr(\rho^2) &= Tr\left(\tfrac{1}{4}\left(I + v_1\sigma_1 + v_2\sigma_2 + v_3\sigma_3\right)\left(I + v_1\sigma_1 + v_2\sigma_2 + v_3\sigma_3\right)\right) \\
&= \tfrac{1}{4}Tr\left(I^2 + 2\left(v_1\sigma_1 + v_2\sigma_2 + v_3\sigma_3\right) + \left(v_1\sigma_1 + v_2\sigma_2 + v_3\sigma_3\right)^2\right) \\
&= \tfrac{1}{4}Tr\left(I + 2\left(v_1\sigma_1 + v_2\sigma_2 + v_3\sigma_3\right) + \left(v_1\sigma_1 + v_2\sigma_2 + v_3\sigma_3\right)^2\right) \\
&= \tfrac{1}{2} + 0 + \tfrac{1}{4}Tr\left(\left(v_1^2\sigma_1^2 + v_2^2\sigma_2^2 + v_3^2\sigma_3^2\right)\right) \\
&+ \ 2\left(v_1v_2\sigma_1\sigma_2 + v_1v_3\sigma_1\sigma_3 + v_2v_3\sigma_2\sigma_3\right)) \\
&= \frac{1 + v_1^2 + v_2^2 + v_3^2}{2} \\
&= \frac{1 + |\vec{v}|^2}{2}.
\end{aligned} \tag{3.115}$$

Now since the amplitude of the Bloch vector is unity for pure state, $Tr(\rho^2) = 1$. This is perfectly consistent with the definition of pure state and we may conclude that the points on the surface of the Bloch sphere depict pure states. Then what happens for the mixed states? We know that the mixed states are characterized by the condition $Tr(\rho^2) < 1$.

Thus using (3.115) we can obtain the condition for mixed state as $\frac{1 + |\vec{v}|^2}{2} < 1$ or $|\vec{v}|^2 < 1$. Thus a mixed state of single qubit would be represented by the points inside the Bloch sphere. In other words, the points inside the Bloch sphere represent mixed states and the points on the surface represent pure states. This can be clarified through a few examples. Before we give examples we would like to mention that if there exists a mixed state $\rho = \sum_i p_i |\psi_i\rangle\langle\psi_i|$ and if the Bloch vector for $|\psi_i\rangle$ is $(v_{x,i}, v_{y,i}, v_{z,i})$, then the Bloch vector for the mixed state would be $(\sum_i p_i v_{x,i}, \sum_i p_i v_{y,i}, \sum_i p_i v_{z,i})$.

Example 3.53: Consider the ensemble $\{(|0\rangle\langle0|, \tfrac{1}{2}), (|1\rangle\langle1|, \tfrac{1}{2})\}$. Since the Bloch vectors of $|0\rangle$ and $|1\rangle$ are $(0, 0, 1)$ and $(0, 0, -1)$, therefore the Bloch vector of the mixed state is $(\frac{0+0}{2}, \frac{0+0}{2}, \frac{1-1}{2}) = (0, 0, 0)$. Thus this ensemble corresponds to the center of the Bloch sphere.

Example 3.54: Consider the ensemble: $\{p_1 = 0.5, |\psi_1\rangle = 0.8|0\rangle + 0.6|1\rangle, p_2 = 0.5, |\psi_2\rangle = 0.8|0\rangle - 0.6|1\rangle\}$. In (3.47) we have shown that the density matrix for this state is

$$\rho = \begin{pmatrix} 0.64 & 0 \\ 0 & 0.36 \end{pmatrix} = \frac{1}{2}\left(\begin{pmatrix} 1 & 0 \\ 0 & 1 \end{pmatrix} + 0.28\begin{pmatrix} 1 & 0 \\ 0 & -1 \end{pmatrix}\right) = \frac{I + 0.28\sigma_z}{2}$$

and consequently, the Bloch vector for this mixed state is $(0, 0, 0.28)$. Therefore, it is represented by a point inside the Bloch sphere.

We have already shown in Subsection 3.2.3.2 that this density operator can represent more than one ensemble. Consequently, a point inside the Bloch sphere represents a set of states.

3.3.11   Nocloning theorem

There are several no-go theorems in quantum information theory. The most important among them is the nocloning theorem, which may be stated as follows.

Theorem: In a complex Hilbert space H, there does not exist a unitary transformation $U : H \otimes H \rightarrow H \otimes H$ such that there exists a state $|s\rangle \in H$ satisfying

$$U\left(|\psi\rangle|s\rangle\right) = |\psi\rangle|\psi\rangle, \quad \forall|\psi\rangle \in H.$$

Proof: This can be proved by reductio-ad-absurdum[13]. Assume that such a unitary operation U exists, then we can have

$$U\left(|\psi\rangle|s\rangle\right) = |\psi\rangle|\psi\rangle \tag{3.116}$$

and

$$U\left(|\phi\rangle|s\rangle\right) = |\phi\rangle|\phi\rangle. \tag{3.117}$$

Now combining (3.116), (3.117) and using the property of unitary operator we may obtain

$$\langle\psi|\phi\rangle = \langle\psi|\langle s|s\rangle|\phi\rangle = \langle\psi, s|U^\dagger U|s, \phi\rangle = \langle\psi, \psi|\phi, \phi\rangle = \left(\langle\psi|\phi\rangle\right)^2. \tag{3.118}$$

This equation can be satisfied only for $\langle\psi|\phi\rangle = 0$ or $\langle\psi|\phi\rangle = 1$. Consequently, our initial assumption is wrong and we cannot copy $\forall|\psi\rangle \in H$. Thus if we choose two nonorthogonal states ($|\psi\rangle$ and $|\phi\rangle$) in H then we obtain a contradiction. $\langle\psi|\phi\rangle = 0$ means the states are orthogonal, and $\langle\psi|\phi\rangle = 1$ means the states are the same. Thus a quantum state chosen from a given set can be perfectly cloned if and only if the distinct states of that set are mutually orthogonal. In that particular case the states are related to each other in a fashion which is similar to their classical counterparts. All the advantages of quantum states are lost in that case. Thus it is apt to say that cloning is possible only if the information being cloned is essentially classical.

---

[13] Reductio-ad-absurdum is an excellent trick. In this we assume something and show that the assumption leads to an absurd conclusion and from that we conclude that either the initial assumption was wrong or the opposite of the assumption is true. A nice example is to prove that the largest prime number is not finite. We assume that p is the largest prime number and then we multiply all the prime numbers up to p and add 1 to generate $N = (3 \times 5 \times 7 \times 11 \times \cdots \times p) + 1$. Now since none of the prime numbers till p is a factor of N, either N itself is a prime number or there exists a prime factor of N which is greater than p. In both the cases our initial assumption is wrong. We have elaborated this technique in detail because it is often used in quantum information theory.

The credit for formal introduction of nocloning theorem normally goes to Wootters and Zurek [16], but almost simultaneously and independently it was introduced by Dieks in 1982 [17]. Interestingly, the idea of nocloning theorem was present in the pioneering paper of Wiesner, which was written in 1970 [11]. Actually nocloning theorem in some form or other was known to many people before 1982, but its relevance was not probably clear to them. In this context it would be apt to quote a relevant comment of Peres [61], "... these things were well known to those who know things well."

3.3.11.1 Conclusions from nocloning theorem

1. Classical fan-out gate is not allowed in quantum computing because it essentially means copying operation. Here we would like to note that if you are working with a classical reversible circuit then you can use a CNOT gate to copy the input state present in the controlled bit. To be precise, $U_{\text{CNOT}}|x\rangle|0\rangle = |x\rangle|x\rangle$ for $x \in \{0,1\}$. It is easy to see that it does not work for an arbitrary superposition state as $U_{\text{CNOT}}(\alpha|0\rangle + \beta|1\rangle)|0\rangle = \alpha|00\rangle + \beta|11\rangle \neq (\alpha|0\rangle + \beta|1\rangle) \otimes (\alpha|0\rangle + \beta|1\rangle)$. This provides an important difference between classical reversible circuit and quantum circuit.

2. Classical fan-in operation is the inverse of fan-out operation. Since quantum computing is reversible, if fan-out is disallowed then fan-in has to be disallowed. Thus nocloning theorem disallows fan-in operations in quantum circuits.

3. Usually classical error correction is performed by majority voting technique. This technique requires copying operation. Keeping this in mind, initially people thought quantum error correction would not be possible because of the nocloning theorem. But later on intelligent schemes for quantum error correction were introduced. Such protocols will be described in Chapter 6.

4. Nocloning theorem is very helpful for quantum cryptography. This is so because if Alice sends a qubit to Bob prepared randomly from a set of states which are not mutually orthogonal, then Eve cannot copy the state of the qubit. Since she cannot copy the state of the qubit, she has to measure it to know the state of the qubit. The moment she does a measurement on the system, the system gets perturbed and following some specific protocols Alice and Bob can detect the existence of Eve. Thus nocloning theorem is essential for unconditionally secure quantum cryptography. This point will be more elaborately described in Chapter 8.

3.3.11.2 Other no-go theorems

A no-go theorem shows that a particular situation (task or result) is not physically possible even though it may appear as if it should be. Thus no-cloning theorem is an example of no-go theorem. The nocloning theorem is not the only no-go theorem in quantum information theory. A classic example of no-go theorem is Heisenberg's uncertainty relation. There are several no-go theorems and they are very important for secure quantum communication and quantum information processing. Here we briefly introduce a few of them.

1. **No-deletion theorem:** Given two copies of an unknown quantum state, it is not possible to delete one copy against the other by any physical operation. The concept of this no-go theorem can be illustrated as follows: Assume that initially we have two copies of an unknown quantum state $|\psi\rangle$ and an ancilla state $|A_i\rangle$, which is prepared in the same dimension as that of $|\psi\rangle$. Now the quantum deleting operation U is an operator that acts on the combined Hilbert space of the input states and ancilla and deletes one of the two copies of $|\psi\rangle$ and keeps the other intact. Thus,

$$U\left(|\psi\rangle|\psi\rangle|A_i\rangle\right) = |\psi\rangle|b\rangle|A_f\rangle, \qquad (3.119)$$

where $|b\rangle$ is a blank state in the same dimension and $|A_f\rangle$ is the final state of the ancilla, which is independent of $|\psi\rangle$. The no-deletion theorem states that there does not exist any quantum mechanically allowed (linear isomorphic) transformation U which satisfies (3.119). It is important to note that $|A_f\rangle$ is required to be independent of $|\psi\rangle$, otherwise the copy ($|\psi\rangle$) can be swapped into the final state of the ancilla. Swapping only hides the quantum information. It does not delete the state.

2. **No-broadcast theorem:** Broadcasting is a slightly weaker version of cloning. Before we state the theorem it is important to understand what we mean by broadcasting a state. To understand that, first assume that we have a state ρ in H and a map E that maps state ρ on H to state ρ^{AB} on $H_A \otimes H_B$. Now E broadcasts the state ρ if $Tr_A\left(\rho^{AB}\right) = \rho$ and $Tr_B\left(\rho^{AB}\right) = \rho$. Note that the final state is not necessarily a product state. The no-broadcast theorem states that no such map E exists, i.e., an arbitrary quantum state cannot be broadcasted. Alternatively, this theorem can be stated as: Given an arbitrary quantum state ρ it is impossible to create a state ρ_{AB} such that $Tr_A\left(\rho^{AB}\right) = Tr_B\left(\rho^{AB}\right) = \rho$.

3. **No-hiding theorem:** The theorem states that if a physical process leads to loss (bleaching) of quantum information from a quantum system, then the lost information must reside in the rest of the universe

with no information being hidden in the correlation between the system and its environment. The no-hiding theorem can be illustrated in a simpler way as follows: Consider that some information is lost from a quantum system due to the interaction of the system with its environment. Now the following questions come to our mind: What happened to the lost information? Is it really lost? Is it hiding somewhere else? If yes, where is it hiding? The no-hiding theorem answers these questions by stating two things: (1) All of the lost information has moved somewhere else in the universe and (2) The lost information cannot remain hidden in the correlations between the system and its environment. It is interesting to note that this theorem has recently been verified experimentally.

4. **No-teleportation theorem:** This theorem states that it is impossible to determine a quantum state by a single measurement. You can visualize the theorem easily if you recall that when we introduced the measurement postulate we clearly stated that a measurement provides one of the possible outcomes. The outcome of the measurement is a classical information. The classical information obtained in a single measurement does not provide us any information about the probabilities of other possible outcomes. Therefore, a quantum state cannot be reconstructed by using the classical information obtained in a single measurement. However, one can determine a quantum state by a large number of measurements, but that would require a large number of copies of the input quantum state. A corollary of this is that a classical information channel cannot transmit a quantum state with unit fidelity.

5. **No-communication theorem:** This theorem essentially states that superluminal (i.e., faster than the speed of light in vacuum) quantum communication is impossible. Thus it is a variant of the postulate of the special theory of relativity. This theorem also implies that the shared entanglement alone cannot be used to transmit any information. One can easily visualize the meaning of the last statement through a simple example. Consider that Alice and Bob share an entangled state $\frac{1}{\sqrt{2}} (|00\rangle + |11\rangle)$. The first qubit is with Alice and the second is with Bob. Alice does a measurement on her qubit and obtains $|0\rangle$. As a consequence, the combined state of Alice and Bob collapses to $|00\rangle$ and consequently any subsequent measurement of Bob will yield $|0\rangle$. For a while, it may appear that the information that Alice has done a measurement is instantaneously communicated to Bob, but unfortunately Bob has no way to distinguish between the possible input states that he measures. In other words, looking at his output ($|0\rangle$) Bob cannot conclude whether he performed a measurement on the second qubit of $|00\rangle$ or that of $\frac{1}{\sqrt{2}} (|00\rangle + |11\rangle)$.

This example shows that shared entanglement alone is not sufficient for quantum communication. For effective quantum communication we need to physically transmit a bit or qubit and that cannot be transmitted superluminally, and as a result superluminal quantum communication is impossible.

In the no-teleportation theorem we have seen that a classical information channel cannot transmit quantum information with unit fidelity, and now we have observed that shared entanglement alone cannot transmit any information. But it is interesting to note that shared entanglement and the classical information channel together can transmit quantum states with unit fidelity. A very exciting example of this process is known as perfect quantum teleportation. Quantum teleportation will be discussed in detail in Chapter 7.

6. **No-programming theorem:** This theorem states that it is impossible to build a programmable quantum computer having an architecture similar to Von Neumann (or Harvard) architecture. This is because distinct unitary operators $U_0, U_1, \cdots, U_n$ require orthogonal programs $|U_0\rangle, |U_1\rangle, \cdots, |U_n\rangle$ [62].

With this introductory knowledge of algebra and quantum mechanics we are now ready to do computation with quantum resources. To start with let us learn about quantum gates and quantum circuits in the next chapter.

3.4   Solved examples

1. Show that the set of all vectors of the form $\vec{V_i} = (1, y_i, z_i)$ do not form a vector space.
   Solution: If we add any two vectors of the given form then we obtain $\vec{V_i} + \vec{V_j} = (1+1, y_{i+j}, z_{i+j}) = (2, y_{i+j}, z_{i+j})$, which is not a vector of the form $(1, y_i, z_i)$. Consequently, closure property of vector space is not satisfied and we may conclude that the set of vectors of the form $\vec{V_i} = (1, y_i, z_i)$ do not form a vector space.

2. Measurement is made on each of the following qubits. What are the probabilities that the qubit is found in state $|0\rangle$ and $|1\rangle$?
   (a) $\frac{i}{\sqrt{2}}|0\rangle + \frac{1}{\sqrt{2}}|1\rangle$, (b) $\frac{1+i}{\sqrt{3}}|0\rangle + \frac{i}{\sqrt{3}}|1\rangle$, (c) $\frac{1}{\sqrt{5}}|0\rangle + \frac{2}{\sqrt{5}}|1\rangle$.
   Solution: (a) Probability of obtaining the state in $|0\rangle$ on measurement is $p(0) = |\frac{i}{\sqrt{2}}|^2 = \frac{1}{2}$ and probability of obtaining the state in $|1\rangle$ on measurement is $p(1) = |\frac{1}{\sqrt{2}}|^2 = \frac{1}{2}$. Similarly, (b) $p(0) = \frac{2}{3}$ and $p(1) = \frac{1}{3}$, (c) $p(0) = \frac{1}{5}$ and $p(1) = \frac{4}{5}$.

3. Assume that the second qubit of the state $\frac{1}{\sqrt{30}}(|00\rangle + 2i|01\rangle - 3|10\rangle - 4i|11\rangle)$ is measured and observed to be in state $|1\rangle$. What is the state of the system after the measurement? What is the probability that

a subsequent measurement of the first qubit will yield $|1\rangle$?

Solution: After the measurement the state reduces (collapses) to a superposition of $|01\rangle$ and $|11\rangle$ as these are the only possibilities where second qubit is in $|1\rangle$. Thus the normalized state after the measurement is $\dfrac{\frac{2i}{\sqrt{30}}}{\sqrt{\left(\frac{2}{\sqrt{30}}\right)^2+\left(\frac{4}{\sqrt{30}}\right)^2}}|01\rangle - \dfrac{\frac{4i}{\sqrt{30}}}{\sqrt{\left(\frac{2}{\sqrt{30}}\right)^2+\left(\frac{4}{\sqrt{30}}\right)^2}}|11\rangle = \frac{1}{\sqrt{5}}|01\rangle - \frac{2}{\sqrt{5}}|11\rangle$.

Note that we have dropped i from both the coefficients. This is allowed as the common factor i is nothing but a global phase $\frac{\pi}{2}$ (as $i = e^{i\frac{\pi}{2}}$). Now the probability of getting the first qubit at state $|1\rangle$ is $\left(\frac{2}{\sqrt{5}}\right)^2 = \frac{4}{5}$.

4. Find out whether the following states are entangled or not:

   (a) $\frac{1}{\sqrt{30}}\left(|00\rangle + 2i|01\rangle - 3|10\rangle - 4i|11\rangle\right)$

   (b) $\frac{1}{5}\left(|00\rangle + 2i|01\rangle - 2|10\rangle - 4i|11\rangle\right)$

   Solution: Consider the states as $a|00\rangle + b|01\rangle + c|10\rangle + d|11\rangle$ and check whether the separability condition $ad = bc$ is satisfied or not. In (a) $ad = -\frac{4i}{30} \neq bc = -\frac{6i}{30}$, so the state is not separable. This implies that the state is entangled. Now in (b) $ad = bc = -\frac{4i}{25}$, consequently, the state is separable. You can easily check that as follows,

   $$\frac{1}{5}\left(|00\rangle + 2i|01\rangle - 2|10\rangle - 4i|11\rangle\right)$$
   $$= \frac{1}{5}\left(|0\rangle(|0\rangle + 2i|1\rangle) - 2|1\rangle(|0\rangle + 2i|1\rangle)\right)$$
   $$= \frac{1}{\sqrt{5}}\left(|0\rangle - 2|1\rangle\right) \otimes \frac{1}{\sqrt{5}}\left(|0\rangle + 2i|1\rangle\right).$$

5. Which of the following state vectors are valid representations of a qubit?

   (a) $0.7|0\rangle + 0.3|1\rangle$

   (b) $\cos^2(x)|0\rangle - \sin^2(x)|1\rangle$

   (c) $0.8|0\rangle + 0.6|1\rangle$

   Solution: The condition for a state vector being valid representation of a qubit $|\psi\rangle = \alpha|0\rangle + \beta|1\rangle$ is $|\alpha|^2 + |\beta|^2 = 1$. Thus the state vector has to be normalized.

   (a) Now $0.7|0\rangle + 0.3|1\rangle$ is not a valid state since $(0.7)^2 + (0.3)^2 = 0.58 \neq 1$,

   (b) $\cos^2(x)|0\rangle - \sin^2(x)|1\rangle$ is a valid representation of a qubit when $\cos^4(x) + \sin^4(x) = 1$, i.e., when $x = 0, \pm\frac{\pi}{2}, \pm\pi$.

   (c) It is a valid representation of a qubit since $(0.8)^2 + (0.6)^2 = 1$.

6. In the last problem if you have found any state which is not a valid representation of a qubit then try to normalize the state and obtain a valid representation of qubit.

   Solution: In order to normalize an arbitrary one qubit state vector $|\psi\rangle = \alpha|0\rangle + \beta|1\rangle$ we need to divide it by its norm $\sqrt{\langle\psi|\psi\rangle} =$

$\sqrt{|\alpha|^2 + |\beta|^2}$. Now the last state at the previous problem is already normalized. The first state can be normalized as $|\psi\rangle = \frac{0.7}{\sqrt{0.58}}|0\rangle + \frac{0.3}{\sqrt{0.58}}|1\rangle$.

The second state vector was normalized for specific values of x. It can be normalized in general for all values of x as $|\psi\rangle = \frac{\cos^2(x)}{\sqrt{\cos^4(x)+\sin^4(x)}}|0\rangle - \frac{\sin^2(x)}{\sqrt{\cos^4(x)+\sin^4(x)}}|1\rangle$.

7. Can we rewrite $|\psi\rangle = \alpha|0\rangle + i\beta|1\rangle$ as $|\psi'\rangle = i\alpha|0\rangle - \beta|1\rangle$?
   Solution: Yes, we can. Since the states are different by a global phase, only so they are equivalent. See

   $$|\psi\rangle = \alpha|0\rangle + i\beta|1\rangle = -i\,(i\alpha|0\rangle - \beta|1\rangle) = e^{-\frac{\pi i}{2}}|\psi'\rangle \equiv |\psi'\rangle.$$

8. Express σ_z in matrix form in diagonal basis. Also provide the spectrum of σ_z.
   Solution: Matrix form of σ_z in the diagonal basis is

   $$\sigma_z = \begin{pmatrix} \langle+|\sigma_z|+\rangle & \langle+|\sigma_z|-\rangle \\ \langle-|\sigma_z|+\rangle & \langle-|\sigma_z|-\rangle \end{pmatrix} = \begin{pmatrix} \langle+|-\rangle & \langle+|+\rangle \\ \langle-|-\rangle & \langle-|+\rangle \end{pmatrix} = \begin{pmatrix} 0 & 1 \\ 1 & 0 \end{pmatrix}.$$

   Now σ_z is already diagonal in computational basis as in that basis $\sigma_z = \begin{pmatrix} 1 & 0 \\ 0 & -1 \end{pmatrix}$. Further, as $\sigma_z|0\rangle = |0\rangle$ and $\sigma_z|1\rangle = -|1\rangle$, therefore, eigenvalues of σ_z are $+1, -1$. This is the spectrum of σ_z. Now we can express it as $\sigma_z = |0\rangle\langle0| - |1\rangle\langle1|$.

9. Find out the tensor product of $\begin{pmatrix} 1 & 4 \\ 3 & 2 \end{pmatrix}$ and $\begin{pmatrix} 2 & 0 \\ 3 & -1 \end{pmatrix}$ and also find the trace of the product.
   Solution: The tensor product of the given matrices are

   $$\begin{pmatrix} 1 & 4 \\ 3 & 2 \end{pmatrix} \otimes \begin{pmatrix} 2 & 0 \\ 3 & -1 \end{pmatrix} = \begin{pmatrix} 2 & 0 & 8 & 0 \\ 3 & -1 & 12 & -4 \\ 6 & 0 & 4 & 0 \\ 9 & -3 & 6 & -2 \end{pmatrix}$$

   and the trace of the product matrix is $2 - 1 + 4 - 2 = 3$ (sum of the diagonal terms).

10. Suppose a unitary operator U is applied to a quantum state. Show that the density matrix evolves as $\rho' = U\rho U^\dagger$.
    Solution: We know that $\rho = \sum_i p_i|\psi_i\rangle\langle\psi_i|$ and U is applied to all $|\psi_i\rangle$. Thus each $|\psi_i\rangle$ is now mapped to $U|\psi_i\rangle$ and as an effect of this transformation the modified density matrix would be

    $$\rho' = \sum_i p_i U|\psi_i\rangle\langle\psi_i|U^\dagger = U\sum_i p_i|\psi_i\rangle\langle\psi_i|U^\dagger = U\rho U^\dagger.$$

11. Prove that $Tr(A \otimes B) = Tr(A)Tr(B)$.
    Solution:

$$Tr(A \otimes B) = \sum_{i=1}^{m}\sum_{j=1}^{n} a_{ii}b_{jj} = \sum_{i=1}^{m} a_{ii}\sum_{j=1}^{n} b_{jj} = Tr(A)Tr(B).$$

12. Show that the Schmidt measure for all separable states is zero. Is it logically expected?
    Solution: As the Schmidt number for all separable states is 1, corresponding Schmidt measure is $E_S = \log_2 1 = 0$. It is also logically expected as the state is separable so its entanglement content must be 0.

13. Show that the trace distance between two single qubit states is equal to half of the ordinary Euclidean distance between them on the Bloch sphere.
    Solution: If ρ_1 and ρ_2 are the quantum states and $\vec{v}_1$ and $\vec{v}_2$ are the corresponding Bloch vectors then using (3.114) we can write $\rho_1 = \frac{I + \vec{v}_1.\vec{\sigma}}{2}$ and $\rho_2 = \frac{I + \vec{v}_2.\vec{\sigma}}{2}$. Therefore, trace distance

$$
\begin{aligned}
D(\rho_1, \rho_2) &= \tfrac{1}{2}Tr|\rho_1 - \rho_2| \\
&= \tfrac{1}{4}Tr|(\vec{v}_1 - \vec{v}_2).\vec{\sigma}| \\
&= \tfrac{1}{2}|(\vec{v}_1 - \vec{v}_2)|,
\end{aligned}
$$

which is clearly half of the Euclidean distance ($|(\vec{v}_1 - \vec{v}_2)|$) between the states in the Bloch sphere. In the last step we have used the fact that $(\vec{v}_1 - \vec{v}_2).\vec{\sigma}$ has eigenvalues $\pm|(\vec{v}_1 - \vec{v}_2)|$, which implies $Tr|(\vec{v}_1 - \vec{v}_2).\vec{\sigma}| = 2|(\vec{v}_1 - \vec{v}_2)|$.

14. Show that if U and V are unitary, then $U \otimes V$ is also unitary.
    Solution: Since U and V are unitary, so $UU^\dagger = U^\dagger U = VV^\dagger = V^\dagger V = I$. Now

$$(U \otimes V)^\dagger (U \otimes V) = (V^\dagger \otimes U^\dagger)(U \otimes V) = I.$$

Similarly, we can show that $(U \otimes V)(U \otimes V)^\dagger = I$ and consequently $U \otimes V$ is unitary.

15. Show that $\{|+0\rangle, |+1\rangle, |-0\rangle, |-1\rangle\}$ forms a basis set in C^4 but $\{|+0\rangle, |+1\rangle, |0-\rangle, |1-\rangle\}$ does not form a basis set.
    Solution: Check that the four vectors of the first set are normalized and they are mutually orthogonal. Consequently, they form a basis set in C^4. However, in the second set all the vectors are not mutually orthogonal. For example, $\langle+0|0-\rangle = \langle+|0\rangle\langle0|-\rangle = \frac{1}{2} \neq 0$. Consequently, the second set of vectors do not form a basis set.

16. What happens to the quantum state $\rho = \begin{pmatrix} 0.8 & 0 \\ 0 & 0.2 \end{pmatrix}$ when a NOT gate is applied on it?

Solution: When a unitary operator U operates on ρ then it transforms to $\rho' = U\rho U^\dagger$. Here $U = U^\dagger = X$. So

$$\rho' = \begin{pmatrix} 0 & 1 \\ 1 & 0 \end{pmatrix}\begin{pmatrix} 0.8 & 0 \\ 0 & 0.2 \end{pmatrix}\begin{pmatrix} 0 & 1 \\ 1 & 0 \end{pmatrix} = \begin{pmatrix} 0.2 & 0 \\ 0 & 0.8 \end{pmatrix}.$$

17. Consider the states $|\psi\rangle = |0\rangle$ and $|\phi\rangle = \cos(\theta)|0\rangle + \sin(\theta)|1\rangle$. Now show that trace distance and fidelity between these two states are related as follows:

$$D(|\psi\rangle\langle\psi|, |\phi\rangle\langle\phi|) = \sqrt{1 - (F(|\psi\rangle, |\phi\rangle))^2}.$$

Solution: Since the given states are pure, fidelity

$$F(|\psi\rangle, |\phi\rangle) = |\langle\psi|\phi\rangle| = |\cos(\theta)|.$$

Now

$$
\begin{aligned}
\rho_1 - \rho_2 &= |\psi\rangle\langle\psi| - |\phi\rangle\langle\phi| \\
&= \begin{pmatrix} 1 & 0 \\ 0 & 0 \end{pmatrix} - \begin{pmatrix} \cos^2(\theta) & \cos(\theta)\sin(\theta) \\ \cos(\theta)\sin(\theta) & \sin^2(\theta) \end{pmatrix} \\
&= -\begin{pmatrix} -\sin^2(\theta) & \cos(\theta)\sin(\theta) \\ \cos(\theta)\sin(\theta) & \sin^2(\theta) \end{pmatrix}.
\end{aligned}
$$

Therefore $(\rho_1 - \rho_2) = (\rho_1 - \rho_2)^\dagger$ and consequently

$$
\begin{aligned}
&(\rho_1 - \rho_2)^\dagger(\rho_1 - \rho_2) \\
&= \begin{pmatrix} -\sin^2(\theta) & \cos(\theta)\sin(\theta) \\ \cos(\theta)\sin(\theta) & \sin^2(\theta) \end{pmatrix} \\
&\quad \begin{pmatrix} -\sin^2(\theta) & \cos(\theta)\sin(\theta) \\ \cos(\theta)\sin(\theta) & \sin^2(\theta) \end{pmatrix} \\
&= \begin{pmatrix} \sin^4(\theta) + \cos^2(\theta)\sin^2(\theta) & 0 \\ 0 & \sin^4(\theta) + \cos^2(\theta)\sin^2(\theta) \end{pmatrix} \\
&= \sin^2(\theta)\begin{pmatrix} 1 & 0 \\ 0 & 1 \end{pmatrix}
\end{aligned}
$$

and

$$\sqrt{(\rho_1 - \rho_2)^\dagger(\rho_1 - \rho_2)} = \sin(\theta)\begin{pmatrix} 1 & 0 \\ 0 & 1 \end{pmatrix}.$$

Therefore,

$$
\begin{aligned}
D\left(|\psi\rangle\langle\psi|,|\phi\rangle\langle\phi|\right) &= \tfrac{1}{2}Tr|\rho_1 - \rho_2| \\
&= \tfrac{1}{2}Tr|\sin(\theta)\begin{pmatrix} 1 & 0 \\ 0 & 1 \end{pmatrix}| \\
&= |\sin(\theta)| \\
&= |\sqrt{1 - \cos^2(\theta)}| \\
&= \sqrt{1 - \left(F\left(|\psi\rangle,|\phi\rangle\right)\right)^2}.
\end{aligned}
$$

18. Show that for a pure bipartite separable state $|\psi_{AB}\rangle = |\psi_A\rangle \otimes |\psi_B\rangle$ implies $\rho^{AB} = \rho^A \otimes \rho^B$.

    Solution: As $|\psi_{AB}\rangle$ is a pure and separable state, so the density matrix of the composite system is

    $$
    \rho^{AB} = |\psi_{AB}\rangle\langle\psi_{AB}| = |\psi_A\rangle|\psi_B\rangle\langle\psi_A|\langle\psi_B| = |\psi_A\rangle\langle\psi_A| \otimes |\psi_B\rangle\langle\psi_B|. \tag{3.120}
    $$

    Now taking partial trace of ρ^{AB} over system B and A respectively, we obtain

    $$
    \begin{aligned}
    \rho^A &= |\psi_A\rangle\langle\psi_A|, \\
    \rho^B &= |\psi_B\rangle\langle\psi_B|.
    \end{aligned} \tag{3.121}
    $$

    Combining (3.120) and (3.121) we obtain $\rho^{AB} = \rho^A \otimes \rho^B$.

3.5 Further reading

1. G. Chen and R. K. Brylinski (Ed.), Mathematics of quantum computation, Chapman and Hall, Boca Raton, Florida, United States (2002).

2. We have provided some solved examples and exercises in this textbook. Interested readers may find more problems in the book: W.-H. Steeb and Y. Hardy, Problems and solutions in quantum computing and quantum information, World Scientific, Singapore (2004).

3. A. K. Ekert and P. L. Knight, Entangled quantum systems and the Schmidt decomposition, Am. J. Phys. **63** (1995) 415-422.

4. An excellent history of origin and development of nocloning theorem may be found in A. Peres's article, "How the no-cloning theorem got its name", Fortschritte der Physik, **51** (2003) 458-461. The article can also be read for free from http://arxiv.org/pdf/quant-ph/0205076.

5. N. D. Mermin, From Cbits to Qbits: Teaching computer scientists quantum mechanics, Am. J. Phys. **71** (2003) 23-30.

6. J. H. Eberly, Bell inequalities and quantum mechanics, Am. J. Phys. **70** (2002) 276-279.

7. K. Jacobs and H. M. Wiseman, An entangled web of crime: Bell's theorem as a short story, Am. J. Phys. **73** (2005) 932-937.

8. Readers who are not familiar with quantum mechanics and wish to learn it in further detail may start with: L. I. Schiff, Quantum mechanics, McGraw-Hill, New York (1968).

3.6 Exercises

1. Compute the eigenvectors of the Pauli matrices, and find the points on the Bloch sphere which correspond to the normalized eigenvectors of the different Pauli matrices.

2. Prove the following properties of unitary operators:
   (a) Unitary operators preserve inner product.
   (b) Unitary operators preserve norm.
   (c) $Tr(U^\dagger AU) = Tr(A)$.

3. Find the inner product and tensor product of the qubits $|\psi\rangle_1 = 0.6|0\rangle + 0.8|1\rangle$ and $|\psi\rangle_2 = |0\rangle$.

4. Assume that the Schmidt rank of a composite state is 4. Is the state entangled? If yes, how much? What are the minimum dimensions of the subsystems of the composite Hilbert space?

5. Assume that you have a molecule which can be used as a 33-qubit quantum register. Can you tell me how many bits of information you can store on this molecule. Compare the storage capacity of this molecule with a standard hard disk and a blue ray disk. What do you conclude from your comparison?

6. Which of the following operators are (a) Hermitian, (b) unitary and (c) normal? Given $\langle i|j\rangle = \delta_{i,j}$, where $i, j \in \{0, 1, 2\}$.
   (i) $|0\rangle\langle 1|$, (ii) $|0\rangle\langle 1| + |1\rangle\langle 0|$, (iii) $|0\rangle\langle 2| + |1\rangle\langle 0| + |2\rangle\langle 1|$.

7. Find out the eigenvalues of the Pauli matrices [Hint: find roots of the equation $(\sigma_i - \lambda_i = 0)$].

8. Consider the normalized states

$$\begin{pmatrix} \alpha_1 \\ \alpha_2 \end{pmatrix}, \begin{pmatrix} \alpha_3 \\ \alpha_4 \end{pmatrix}.$$

Assume that α_i's are real and find the condition on α_i's such that

$$\begin{pmatrix} \alpha_1 \\ \alpha_3 \end{pmatrix} - \begin{pmatrix} \alpha_2 \\ \alpha_4 \end{pmatrix}$$

is also normalized. If the probability of finding the state $\begin{pmatrix} \alpha_3 \\ \alpha_4 \end{pmatrix}$
in $|0\rangle$ is 0.36 then what will be the probability of finding the state
$\begin{pmatrix} \alpha_1 \\ \alpha_3 \end{pmatrix} - \begin{pmatrix} \alpha_2 \\ \alpha_4 \end{pmatrix}$ in $|1\rangle$? Assume that we are working in $\{|0\rangle, |1\rangle\}$
basis.

9. Find the density matrices for the following states/ ensembles:
   (a) $\frac{4}{5}|0\rangle + \frac{3}{5}|1\rangle$, (b) $\{\frac{1}{2}, \frac{4}{5}|0\rangle + \frac{3}{5}|1\rangle, \frac{1}{4}, \frac{4}{5}|0\rangle - \frac{3}{5}|1\rangle, \frac{1}{4}, \frac{3}{5}|0\rangle - \frac{4}{5}|1\rangle\}$.

10. We know that the basis states used in computational basis are separable, and in contrary to that the basis states used in Bell basis are maximally entangled. Can you provide two examples of basis sets in C^4 where 50% of the elements of the basis set are separable and the rest are maximally entangled?

11. Show that for any density operator ρ, $Tr\left(\rho^2\right) \leq 1$ with $Tr\left(\rho^2\right) = 1$ iff the state is pure.

12. Compute trace distance $D(\rho_1, \rho_2)$ between $|0\rangle$ and $|+\rangle$ states.

13. Given a quantum system $\rho = \frac{1}{3}\begin{pmatrix} 2 & 1 \\ 1 & 1 \end{pmatrix}$, determine the Von Neumann entropy. Assume that the qubits of an arbitrary long length quantum message are generated by this density operator. Find out the maximum possible compression factor for the quantum message.

14. Explicitly describe the measurement operators in
$$\left\{ \frac{|01\rangle + |10\rangle}{\sqrt{2}}, \frac{|01\rangle - |10\rangle}{\sqrt{2}}, |00\rangle, |11\rangle \right\}$$
basis. Also express them in matrix form.

15. Show that the state $\frac{1}{\sqrt{3}}(|00\rangle + |01\rangle + |11\rangle)$ is entangled. Use Von Neumann entropy of the subsystem to provide a quantitative measure of the amount of entanglement.

16. If $|u\rangle$ is an eigen state of a square matrix A with eigen value λ then show that $Det(A - \lambda I) = 0$, where I is the identity matrix of the same dimension as that of A.

17. Compute the fidelity and Bures distance between $\frac{4}{7}|0\rangle\langle 0| + \frac{3}{7}|1\rangle\langle 1|$ and $\frac{2}{9}|0\rangle\langle 0| + \frac{7}{9}|1\rangle\langle 1|$.

18. Show that X and Z do not commute.

19. Show that the reduced density operator $\rho^A \equiv Tr_B(\rho^{AB})$ is Hermitian, nonnegative and it has unit trace.

20. Show that the singlet state $\left(\frac{|01\rangle - |10\rangle}{\sqrt{2}}\right)$ has nonpositive partial transposition. Is it expected?

21. Check whether the following set of three vectors is linearly independent or not: $|e_1\rangle = (1 - i)\hat{i} + 2\hat{j} + i\hat{k}$, $|e_2\rangle = (2i)\hat{i} + 3\hat{j} + 2\hat{k}$ and $|e_3\rangle = (1 - i)\hat{i} + \hat{j} + (1 + i)\hat{k}$. If they are found to be linearly independent then use Gram-Schmidt procedure to obtain an orthonormalized basis set using these set of vectors.

22. Compute the concurrence and entanglement of formation for the Werner state $\rho = \frac{5}{6}|\psi^+\rangle\langle\psi^+| + \frac{1}{24}I_4$.

23. Use the following techniques/criteria to show that $|\phi^-\rangle$ is entangled:
    (a) Criterion described in Equation (3.57),
    (b) Von Neumann entropy of the subsystem,
    (c) Schmidt measure,
    (d) PPT criterion of Peres-Horodecki
    (e) Negativity,
    (f) Concurrence,
    (g) Entanglement of formation.

24. Find fidelity and trace distance between $\rho_1 = \frac{3}{5}|0\rangle\langle0| + \frac{2}{5}|1\rangle\langle1|$ and $\rho_2 = \frac{5}{7}|0\rangle\langle0| + \frac{2}{7}|1\rangle\langle1|$. Also obtain Bures distance.

25. Check whether the following states are valid or not: $\rho_1 = \frac{1}{5}|0\rangle\langle0| + \frac{1}{5}|0\rangle\langle1| + \frac{1}{5}|1\rangle\langle0| + \frac{2}{5}|1\rangle\langle1|$ and $\rho_2 = \frac{1}{4}|0\rangle\langle0| + \frac{1}{4}|0\rangle\langle1| + \frac{1}{4}|1\rangle\langle0| + \frac{1}{4}|1\rangle\langle1|$. If the above are valid density matrices, then find the fidelity and trace distance between them. Also obtain Bures distance.

Chapter 4

Quantum gates and quantum circuits

Except for the NOT gate, all the other familiar classical gates are irreversible in the sense that we cannot uniquely reconstruct the input states from the output states. For example, OR, AND, NAND, NOR, etc. are irreversible. They all map a two-bit input state into a single bit output state. Thus one bit is erased during operation of each of these gates, and according to Landauer's principle that requires dissipation of a minimum amount of energy. Now a simple question arises in our curious mind: Is it possible to avoid this loss of energy? The answer is yes. If we don't erase any bit then we can circumvent this energy loss. Thus we need to map n bit input states to n bit output states. In addition, if a one-to-one correspondence exists between the input states and the output states then only we will be able to uniquely reconstruct the input states from the output states. In such a case the gate is called reversible. For example, if we have $f(00) = 00, f(01) = 10, f(10) = 01, f(11) = 11$ then f represents a reversible gate. Quantum evolution operators are unitary so for every operator U we have an inverse operator $U^{-1} = U^\dagger$. Therefore, quantum evolution operators are the natural choice for the construction of energy efficient reversible gates. However, it is not the only choice. We can have classical reversible gates, too. Usually by reversible gates we refer to classical reversible gates, and quantum gates are specifically referred to as quantum gates. Reversible gates and quantum gates are similar but there exists a fundamental difference that reversible gates cannot accept superposition states (e.g. $\alpha|0\rangle + \beta|1\rangle$) as input states, whereas quantum gates can. Thus all quantum gates are essentially reversible but the converse is not true.

In brief, simple unitary operations on qubits are called quantum logic gates. If a gate acts on a single qubit then it is called a single qubit gate.

Similarly, we can define a two qubit gate, three qubit gate and so on. We know that gates are combined together to form circuits, so in this chapter we will first describe single qubit gates, two qubit gates and three qubit gates. Then we will provide a few examples of quantum circuits and briefly describe the quantitative measures of the quality of the quantum circuits. We will also describe a few simple tricks that are usually used to improve the quality of quantum circuits. This chapter is focused on quantum gates and quantum circuits, but the techniques described here are also valid for reversible gates and reversible circuits.

4.1   Single qubit gates

A general structure of single qubit gates is shown in Fig. 4.1. Here the single qubit gate is a unitary operator which transforms a single qubit state $|\psi\rangle_{\text{in}}$ to another single qubit state $|\psi\rangle_{\text{out}} = U|\psi\rangle_{\text{in}}$. In this figure and in all the subsequent figures that depict quantum gates and quantum circuits, time moves from left to right, and each horizontal line represents a qubit. The horizontal lines are often referred to as qubit lines. Single qubit gates are represented by 2×2 unitary matrices. Every unitary operator (U), which is represented by a 2×2 matrix, is a valid single qubit gate. In principle, we can construct an infinite number of 2×2 unitary matrices. Consequently, there are an infinite number of possible single qubit quantum gates. However, in the conventional classical circuit theory, only two single bit logic gates are possible, namely the Identity gate and the logical NOT gate. Among this infinite number of possible single qubit quantum gates, some have special importance as they are used most frequently, and as they can be used as elements of a set of gates, which form a universal gate library. In this section we will briefly introduce these important and useful single qubit quantum gates, which are nothing but single qubit quantum state transformations. Since these transformations are linear, they are completely specified by their effect on the basis vectors. For instance, if we know that a single qubit quantum gate A maps $|0\rangle$ to $|\psi_0\rangle$ and $|1\rangle$ to $|\psi_1\rangle$ then linearity implies that the gate maps an arbitrary single qubit state $\alpha|0\rangle + \beta|1\rangle$ to $\alpha|\psi_0\rangle + \beta|\psi_1\rangle$. Keeping this in mind, we will now describe the effect of important single qubit gates on the basis vectors $|0\rangle$ and $|1\rangle$ and will also provide the corresponding 2×2 unitary matrices that represent the gates.

Figure 4.1: An arbitrary single qubit gate U.

1. **Pauli gates:** The quantum NOT gate transforms $|0\rangle$ to $|1\rangle$ and vice versa, so it is analogous to a classical NOT gate. But there is a fundamental difference with a classical NOT gate. A quantum NOT gate can accept a superposition state $\alpha|0\rangle + \beta|1\rangle$ as an input state, but a classical NOT gate cannot accept it as an input. The NOT gate is also called X gate since the unitary matrix that represents this gate is given by

$$X = \begin{pmatrix} 0 & 1 \\ 1 & 0 \end{pmatrix}, \tag{4.1}$$

which is the same as the Pauli matrix σ_x. Consequently, this gate is also called the Pauli gate. It is easy to obtain the matrix (4.1). Suppose we don't know the matrix of the NOT gate, but we know that it transforms $|0\rangle$ to $|1\rangle$ and $|1\rangle$ to $|0\rangle$. A single qubit operation must be a 2×2 matrix since the states $|0\rangle$ and $|1\rangle$ are represented by column matrices of dimension 2×1. Now if we assume that the matrix for the NOT gate is $\text{NOT} = \begin{pmatrix} a & b \\ c & d \end{pmatrix}$ then using $\text{NOT}|0\rangle = |1\rangle$ we obtain

$$\begin{pmatrix} a & b \\ c & d \end{pmatrix} \begin{pmatrix} 1 \\ 0 \end{pmatrix} = \begin{pmatrix} 0 \\ 1 \end{pmatrix}$$
$$\Rightarrow \begin{pmatrix} a \\ c \end{pmatrix} = \begin{pmatrix} 0 \\ 1 \end{pmatrix}$$
$$\Rightarrow a = 0, c = 1$$

and similarly, $\text{NOT}|1\rangle = |0\rangle$ implies $b = 1$ and $d = 0$. Thus we have obtained the matrix of the NOT gate. We can also express it in bra-ket notation as follows:

$$X = |0\rangle\langle 1| + |1\rangle\langle 0|. \tag{4.2}$$

One can quickly check that $X|0\rangle = |0\rangle\langle 1|0\rangle + |1\rangle\langle 0|0\rangle = |1\rangle$ and $X|1\rangle = |0\rangle\langle 1|1\rangle + |1\rangle\langle 0|1\rangle = |0\rangle$. We can also obtain the matrix of NOT gate from (4.2) by replacing $|0\rangle$, $|1\rangle$, $\langle 0|$ and $\langle 1|$ by their equivalent matrices. The other two Pauli matrices (σ_y, σ_z) represent two more single qubit gates

$$Y = \sigma_y = \begin{pmatrix} 0 & -i \\ i & 0 \end{pmatrix} \text{ and } Z = \sigma_z = \begin{pmatrix} 1 & 0 \\ 0 & -1 \end{pmatrix}. \tag{4.3}$$

In general X, Y, Z are known as Pauli gates. The quantum NOT or X gate just flips the bits. In analogy to X we can understand the effect of Y and Z on the single qubit state. For example Z will map an arbitrary quantum state $\alpha|0\rangle + \beta|1\rangle$ to $\alpha|0\rangle - \beta|1\rangle$. Thus the sole effect of Z is to flip the relative phase. Y is actually a combination of bit flip and phase flip. To visualize this feature we may note that $Y = iXZ \equiv XZ$ (as the global phase does not have any meaning). Now we provide a simple rule which will be found useful in the rest of the book. If an arbitrary single qubit gate A maps $|0\rangle$ to a single qubit state $|\psi_0\rangle$ and $|1\rangle$ to another single qubit state $|\psi_1\rangle$ then we can write

$$A = |\psi_0\rangle\langle 0| + |\psi_1\rangle\langle 1|. \tag{4.4}$$

Similarly, if an arbitrary two qubit gate B maps $|00\rangle, |01\rangle, |10\rangle$ and $|11\rangle$ to two qubit states $|\psi_{00}\rangle, |\psi_{01}\rangle, |\psi_{10}\rangle$ and $|\psi_{11}\rangle$ respectively then

$$B = |\psi_{00}\rangle\langle 00| + |\psi_{01}\rangle\langle 01| + |\psi_{10}\rangle\langle 10| + |\psi_{11}\rangle\langle 11|. \tag{4.5}$$

One can easily extend this simple rule to n−qubit gates and to other basis sets.

2. **Hadamard gate:** The Hadamard transformation is defined by the operation

$$H|0\rangle = \frac{1}{\sqrt{2}} \left(|0\rangle + |1\rangle \right)$$

and

$$H|1\rangle = \frac{1}{\sqrt{2}} \left(|0\rangle - |1\rangle \right).$$

Thus the unitary matrix corresponding to this gate can be given as

$$H = \frac{1}{\sqrt{2}} \begin{pmatrix} 1 & 1 \\ 1 & -1 \end{pmatrix} \tag{4.6}$$

and following (4.4) we can express it in bra-ket notation as

$$
\begin{aligned}
H &= \frac{1}{\sqrt{2}} \left(|0\rangle + |1\rangle \right) \langle 0| + \frac{1}{\sqrt{2}} \left(|0\rangle - |1\rangle \right) \langle 1| \\
&= \frac{1}{\sqrt{2}} \left(|0\rangle\langle 0| + |1\rangle\langle 0| + |0\rangle\langle 1| - |1\rangle\langle 1| \right) \\
&= \sum_{x,y=0}^{1} (-1)^{xy} |x\rangle\langle y|.
\end{aligned}
$$

Here we can easily observe that H is self-inverse as

$$HH = \frac{1}{2} \begin{pmatrix} 1 & 1 \\ 1 & -1 \end{pmatrix} \begin{pmatrix} 1 & 1 \\ 1 & -1 \end{pmatrix} = \begin{pmatrix} 1 & 0 \\ 0 & 1 \end{pmatrix}.$$

A self-inverse operator always maps the input states into mutually orthogonal output states and we can check that H maps $|0\rangle$ and $|1\rangle$ to mutually orthogonal states $|+\rangle$ and $|-\rangle$. A Hadamard operation can be described in a compact notation as

$$H|a\rangle = \frac{(-1)^a |a\rangle + |\bar{a}\rangle}{\sqrt{2}}, \tag{4.7}$$

where $a \in \{0, 1\}$.

3. **Phase gate or phase shift gate:** Consider a unitary operation that acts as $|0\rangle \rightarrow |0\rangle$, $|1\rangle \rightarrow \exp(i\phi)|1\rangle$. Thus it keeps $|0\rangle$ unchanged and changes the phase of $|1\rangle$ by $\exp(i\phi)$. We can represent this gate as

$$P(\phi) = |0\rangle\langle 0| + \exp(i\phi)|1\rangle\langle 1| = \begin{pmatrix} 1 & 0 \\ 0 & \exp(i\phi) \end{pmatrix}. \tag{4.8}$$

Since ϕ can have infinitely many values, we can have infinitely many different single qubit gates. This fact is in sharp contrast to the classical case where only one non-trivial single bit operation (NOT) is possible. Again among these infinitely many possible values of ϕ, particular values have drawn special attention. For example,

$$S = P\left(\frac{\pi}{2}\right) = \begin{pmatrix} 1 & 0 \\ 0 & i \end{pmatrix} \tag{4.9}$$

is often called a phase gate. $P(\frac{\pi}{4})$ is also a very popular gate and is often referred to as T gate or $\frac{\pi}{8}$ gate. Here a question is expected to arise in the reader's mind: Why is $P(\frac{\pi}{4})$ called $\frac{\pi}{8}$ gate? This unfortunate nomenclature arises from the fact that historically this gate was referred to as $\frac{\pi}{8}$ gate as up to a global phase this gate is equivalent to a gate which has $\exp\left(\pm\frac{i\pi}{8}\right)$ in its diagonals. To clarify this, we may note

$$T = \begin{pmatrix} 1 & 0 \\ 0 & \exp(i\frac{\pi}{4}) \end{pmatrix} = \exp(i\frac{\pi}{8}) \begin{pmatrix} \exp(-i\frac{\pi}{8}) & 0 \\ 0 & \exp(i\frac{\pi}{8}) \end{pmatrix}. \tag{4.10}$$

Now we may show that $T^2 = S$ as follows:

$$T^2 = \begin{pmatrix} 1 & 0 \\ 0 & \exp(i\frac{\pi}{4}) \end{pmatrix} \begin{pmatrix} 1 & 0 \\ 0 & \exp(i\frac{\pi}{4}) \end{pmatrix} = \begin{pmatrix} 1 & 0 \\ 0 & \exp(i\frac{\pi}{2}) \end{pmatrix} = S. \tag{4.11}$$

Another important point is that the $P(\phi)$ gates are not self-inverse gates in general. For example, $T^2 = S \neq I$ is the manifestation of the fact that $P(\phi)$ gates are not self-inverse. There is a special case $P(\pi) = Z$, which is self-inverse.

A very special case of $P(\phi)$ gate is

$$R_k = P\left(\frac{2\pi}{2^k}\right) = \begin{pmatrix} 1 & 0 \\ 0 & \exp\left(\frac{2\pi i}{2^k}\right) \end{pmatrix}. \tag{4.12}$$

In Chapter 5 we show that R_k gates are very useful for the implementation of quantum Fourier transform operation which is required for implementation of Shor's algorithm.

4. **Rotation gates:** Rotation gates are defined as follows:

$$\begin{aligned} R_x(\theta) &= e^{-\frac{i\theta X}{2}}, \\ R_y(\theta) &= e^{-\frac{i\theta Y}{2}}, \\ R_z(\theta) &= e^{-\frac{i\theta Z}{2}}. \end{aligned} \tag{4.13}$$

Now to obtain the matrix forms of these useful single qubit gates we have to prove a simple identity which is stated as:

Identity: If x is a real number and A is a matrix such that $A^2 = I$ then $e^{iAx} = I\cos(x) + iA\sin(x)$.

Proof:

$$
\begin{aligned}
e^{iAx} &= I + iAx - \frac{AAx^2}{2!} - \frac{iAAAx^3}{3!} + \frac{AAAAx^4}{4!} + \frac{iAAAAAx^5}{5!} \\
&= I\left(1 - \frac{x^2}{2!} + \frac{x^4}{4!} + \cdots\right) + iA\left(x - \frac{x^3}{3!} + \frac{x^5}{5!} + \cdots\right) \\
&= I\cos(x) + iA\sin(x).
\end{aligned}
$$

$$(4.14)$$

We have already seen in Subsection 3.1.3 that the squares of Pauli matrices are Identity (i.e., $\sigma_i^2 = I$). Now we can use this property of Pauli matrices and the above identity (4.14) to provide matrices for rotation gates as follows:

$$
\begin{aligned}
R_x(\theta) &= I\cos(\tfrac{\theta}{2}) - iX\sin(\tfrac{\theta}{2}) \\
&= \begin{pmatrix} 1 & 0 \\ 0 & 1 \end{pmatrix}\cos(\tfrac{\theta}{2}) - i\begin{pmatrix} 0 & 1 \\ 1 & 0 \end{pmatrix}\sin(\tfrac{\theta}{2}) \\
&= \begin{pmatrix} \cos(\tfrac{\theta}{2}) & -i\sin(\tfrac{\theta}{2}) \\ -i\sin(\tfrac{\theta}{2}) & \cos(\tfrac{\theta}{2}) \end{pmatrix}, \\
R_y(\theta) &= I\cos(\tfrac{\theta}{2}) - iY\sin(\tfrac{\theta}{2}) \\
&= \begin{pmatrix} 1 & 0 \\ 0 & 1 \end{pmatrix}\cos(\tfrac{\theta}{2}) - i\begin{pmatrix} 0 & -i \\ i & 0 \end{pmatrix}\sin(\tfrac{\theta}{2}) \\
&= \begin{pmatrix} \cos(\tfrac{\theta}{2}) & -\sin(\tfrac{\theta}{2}) \\ \sin(\tfrac{\theta}{2}) & \cos(\tfrac{\theta}{2}) \end{pmatrix}, \\
R_z(\theta) &= I\cos(\tfrac{\theta}{2}) - iZ\sin(\tfrac{\theta}{2}) \\
&= \begin{pmatrix} 1 & 0 \\ 0 & 1 \end{pmatrix}\cos(\tfrac{\theta}{2}) - i\begin{pmatrix} 1 & 0 \\ 0 & -1 \end{pmatrix}\sin(\tfrac{\theta}{2}) \\
&= \begin{pmatrix} e^{-i\frac{\theta}{2}} & 0 \\ 0 & e^{i\frac{\theta}{2}} \end{pmatrix}.
\end{aligned}
$$

$$(4.15)$$

Now we may recall Equation (3.108) of the previous chapter, where we had described an arbitrary qubit in a Bloch sphere as $|\psi\rangle = \cos\left(\frac{\theta}{2}\right)|0\rangle + \sin\left(\frac{\theta}{2}\right)e^{i\phi}|1\rangle = \begin{pmatrix} \cos\left(\frac{\theta}{2}\right) \\ \sin\left(\frac{\theta}{2}\right)e^{i\phi} \end{pmatrix}$. If we apply the rotational gate $R_z(\theta_1)$ on this Bloch state then the state will be transformed to

$$
\begin{aligned}
R_z(\theta_1)|\psi\rangle &= \begin{pmatrix} e^{-i\frac{\theta_1}{2}} & 0 \\ 0 & e^{i\frac{\theta_1}{2}} \end{pmatrix}\begin{pmatrix} \cos\left(\frac{\theta}{2}\right) \\ \sin\left(\frac{\theta}{2}\right)e^{i\phi} \end{pmatrix} \\
&= e^{-i\frac{\theta_1}{2}}\begin{pmatrix} \cos\left(\frac{\theta}{2}\right) \\ \sin\left(\frac{\theta}{2}\right)e^{i(\phi+\theta_1)} \end{pmatrix} \\
&= \begin{pmatrix} \cos\left(\frac{\theta}{2}\right) \\ \sin\left(\frac{\theta}{2}\right)e^{i(\phi+\theta_1)} \end{pmatrix}.
\end{aligned}
$$

$$(4.16)$$

In the last step of the previous equation we ignored the global phase. Now one can easily see that $R_z(\theta_1)$ has rotated the Bloch vector by an angle θ_1 about the z axis of the Bloch sphere. Similarly, $R_x(\theta_1)$ and $R_y(\theta_1)$ gates rotate the Bloch vector by an angle θ_1 with respect to x and y axes respectively.

5. **Square-root-of-NOT gates:** This interesting single qubit gate is usually denoted as V gate or $\sqrt{\text{NOT}}$ gate. In matrix representation

$$V = \sqrt{\text{NOT}} = \frac{(1+i)}{2}\begin{pmatrix} 1 & -i \\ -i & 1 \end{pmatrix} = \frac{1}{2}\begin{pmatrix} 1+i & 1-i \\ 1-i & 1+i \end{pmatrix}.$$

Now it is easy to see that

$$VV = \frac{1}{4}\begin{pmatrix} 1+i & 1-i \\ 1-i & 1+i \end{pmatrix}\begin{pmatrix} 1+i & 1-i \\ 1-i & 1+i \end{pmatrix} = \begin{pmatrix} 0 & 1 \\ 1 & 0 \end{pmatrix} = \text{NOT}.$$

This is why V gate is called $\sqrt{\text{NOT}}$ gate. Now we can quickly check that $V^\dagger = \frac{1}{2}\begin{pmatrix} 1-i & 1+i \\ 1+i & 1-i \end{pmatrix}$ satisfies the following relations: $VV^\dagger = V^\dagger V = I$ and $V^\dagger V^\dagger = \text{NOT}$. Therefore, $V^\dagger = V^{-1} = \sqrt{\text{NOT}}$. In other words, $\sqrt{\text{NOT}}$ operation can be described by both V and $V^\dagger$ but usually we use V to represent $\sqrt{\text{NOT}}$ gate.

4.2 Two qubit gates

1. **Controlled-NOT gate:** The most popular and the most important example of a two qubit gate is the controlled $-$ NOT or CNOT gate, which complements the second qubit if the first qubit is in the state $|1\rangle$ and leaves the second qubit unchanged otherwise. Thus the matrix representation for this gate is

$$\text{CNOT} = \begin{pmatrix} 1 & 0 & 0 & 0 \\ 0 & 1 & 0 & 0 \\ 0 & 0 & 0 & 1 \\ 0 & 0 & 1 & 0 \end{pmatrix}. \tag{4.17}$$

Alternatively, it can also be represented in bra-ket notation (Dirac notation) as

$$\text{CNOT} = |00\rangle\langle 00| + |01\rangle\langle 01| + |11\rangle\langle 10| + |10\rangle\langle 11|.$$

As the state of the first qubit controls whether the second qubit will be flipped or not, the first qubit is referred to as control qubit and the second qubit is called target qubit. The same convention is used in other controlled gates. In multiqubit gates there may exist more than one control qubit. The basic reason of the popularity of the CNOT gate is twofold, firstly it is used in many quantum circuits of practical importance and secondly if we can construct all single qubit gates and CNOT gate then we can construct any other unitary quantum operation with suitable combination of these gates. Specifically, if we can physically realize all single qubit gates and any two qubit entangled quantum gate[1] then in principle we can construct all possible quantum circuits. In most of the physical realization of two qubit gates, CNOT is reported and gradually it has become almost synonymous to universal gate (a quantum correspondent of classical NAND). A symbolic representation of CNOT is shown in Fig. 4.2.

2. **Swap gate:** Another popular two qubit gate is the SWAP gate. It swaps the states of the two qubits; thus it maps $|ab\rangle \to |ba\rangle$ (i.e., $|00\rangle \to |00\rangle$, $|01\rangle \to |10\rangle$, $|10\rangle \to |01\rangle$ and $|11\rangle \to |11\rangle$) and it can be represented as

$$\text{SWAP} = \begin{pmatrix} 1 & 0 & 0 & 0 \\ 0 & 0 & 1 & 0 \\ 0 & 1 & 0 & 0 \\ 0 & 0 & 0 & 1 \end{pmatrix}. \tag{4.18}$$

Following (4.5) we can write it in bra-ket notation as

$$\text{SWAP} = |00\rangle\langle00| + |10\rangle\langle01| + |01\rangle\langle10| + |11\rangle\langle11|.$$

Symbolic representation of the SWAP gate is shown in Fig. 4.2, and in Fig. 4.3 we have shown that the SWAP gate can be constructed by using three CNOT gates.

3. **Controlled-U gates:** If U is an arbitrary single qubit gate (unitary operation) then we may construct a two qubit Controlled$-U$ gate, such that the single qubit operation U operates on the second (target) qubit if the first (control) qubit is in state $|1\rangle$, otherwise the input state remains unchanged. Thus the Controlled$-U$ gate maps $|00\rangle \to |00\rangle$, $|01\rangle \to |01\rangle$, $|10\rangle \to |1\rangle \otimes U|0\rangle$, $|11\rangle \to |1\rangle \otimes U|1\rangle$. In Dirac notation such a gate is described in general as

$$\text{Controlled}-U = |00\rangle\langle00| + |01\rangle\langle01| + |1\rangle \otimes U|0\rangle\langle10| + |1\rangle \otimes U|1\rangle\langle11|$$

and the corresponding matrix is

$$\text{Controlled}-U = \begin{pmatrix} I & O \\ O & U \end{pmatrix}, \tag{4.19}$$

where $I = \begin{pmatrix} 1 & 0 \\ 0 & 1 \end{pmatrix}$, $O = \begin{pmatrix} 0 & 0 \\ 0 & 0 \end{pmatrix}$ are respectively the Identity and Null operations (gate) in C^2. Now we can see that CNOT is just

---

[1]By an entangled two qubit gate we mean a two qubit gate which cannot be achieved as tensor product of two single qubit gates.

a special case of Controlled$-U$ where $U = X$. In a similar fashion we may construct Controlled$-T$, Controlled$-P(\phi)$, Controlled$-R_x(\theta)$, etc. and it is a straightforward job to write the corresponding matrices. As examples, we may explicitly write the matrices for Controlled$-V$ and Controlled$-V^\dagger$gates as follows:

$$\text{Controlled}-V = \begin{pmatrix} I & O \\ O & V \end{pmatrix} = \begin{pmatrix} 1 & 0 & 0 & 0 \\ 0 & 1 & 0 & 0 \\ 0 & 0 & \frac{1+i}{2} & \frac{1-i}{2} \\ 0 & 0 & \frac{1-i}{2} & \frac{1+i}{2} \end{pmatrix} \quad (4.20)$$

$$\text{Controlled}-V^\dagger = \begin{pmatrix} I & O \\ O & V^\dagger \end{pmatrix} = \begin{pmatrix} 1 & 0 & 0 & 0 \\ 0 & 1 & 0 & 0 \\ 0 & 0 & \frac{1-i}{2} & \frac{1+i}{2} \\ 0 & 0 & \frac{1+i}{2} & \frac{1-i}{2} \end{pmatrix}. \quad (4.21)$$

Controlled$-V$ and Controlled$-V^\dagger$ are very often used in construction of more complex quantum gates and quantum circuits. Actually {NOT, CNOT, Controlled$-V$, Controlled$-V^\dagger$} forms a universal gate library for reversible circuits. Such a gate library is referred to as *NCV* gate library. As $\sqrt{\text{NOT}}$ or Controlled $- \sqrt{\text{NOT}}$ operations cannot be achieved classically so *NCV* is actually a quantum gate library which is universal for classical reversible operations.

Figure 4.2: CNOT (left) and SWAP (right) gates.

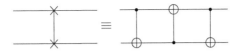

Figure 4.3: SWAP gate as composition of CNOTs.

4.3   Three qubit gates

1. **Toffoli gate and Fredkin gate:** The three qubit gates used in the quantum circuits are generally either controlled-controlled single qubit gate or controlled two qubit gate. The most popular three

qubit gates are the Toffoli gate and Fredkin gate, which are CCNOT and CSWAP gates respectively. Thus in the three qubit Toffoli gate the first two qubits are control qubits and the third one is the target qubit. So the third qubit will be flipped only if both the first and second qubits are in state $|1\rangle$. Similarly, in the case of the Fredkin gate the swap operation between the second and third qubit is done only if the first qubit (i.e., the control qubit) is at state $|1\rangle$. Now we can write the unitary matrices representing these two quantum gates as

$$\text{Toffoli} = \begin{pmatrix} 1 & 0 & 0 & 0 & 0 & 0 & 0 & 0 \\ 0 & 1 & 0 & 0 & 0 & 0 & 0 & 0 \\ 0 & 0 & 1 & 0 & 0 & 0 & 0 & 0 \\ 0 & 0 & 0 & 1 & 0 & 0 & 0 & 0 \\ 0 & 0 & 0 & 0 & 1 & 0 & 0 & 0 \\ 0 & 0 & 0 & 0 & 0 & 1 & 0 & 0 \\ 0 & 0 & 0 & 0 & 0 & 0 & 0 & 1 \\ 0 & 0 & 0 & 0 & 0 & 0 & 1 & 0 \end{pmatrix} \qquad (4.22)$$

and

$$\text{Fredkin} = \begin{pmatrix} 1 & 0 & 0 & 0 & 0 & 0 & 0 & 0 \\ 0 & 1 & 0 & 0 & 0 & 0 & 0 & 0 \\ 0 & 0 & 1 & 0 & 0 & 0 & 0 & 0 \\ 0 & 0 & 0 & 1 & 0 & 0 & 0 & 0 \\ 0 & 0 & 0 & 0 & 1 & 0 & 0 & 0 \\ 0 & 0 & 0 & 0 & 0 & 0 & 1 & 0 \\ 0 & 0 & 0 & 0 & 0 & 1 & 0 & 0 \\ 0 & 0 & 0 & 0 & 0 & 0 & 0 & 1 \end{pmatrix} \qquad (4.23)$$

respectively. Here we would like to note that a quantum Toffoli gate can be built up from CNOT and single qubit gates but a classical (reversible) Toffoli gate cannot be built using two bit and one bit classical reversible gates. Symbolic representations of Toffoli and Fredkin gates are provided in Fig. 4.4.

Figure 4.4: Toffoli (left) and Fredkin (right) gates.

2. **Deutsch gate:** TheDeutsch gate was introduced by David Deutsch

in 1989 as a universal quantum gate. This gate is defined as

$$
\text{Deutsch}(\theta) = \begin{pmatrix}
1 & 0 & 0 & 0 & 0 & 0 & 0 & 0 \\
0 & 1 & 0 & 0 & 0 & 0 & 0 & 0 \\
0 & 0 & 1 & 0 & 0 & 0 & 0 & 0 \\
0 & 0 & 0 & 1 & 0 & 0 & 0 & 0 \\
0 & 0 & 0 & 0 & 1 & 0 & 0 & 0 \\
0 & 0 & 0 & 0 & 0 & 1 & 0 & 0 \\
0 & 0 & 0 & 0 & 0 & 0 & i\cos(\theta) & \sin(\theta) \\
0 & 0 & 0 & 0 & 0 & 0 & \sin(\theta) & i\cos(\theta)
\end{pmatrix},
$$

where θ is a constant angle such that $\frac{2\theta}{\pi}$ is an irrational number. This choice isolates Toffoli gate as $D\left(\frac{\pi}{2}\right) \equiv \text{Toffoli}$. $\{\text{Deutsch}(\theta)\}$ forms a universal quantum gate library. However, circuits built using this gate library are usually inefficient and this is why this particular gate library is not used very frequently [63].

4.4 A little more on quantum gates

Here we would like to note an interesting feature of quantum gates. If a quantum gate is self-inverse then it will always map input states to mutually orthogonal output states. For example, consider a single qubit gate A, which maps $|0\rangle$ to $|\psi_0\rangle$ and $|1\rangle$ to $|\psi_1\rangle$ (i.e., $A = |\psi_0\rangle\langle 0| + |\psi_1\rangle\langle 1|$). If it is self-inverse (i.e., $A = A^{-1} = A^\dagger$) then $\langle 0|A = \langle\psi_0| \Rightarrow \langle 0| = \langle\psi_0|A^{-1} = \langle\psi_0|A = \langle\psi_0|\psi_0\rangle\langle 0| + \langle\psi_0|\psi_1\rangle\langle 1| \Rightarrow \langle\psi_0|\psi_1\rangle = 0, \langle\psi_0|\psi_0\rangle = 1$. The idea can be easily extended to n-qubit systems. In general outputs of self-inverse gates are mutually orthogonal. But the converse is not true. That means there exist quantum gates that are not self-inverse but that map input states to mutually orthogonal output states.

Now we can note another interesting point. A unitary operator A must satisfy $A^{-1} = A^\dagger$ and a Hermitian operator A must satisfy $A = A^\dagger$. Consequently, all unitary operators are not Hermitian and all Hermitian operators are not unitary. If a unitary operator is Hermitian then $A^{-1} = A^\dagger = A$, i.e., $A = A^{-1}$, so the operator is self-inverse. We can easily show the converse (i.e., self-inverse unitary operators are Hermitian). This implies that quantum gates represented by Hermitian unitary operators always map input states to mutually orthogonal states. Now we can observe that the CNOT gate $|00\rangle\langle 00| + |01\rangle\langle 01| + |11\rangle\langle 10| + |10\rangle\langle 11|$ is self-inverse, so it is Hermitian. We can easily drop the symmetry required for the gate to be self-inverse and modify it to another quantum gate $B = |01\rangle\langle 00| + |11\rangle\langle 01| + |10\rangle\langle 10| + |10\rangle\langle 11|$. You can easily check that this gate is not self-inverse as $BB|00\rangle = B|01\rangle = |11\rangle \neq |00\rangle$. Thus this gate is not Hermitian but interestingly it is unitary and it maps the input states into a set of mutually orthogonal output states.

Now we may ask some simple questions: (1) How many such quantum gates are possible that can map input states to mutually orthogonal states? This question is relevant from several perspectives. Especially in the secure quantum communication we need the output states to be mutually orthogonal. (2) How many of these gates are non-Hermitian (non-self-inverse)? (3) Can we physically construct a unitary non-Hermitian gate with the help of a Hermitian Hamiltonian?

Assume that we are working in M dimension and the input states are $\{|a_1\rangle, |a_2\rangle, |a_3\rangle, \cdots, |a_M\rangle\}$. Thus $\{|a_j\rangle\}$ forms our input basis set. Similarly, assume that $\{|b_j\rangle\}$ represent a new basis set in the same dimension and $\{|b_j\rangle\}$ is our output basis set. For quantum gates $\{|b_j\rangle\}$ may or may not be a permutation of $\{|a_j\rangle\}$. Now we may introduce the operators $U_J = \sum_j |b_j\rangle\langle a_j|$, which are unitary, as is easily verified to satisfy

$$U_J U_J^\dagger = U_J^\dagger U_J = \left(\sum_p |a_p\rangle\langle b_p|\right)\left(\sum_q |b_q\rangle\langle a_q|\right) = \left(\sum_j |a_j\rangle\langle a_j|\right) = I_M,$$

where I_M is the identity operation in M dimension. We can elaborate the idea with a few examples.

Example 4.1: Consider SWAP gate. Here $\{|a_j\rangle\} = \{|00\rangle, |01\rangle, |10\rangle, |11\rangle\}$ and $\{|b_j\rangle\} = \{|00\rangle, |10\rangle, |01\rangle, |11\rangle\}$ and so

$$U_J = |00\rangle\langle 00| + |01\rangle\langle 10| + |10\rangle\langle 01| + |11\rangle\langle 11| = \text{SWAP}.$$

Example 4.2: Consider a gate where $\{|a_j\rangle\} = \{|00\rangle, |01\rangle, |10\rangle, |11\rangle\}$ and $\{|b_j\rangle\} = \{|00\rangle, |01\rangle, |11\rangle, |10\rangle\}$. In this case

$$U_J = |00\rangle\langle 00| + |01\rangle\langle 01| + |10\rangle\langle 11| + |11\rangle\langle 10| = \text{CNOT}.$$

Example 4.3: In the last two examples output basis $\{|b_j\rangle\}$ is just a permutation of input basis $\{|a_j\rangle\}$. How many such gates are possible where output basis is a permutation of input basis?

Solution: Arbitrary permutation on M basis vectors can be achieved in $M!$ ways; each of these permutations will provide us with a unitary operator. Thus each of these $M!$ unitary operators U_J is a quantum gate that always maps input states into mutually orthogonal output states. We may call these quantum gates as permutation gates. In brief, we have $M!$ permutation gates of the given type.

Example 4.4: How many reversible n-bit gates are possible?

Solution: As for reversible gates superposition states are not allowed in input and output, so $\{|a_j\rangle\}$ is always in computational basis and $\{|b_j\rangle\}$ is just a permutation of input basis $\{|a_j\rangle\}$. Consequently, only $M!$ gates are possible, where $M = 2^n$. Thus the number of possible 2-bit reversible gates is $4! = 24$.

Example 4.5: How many quantum permutation gates are possible?

Solution: Infinite. We have already mentioned that $\{|b_j\rangle\}$ is not essentially a permutation of input basis $\{|a_j\rangle\}$ and the basis vectors of $\{|b_j\rangle\}$ can be in superposition states. Consider, $\{|a_j\rangle\} = \{|0\rangle, |1\rangle\}$ and $\{|b_j\rangle\} =$

$\{|+\rangle, |-\rangle\}$, then U_J = Hadamard. In general, for each combination of $\{|a_j\rangle\}$ and $\{|b_j\rangle\}$ we can have $M!$ gates and there exist infinitely many possible combinations of $\{|a_j\rangle\}$ and $\{|b_j\rangle\}$. For example, in single qubit cases we can think of $\{|b_j\rangle\} = \{\sin(\theta)|0\rangle + \cos(\theta)|1\rangle, \cos(\theta)|0\rangle - \sin(\theta)|1\rangle\}$. For each choice of θ we have a new basis set. So it is straightforward to see that the number of quantum permutation gates is infinite. But for a specific choice of combination of input basis set and output basis set it is finite ($M!$). In this discussion we have considered that $\{|a_j\rangle\}$ is fixed and each permutation of $\{|b_j\rangle\}$ provides a new quantum gate.

Example 4.6: Are these $M!$ gates self-inverse/Hermitian?

Solution: These unitary gates are not essentially self-inverse/Hermitian. Our task is to find out the number of self-inverse permutations on M letters which is known as involutions. For $M = 1, 2, 3, \cdots$ number of alternating permutations are $1, 2, 4, 10, 26, 76, 232, 764, \cdots$. This implies that in C^{2^2} (i.e., for $M = 4$) we can have 10 self-inverse gates for a specific choice of $\{|a_i\rangle\}$ and $\{|b_j\rangle\}$, which implies that $4! - 10 = 14$ two qubit gates are not self-inverse. Similarly, in C^{2^3} we have $8! - 764 = 39556$ non-Hermitian permutation gates and only 764 Hermitian permutation gates. Therefore, 98.10% of the gates are non-Hermitian (nonself-inverse).

We elaborate on this point just to show that most of the classical reversible gates are not self-inverse and most of the quantum gates are not Hermitian. Here it would be apt to note that this apparent non-Hermiticity does not contradict standard quantum mechanics. Quantum mechanics demands that the Hamiltonian of a quantum system should be Hermitian and that leads to the unitary operators. The evolution operators (quantum gates) are required to be unitary only, they are not bound to be Hermitian as they do not represent physical observables. Thus all the unitary but non-Hermitian quantum gates constructed here are physically realizable and perfectly consistent with quantum mechanics.

We have already learned about single qubit, two qubit and three qubit quantum gates, now we may combine them to form some simple quantum circuits. This is what we do in the next section.

4.5 Quantum circuits

The quantum gates are combined to form quantum circuits. In the subsequent chapters we describe several useful quantum circuits in relation to quantum teleportation, dense coding and quantum algorithms. Here we describe a few simple quantum circuits. To begin with, let us consider a simple circuit comprised of a Hadamard gate followed by a CNOT gate as shown in Fig. 4.5. The operation of this circuit can be mathematically understood as follows. We start with a separable state $|00\rangle$. Thus the input state of the circuit is $|00\rangle$. Then a Hadamard gate operates on the first (upper) qubit and transforms the state of the system to another separable state $\frac{1}{\sqrt{2}}(|00\rangle + |10\rangle)$. Now the CNOT gate operates with the first qubit working as the control qubit and the second qubit working as the target qubit. We have already learned that the CNOT gate flips the target qubit, when the control qubit is $|1\rangle$. Therefore, after the CNOT operation the output state of the circuit is $\frac{1}{\sqrt{2}}(|00\rangle + |11\rangle)$ which is maximally entangled.

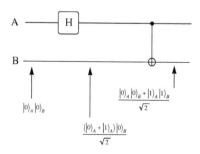

Figure 4.5: EPR circuit.

A computational task can be performed by using more than one quantum circuit. Such an example is shown in Fig. 4.6a and Fig. 4.6b, where both the circuits represent quantum half adder. Now a question naturally appears in our mind: Which of these two circuits is better? To answer this question we need some quantitative measure of the quality of circuits. Normally such a quantitative measure is referred to as cost or cost metric of the circuit. Different quantitative measures of quality of quantum circuits exist and a few of them are described in the following subsection. The circuit shown in Fig. 4.5 is known as an EPR circuit and we can easily justify this nomenclature if we note that when the input state of the EPR circuit is $|1\rangle_A|1\rangle_B$ then the output of the EPR circuit is the singlet state $\left(\frac{|01\rangle - |10\rangle}{\sqrt{2}}\right)_{AB}$.

4.5.1 Quantitative measures of quality of a circuit

Some of the important quantitative measures of the quality of a quantum circuit are gate count (circuit cost), number of garbage bits, quantum cost and delay. We will briefly introduce each of them in this subsection. Let us begin with gate count, which is usually referred to as circuit cost.

4.5.1.1 Gate count or circuit cost

Gate count or circuit cost is the total number of gates present in a circuit. This is an important measure of quality of circuit. The lesser the circuit cost is the better the circuit is. But unfortunately, circuit cost is not unique. One may substantially reduce the circuit cost by using complex gate library and/or new gates. For example, let us consider the EPR circuit shown in Fig. 4.5. The circuit is composed of two gates from the universal gate library $\{H, S, T, \text{CNOT}\}$. Therefore, its gate count (circuit cost) is 2. Now we can put the two gates in a box and call it a new gate. The matrix of the new gate would be

$$
\text{CNOT}\,(H \otimes I_2) = \begin{pmatrix} 1 & 0 & 0 & 0 \\ 0 & 1 & 0 & 0 \\ 0 & 0 & 0 & 1 \\ 0 & 0 & 1 & 0 \end{pmatrix} \frac{1}{\sqrt{2}} \begin{pmatrix} 1 & 1 \\ 1 & -1 \end{pmatrix} \otimes \begin{pmatrix} 1 & 0 \\ 0 & 1 \end{pmatrix}
$$

$$
= \frac{1}{\sqrt{2}} \begin{pmatrix} 1 & 0 & 0 & 0 \\ 0 & 1 & 0 & 0 \\ 0 & 0 & 0 & 1 \\ 0 & 0 & 1 & 0 \end{pmatrix} \begin{pmatrix} 1 & 0 & 1 & 0 \\ 0 & 1 & 0 & 1 \\ 1 & 0 & -1 & 0 \\ 0 & 1 & 0 & -1 \end{pmatrix}
$$

$$
= \frac{1}{\sqrt{2}} \begin{pmatrix} 1 & 0 & 1 & 0 \\ 0 & 1 & 0 & 1 \\ 0 & 1 & 0 & -1 \\ 1 & 0 & -1 & 0 \end{pmatrix}.
$$

Now if we consider it as a new gate

$$
\text{NEWG} = \text{CNOT}\,(H \otimes I_2) = \frac{1}{\sqrt{2}} \begin{pmatrix} 1 & 0 & 1 & 0 \\ 0 & 1 & 0 & 1 \\ 0 & 1 & 0 & -1 \\ 1 & 0 & -1 & 0 \end{pmatrix}
$$

then we can easily find that this gate maps $|00\rangle$ to $\frac{|00\rangle+|11\rangle}{\sqrt{2}}$, $|01\rangle$ to $\frac{|01\rangle+|10\rangle}{\sqrt{2}}$, $|10\rangle$ to $\frac{|00\rangle-|11\rangle}{\sqrt{2}}$ and $|11\rangle$ to $\frac{|01\rangle-|10\rangle}{\sqrt{2}}$. Thus this NEWG is a new gate which is equivalent to the EPR circuit and which reduces the circuit cost of the EPR circuit to 1. It may appear advantageous to use such new and complex gates to reduce the circuit cost. However, it is not allowed. If it is allowed then every quantum circuit block without measurement would have reduced to a single gate. This is so because the product of an arbitrary number of unitary operations is always unitary. Circuits are constructed using standard universal gate libraries and circuit cost of a circuit A can be compared with the circuit cost of another circuit B if and only if the circuit costs are calculated using the same gate library. Often we need to compare circuits designed using different gate libraries. In that case we need to transform one of the circuits into a logically equivalent and optimized circuit prepared in the other gate library. In brief, it is important to define a unique gate library for comparison of circuit costs. See the circuits shown in Fig. 4.6. Both represent half adder and both are constructed using the gates from *NCV* gate library, but the gate count of the circuit shown in Fig. 4.6b is less than the gate count of the circuit shown in Fig. 4.6a.

Irreversible		Reversible	
Input	Output	Input	Output
AB	Z	ABC	XYZ
00	0	000	000
01	0	010	010
10	0	100	100
11	1	110	111

Table 4.1: Irreversible and reversible AND gate. X and Y bits in the output of reversible AND gate are the garbage bits.

Hence the circuit shown in Fig. 4.6b is better than the circuit shown in Fig. 4.6a as far as the circuit cost is concerned.

4.5.1.2   Garbage bit

A garbage bit is the additional output that is used either to make a function reversible or to reduce the gate count. Garbage bits are not used for further computations. Large numbers of garbage bits are undesirable in a quantum/reversible circuit as it increases the width of the circuit. As an example in Table 4.1 we have described the truth table of an irreversible and a reversible AND gate. It is evident that the output Z gives us the required output of AND gate in both the cases. To be precise, we may use a Fredkin gate as AND gate if we use $|AB0\rangle$ as input (i.e., we keep the third input bit at a constant value of 0). In that case we will obtain the truth table of reversible AND gate described in Table 4.1. Here the third output bit of the Fredkin gate will be $Z = AB$, which is the required output of AND gate. As we are interested to implement an AND gate only, the other two outputs of the Fredkin (reversible AND) gate (i.e., X and Y) will not be used for further computation and they are the garbage bits. Similarly, if we use $|A1B\rangle$ as input in a Fredkin gate then we obtain a reversible OR gate. Here the third bit of output is $Z = A + B$, which is the desired output of OR gate and the other two outputs are garbage bits. We can also note that the first two output bits in the half adder circuits shown in Fig. 4.6a and Fig. 4.6b are garbage bits. It is important to note that the garbage bits are often introduced to make a function reversible and there can be many ways in which garbage values can be assigned. A circuit having a lesser number of garbage bits is better than the one performing the same task with a higher number of garbage bits. This is an important quantitative measure of quality of quantum circuit.

4.5.1.3   Quantum cost

The quantum cost of a reversible/quantum gate/circuit is the number of primitive quantum gates needed to implement the gate/circuit. Primitive

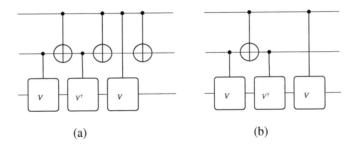

(a)   (b)

Figure 4.6: Quantum circuit of half adder is presented in two different ways: (a) Circuit cost and quantum cost of this quantum half adder circuit is 6 and width is 3. (b) Circuit cost and quantum cost of this quantum half adder circuit is 4 and width is 3. Circuit (b) is better than (a) as far as circuit cost and quantum cost are concerned. Local optimization tools like moving rule and deletion rule can be used to obtain circuit (b) from circuit (a).

quantum gates are the elementary building blocks, like NOT gate, CNOT gate, controlled-V, controlled-$V^\dagger$, rotation gates, etc. In fact all single qubit and 2-qubit gates are considered as quantum primitive gates and the cost of all quantum primitive gates are considered as one. Since Toffoli gate is not a quantum primitive gate, an NCT circuit (i.e., a circuit built using gates from the gate library $\{NOT, CNOT, Toffoli\}$)[2] cannot be used directly to determine the quantum cost. But Toffoli can be constructed using CNOT gate, controlled-V and controlled-$V^\dagger$ and that may reveal that the quantum cost of Toffoli gate is five. This helps us to convert an NCT circuit into an equivalent NCV circuit and then to optimize the NCV circuit to obtain the quantum cost. It is important to note that while computing the quantum cost we are allowed to combine gates and form new two qubit gates. This important feature isolates quantum cost from the circuit cost obtained in NCV or any other gate library where all the elements of the universal gate set are the quantum primitive gates. Further, quantum cost is often classified as linear quantum cost and nonlinear quantum cost. Given a circuit, if we just add the quantum costs of each of the gates used in the circuit then we obtain the linear quantum cost. However, if we replace all the 3-qubit and larger gates of the given circuit by their equivalent circuits built using quantum primitive gates and subsequently optimize the circuit without restricting us to any gate library then the gate count of the optimized circuit is called nonlinear quantum cost. The lesser the quantum cost, the better the circuit.

Example 4.7: The quantum circuit of a half adder is presented in two

---

[2]$\{NOT, CNOT, Toffoli\}$ forms a universal gate library for reversible circuits. It is known as the NCT gate library.

different ways in Fig. 4.6. The circuit shown in Fig. 4.6b is better than the circuit shown in Fig. 4.6a as its quantum cost is less.

4.5.1.4 Depth and width of a circuit

In addition to the measures discussed above, Kaye, Laflamme and Mosc [54] have prescribed two more important measures of complexity of a circuit (i) width (number of qubit lines) and (ii) depth (total number of time slices) of a circuit. These two measures are already introduced in Section 2.3. Mohammadi and Eshghi [64] have described another measure called delay which is closely related to the depth of the circuit. Delay is a technology dependent parameter and it provides a measure of how much time is required to evaluate a function. If we approximate that the evolution of each single qubit and two qubit gates require Δ amount of time then delay of a circuit = (depth of the circuit using primitive quantum gates)Δ. Delay is proportional to depth and we may consider them as the same measure.

Now it is easy to note that the more is the number of garbage bits in a circuit the more will be the width of the circuit. Further, a circuit having lesser width requires lesser space. Consequently, if we have two circuits for the same task then the circuit having lesser width is better. Similarly, the lesser the depth of a circuit, the better it is as it would take less time to perform the computational task. Now look at Fig. 4.7, where a schematic diagram of a general quantum circuit is shown. Each rectangular box represents a quantum gate and each vertical line is a separator between two time slices. If we consider Fig. 4.7 as an optimized circuit built using quantum gates from a well defined gate library, then there are 8 quantum gates, so its circuit cost is 8. But there are disjoint quantum gates which can be applied simultaneously as shown in the first, second and the last time slices of Fig. 4.7. So the depth of the circuit is 5. Further, there are 6 qubit lines so the width of the circuit is 6. However, we cannot directly compute the delay as three nonprimitive quantum gates are used here. We may now look back to Fig. 4.6a and Fig. 4.6b. It is easy to observe that width of both the circuits is 3. Depth and delay of the circuit shown in Fig. 4.6a are 6 and 6Δ respectively. Similarly, depth and delay of the circuit shown in Fig. 4.6b are 4 and 4Δ respectively. Consequently, as far as the depth and delay are concerned circuit shown in Fig. 4.6b is better than the circuit shown in Fig. 4.6a.

4.5.1.5 Total cost

A circuit is better if it has a lesser number of garbage bits, circuit cost and quantum cost. But it is often observed that reduction of circuit cost leads to increase in garbage bits and reduction of quantum cost leads to increase in circuit cost [65]. A new parameter called "total cost" (TC), which is the sum of gate count of an optimized circuit, number of garbage

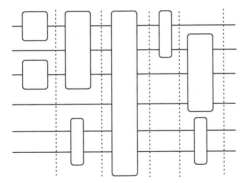

Figure 4.7: A schematic diagram of quantum circuit. Here each vertical line separates two time slices and each rectangular box represents a quantum gate. So depth of the circuit is 5. We cannot directly compute delay as the circuit uses three nonprimitive quantum gates.

bits and quantum cost, is recently introduced [66]. For reduction of TC, we need to simultaneously reduce the circuit cost, garbage count (number of garbage bits) and quantum cost. This is an open problem and at present neither an algorithm for simultaneous reduction of all these measures nor a satisfactory algorithm for reduction of quantum cost exists. Here it would be apt to note that TC is a weak measure of quality of a circuit.

4.5.2 Circuit optimization rules

In the previous section we observed that there exist several measures of quality of a quantum circuit and the lesser the value of a particular measure, the better is the circuit. So we need to optimize quantum circuits. Global optimization of a large quantum circuit is difficult. Most of the tools used for optimization of quantum circuits are actually local optimization tools. To be precise, instead of optimizing the whole circuit, they locally optimize comparatively smaller circuit blocks of the large quantum circuit. Some interesting techniques of circuit optimization are discussed below.

4.5.2.1 Moving rule

Moving rule is a consequence of commutation rule. Let quantum gates U_1, U_2 and U_3 be placed in sequence in a circuit. Thus our circuit is $U_1U_2U_3$. Now if $U_1U_2 = U_2U_1$ (i.e., if $[U_1, U_2] = 0$) then we can interchange the first two gates (equivalently we can move the first (second) gate in right (left) by one step) to obtain an equivalent circuit $U_2U_1U_3$. A simple example that demonstrates the moving rule is shown in Fig. 4.8. Here it would be apt to note that a specific application of moving rule in the optimization of

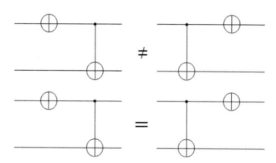

Figure 4.8: An example of commutation rule in context of quantum circuits. If a NOT gate is placed at the control qubit line of a CNOT gate it does not commute with the CNOT gate but if a NOT gate is placed at the target qubit line then it commutes with the CNOT gate.

a quantum circuit can be visualized through deletion rule. If moving rule brings two inverse gates (two CNOT gates or controlled-V and controlled-$V^\dagger$) in two consecutive positions in the same qubit lines then the pair of gates would be deleted and that would lead to optimization of the circuit. Let us now provide two explicit examples to show that moving rule and deletion rule are useful for optimization of the quantum circuits.

Example 4.8: Consider the circuit shown in Fig. 4.6a. Here the last two gates commute. So we can move the last CNOT gate to the left of the adjacent controlled-V gate. This movement brings two CNOT gates in consecutive positions. As CNOT is self-inverse so CNOT(CNOT) = I and we can delete the consecutive CNOTs (here we are applying deletion rule) to obtain the optimized circuit shown in Fig. 4.6b. We have already seen that the circuit shown in Fig. 4.6b is better than the circuit shown in Fig. 4.6a with respect to different cost metrics. Consequently, moving rule and deletion rule are useful for reduction of different quantitative measures of quality of quantum circuits.

Example 4.9: Consider the circuit shown on the left hand side of Fig. 4.9, where the first two gates commute. After application of moving rule we obtain the equivalent circuit where two V gates appear in consecutive positions in the second qubit line. As $VV = $ NOT so we can replace these two V gates by a NOT gate and obtain an optimized circuit comprised of two gates. The optimized circuit is shown on the right hand side of the circuit identity shown in Fig. 4.9.

4.5.2.2 Template matching

A template is a circuit that makes an Identity. A trivial example is $VV\,X=$ I. Now in the previous example, two VV gates appeared in consecutive positions on the same qubit line. In that case, we can use the template

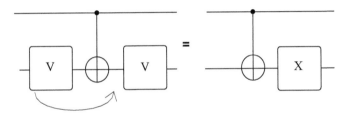

Figure 4.9: Local optimization of a circuit using moving rule and template matching. Here the first V gate commutes with the CNOT gate. The movement of the V gate is shown with the arrow. Subsequently $VVX = I$ template is used to replace two consecutive V gates by a X gate.

$VVX = I$ to obtain $VV = X^{-1} = X$ and replace VV by X gate to reduce the gate count. This is a simple example. In general, if more than half of the consecutive quantum gates present in a given template appear in consecutive positions in a quantum circuit then we can use the template to reduce the gate count. For example, look at the schematic diagram of a general quantum circuit shown in Fig. 4.10a. In Fig. 4.10b we show a template. We observe that the three consecutive gates which are shown in a rectangular box in Fig. 4.10a are the same as the first three gates of the template shown in Fig. 4.10b. Now the template shown in Fig. 4.10b implies the circuit identity shown in Fig. 4.10c. We can now use the right hand side of the circuit equation shown in Fig. 4.10c to replace the three gates shown in the rectangular box in Fig. 4.10a. This will lead to reduction of gate count and we will obtain the circuit shown in Fig. 4.10d. This is how template matching works. The template matching is of two types: forward matching and backward matching.

Forward matching and backward matching

Let us consider a template $U_1U_2U_3U_4U_5U_6 = I$. For simplification we assume that $U_i = U_i^{-1}$. Now if a sequence of gates $U_3U_4U_5U_6$ appears in a circuit, then this sequence of gates can be replaced by $U_2^{-1}U_1^{-1}$. If quantum gates are self-inverse then $U_2^{-1}U_1^{-1} = U_2U_1$. This substitution is called forward matching. This type of substitution by template matching reduces the circuit cost provided the number of gates present in the sequence of gates to be replaced is more than half of the template size. Similarly, when a sequence of gates $U_4U_3U_2U_1$ is substituted by U_5U_6 then the template matching is referred to as backward matching.

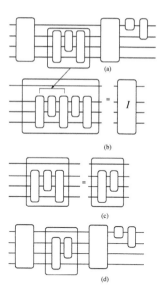

Figure 4.10: Application of template matching tool: (a) An arbitrary quantum circuit, (b) a template, (c) matched sequence that will be substituted by smaller sequence, (d) optimized circuit.

4.5.3   Let us visualize the quantum gate

Time dependent Schrodinger equation is written as

$$H\psi = i\hbar \frac{d\psi}{dt}.$$

For convenience, if we consider $\hbar = 1$ then we can write the time evolution of wave function ψ as

$$\psi(t) = e^{-iHt}\psi(0).$$

If we wish to visualize it as a gate then we have to visualize $\psi(0)$ as an input state and $\psi(t)$ as an output state. In that case we have $\psi_{\text{out}} = U\psi_{\text{in}}$ where $U = e^{-iHt}$ is an operator which may be visualized as a quantum gate or a quantum circuit that maps the input states $\psi_{\text{in}} = \psi(0)$ into the output state $\psi_{\text{out}} = \psi(t)$. Now to achieve a complete circuit we may need to use the output of one operation as the input of the next operation.

4.6   Discussion

In this chapter we have learned about quantum circuits and quantitative measures of their quality. In the next chapter, we will learn about quantum algorithms and will show that quantum circuits play a crucial role in the realization of quantum algorithms. Before we discuss quantum algorithms it would be appropriate to note that the quantum gates described in the present chapter are experimentally realizable and physical realization of quantum gates and circuits are reported by several groups using different implementations of qubits. For example, in 2003 J. L. O'Brien *et al.* [67] provided a completely optical implementation of a CNOT gate. A CNOT gate is also demonstrated using quantum dot, NMR, ion trap, superconductivity, Rydberg blockade interactions between neutral atoms held in optical traps, etc. (see [68] and references therein). Even larger quantum gates (three qubit gates) are experimentally realized in many different ways. For example, experimental realization of a Toffoli gate is reported using ion trap [69] and superconducting circuits [70]. Further, an optical Fredkin gate [71] is also reported. The point we are trying to make is that at the moment there are several implementations of qubit and each of them may be used to build quantum gates. For example, we may make quantum gates using NMR, optical, ion trap, superconductive techniques. More or less all of the popular quantum gates are successfully constructed using each of these techniques. So in principle, we can construct any quantum circuit and demonstrate quantum algorithms and implement quantum teleportation protocol. However, several problems arise when we try to build a large quantum circuit (quantum computer). The problems associated with the implementation of large quantum circuits and the possible ways to circumvent them are discussed in Chapter 6. Before we describe that, it is tempting to see how the quantum gates introduced in this chapter can be used to implement quantum algorithms. In the next chapter we will discuss quantum algorithms.

4.7 Solved examples

1. Suppose a two-qubit system is in the state $|\psi\rangle = 0.8|00\rangle + 0.6|11\rangle$. A NOT gate is applied to the second qubit and a two qubit measurement is performed in the computational basis. What are the probabilities of the possible measurement outcomes?
   Solution: After application of the NOT gate on the second qubit the state becomes $|\psi\rangle_1 = 0.8|01\rangle + 0.6|10\rangle$. Thus the probability of obtaining $|01\rangle$ as outcome is $(0.8)^2 = 0.64$ and the probability of obtaining $|10\rangle$ is $(0.6)^2 = 0.36$.

2. Show that $P(\pi) = Z$.
   Solution: $P(\pi) = \begin{pmatrix} 1 & 0 \\ 0 & \exp(i\pi) \end{pmatrix} = \begin{pmatrix} 1 & 0 \\ 0 & -1 \end{pmatrix} = Z.$

3. Write down the truth table and matrix for Fredkin (CSWAP) gate.
   Solution: Fredkin is Controlled-SWAP gate. If we consider the first bit as the control bit and the last two bits as target bits then the last two bits swap only when the first one is 1. Thus the truth table of

Input state	Output state
$\lvert 000\rangle$	$\lvert 000\rangle$
$\lvert 001\rangle$	$\lvert 001\rangle$
$\lvert 010\rangle$	$\lvert 010\rangle$
$\lvert 011\rangle$	$\lvert 011\rangle$
$\lvert 100\rangle$	$\lvert 100\rangle$
$\lvert 101\rangle$	$\lvert 110\rangle$
$\lvert 110\rangle$	$\lvert 101\rangle$
$\lvert 111\rangle$	$\lvert 111\rangle$

Table 4.2: Truth table of Fredkin gate.

Fredkin is as shown in Table 4.2.

The matrix representation of the gate is already given in the text (see Eqn. (4.23)).

4. Show that it is impossible to define a classical $\sqrt{\text{NOT}}_{\text{cl}}$ gate using binary logic.

Solution: As two consecutive operations of $\sqrt{\text{NOT}}_{\text{cl}}$ should be equivalent to a classical NOT gate so we must have $\sqrt{\text{NOT}}_{\text{cl}}\sqrt{\text{NOT}}_{\text{cl}}\lvert 0\rangle = \lvert 1\rangle$ and $\sqrt{\text{NOT}}_{\text{cl}}\sqrt{\text{NOT}}_{\text{cl}}\lvert 1\rangle = \lvert 0\rangle$. Now we ask: What is the output of $\sqrt{\text{NOT}}_{\text{cl}}\lvert 0\rangle$? In binary logic there are only two possibilities, $\sqrt{\text{NOT}}_{\text{cl}}\lvert 0\rangle = \lvert 1\rangle$ and $\sqrt{\text{NOT}}_{\text{cl}}\lvert 0\rangle = \lvert 0\rangle$. We can analyze these two possibilities separately. (i) If $\sqrt{\text{NOT}}_{\text{cl}}\lvert 0\rangle = \lvert 1\rangle$ and $\sqrt{\text{NOT}}_{\text{cl}}\sqrt{\text{NOT}}_{\text{cl}}\lvert 0\rangle = \lvert 1\rangle$ then we must have $\sqrt{\text{NOT}}_{\text{cl}}\lvert 1\rangle = \lvert 1\rangle$ and consequently we will get $\sqrt{\text{NOT}}_{\text{cl}}\sqrt{\text{NOT}}_{\text{cl}}\lvert 1\rangle = \lvert 1\rangle \neq \lvert 0\rangle$. It cannot implement $\sqrt{\text{NOT}}_{\text{cl}}$ as two consecutive applications of it cannot invert $\lvert 1\rangle$. (ii) Now look at the second possibility, i.e., if $\sqrt{\text{NOT}}_{\text{cl}}\lvert 0\rangle = \lvert 0\rangle$ then $\sqrt{\text{NOT}}_{\text{cl}}\sqrt{\text{NOT}}_{\text{cl}}\lvert 0\rangle = \lvert 0\rangle \neq 1$. So two consecutive application of $\sqrt{\text{NOT}}_{\text{cl}}$ cannot invert $\lvert 0\rangle$. Since there were only two possibilities, we conclude that it is impossible to define a classical $\sqrt{\text{NOT}}_{\text{cl}}$ gate using binary logic.

5. In some books [1] $\sqrt{\text{NOT}}$ gate is given as

$$\sqrt{\text{NOT}} = \frac{1}{\sqrt{2}}\begin{pmatrix} 1 & -1 \\ 1 & 1 \end{pmatrix}. \tag{4.24}$$

In other books [63] it is given as

$$\sqrt{\text{NOT}} = \begin{pmatrix} \frac{1+i}{2} & \frac{1-i}{2} \\ \frac{1-i}{2} & \frac{1+i}{2} \end{pmatrix} = \frac{\exp\left(i\frac{\pi}{4}\right)}{\sqrt{2}}\begin{pmatrix} 1 & -i \\ -i & 1 \end{pmatrix}. \tag{4.25}$$

What is the difference between the two? Which one is a better representation in your opinion?

Solution: Differences are here: (i) $\sqrt{\text{NOT}} = \frac{1}{\sqrt{2}}\begin{pmatrix} 1 & -1 \\ 1 & 1 \end{pmatrix}$ maps

$|0\rangle$ and $|1\rangle$ as follows: $\sqrt{\text{NOT}}|0\rangle = \frac{1}{\sqrt{2}}(|0\rangle+|1\rangle) = |+\rangle$ and $\sqrt{\text{NOT}}|1\rangle =$ $\frac{1}{\sqrt{2}}(|1\rangle - |0\rangle) = |-\rangle$ whereas $\sqrt{\text{NOT}} = \begin{pmatrix} \frac{1+i}{2} & \frac{1-i}{2} \\ \frac{1-i}{2} & \frac{1+i}{2} \end{pmatrix}$ maps $|0\rangle$ and $|1\rangle$ as follows: $\sqrt{\text{NOT}}|0\rangle = \frac{1+i}{2}|0\rangle + \frac{1-i}{2}|1\rangle = \frac{1}{\sqrt{2}}(|+\rangle + i|-\rangle)$ and $\sqrt{\text{NOT}}|1\rangle = \frac{1-i}{2}|0\rangle + \frac{1+i}{2}|1\rangle = \frac{1}{\sqrt{2}}(|+\rangle - i|-\rangle)$. (ii) When $\sqrt{\text{NOT}} = \frac{1}{\sqrt{2}}\begin{pmatrix} 1 & -1 \\ 1 & 1 \end{pmatrix}$ then $\sqrt{\text{NOT}}\sqrt{\text{NOT}} = \begin{pmatrix} 0 & -1 \\ 1 & 0 \end{pmatrix}$ which is equivalent to NOT as far as its action on $|0\rangle$ and $|1\rangle$ are concerned but when it works on $|+\rangle$ then we obtain $\sqrt{\text{NOT}}\sqrt{\text{NOT}}|+\rangle = -|-\rangle \equiv |-\rangle$ but ideally $\text{NOT}|+\rangle = |+\rangle$. Thus $\sqrt{\text{NOT}} = \frac{1}{\sqrt{2}}\begin{pmatrix} 1 & -1 \\ 1 & 1 \end{pmatrix}$ does not really work as square root of NOT for superposition states. Now for $\sqrt{\text{NOT}} = \begin{pmatrix} \frac{1+i}{2} & \frac{1-i}{2} \\ \frac{1-i}{2} & \frac{1+i}{2} \end{pmatrix}$ we obtain $\sqrt{\text{NOT}}\sqrt{\text{NOT}} = \begin{pmatrix} 0 & 1 \\ 1 & 0 \end{pmatrix} = X$ which is exactly the matrix of NOT gate, hence its name is even justified for superposition states. Therefore, in our opinion $\sqrt{\text{NOT}} = \begin{pmatrix} \frac{1+i}{2} & \frac{1-i}{2} \\ \frac{1-i}{2} & \frac{1+i}{2} \end{pmatrix}$ is a better representation.

6. Prove that matrix representation of $\sqrt{\text{NOT}}$ in (4.25) is unitary.

   Solution: It is easy to check that

$$\sqrt{\text{NOT}}\sqrt{\text{NOT}}^\dagger = \begin{pmatrix} \frac{1+i}{2} & \frac{1-i}{2} \\ \frac{1-i}{2} & \frac{1+i}{2} \end{pmatrix}\begin{pmatrix} \frac{1-i}{2} & \frac{1+i}{2} \\ \frac{1+i}{2} & \frac{1-i}{2} \end{pmatrix} = \begin{pmatrix} 1 & 0 \\ 0 & 1 \end{pmatrix} = I,$$

   and

$$\sqrt{\text{NOT}}^\dagger \sqrt{\text{NOT}} = \begin{pmatrix} \frac{1-i}{2} & \frac{1+i}{2} \\ \frac{1+i}{2} & \frac{1-i}{2} \end{pmatrix}\begin{pmatrix} \frac{1+i}{2} & \frac{1-i}{2} \\ \frac{1-i}{2} & \frac{1+i}{2} \end{pmatrix} = \begin{pmatrix} 1 & 0 \\ 0 & 1 \end{pmatrix} = I.$$

   Therefore, $\sqrt{\text{NOT}}$ described in (4.25) is unitary.

7. Is $\sqrt{\text{NOT}}$ described in (4.25) Hermitian operator? Is it a normal operator?

   Solution: No it is not Hermitian as $\sqrt{\text{NOT}} = \begin{pmatrix} \frac{1+i}{2} & \frac{1-i}{2} \\ \frac{1-i}{2} & \frac{1+i}{2} \end{pmatrix} \neq \begin{pmatrix} \frac{1-i}{2} & \frac{1+i}{2} \\ \frac{1+i}{2} & \frac{1-i}{2} \end{pmatrix} = \left(\sqrt{\text{NOT}}\right)^\dagger$. However, it is a normal operator. In the previous problem we have already shown that it is unitary and all unitary operators are normal.

8. Construct a quantum circuit using only Hadamard gates to create the state

$$|\psi\rangle = \frac{(|000\rangle + |001\rangle + |010\rangle + |011\rangle + |100\rangle + |101\rangle + |110\rangle + |111\rangle)}{\sqrt{8}},$$

from three qubits, which are initially in the state $|000\rangle$.

Solution: Just apply three Hadamard gates in parallel on the three input qubits.

9. Show that applying Hadamard transformations individually to N qubits each in the state $|0\rangle$ puts them into an equal superposition of the 2^N possible logical states.

Solution:

$$
\begin{aligned}
& H^{\otimes N}|0\rangle^{\otimes N} \\
=\ & H|0_1\rangle H|0_2\rangle \cdots H|0_N\rangle \\
=\ & \tfrac{1}{\sqrt{2}}(|0_1\rangle + |1_1\rangle)\tfrac{1}{\sqrt{2}}(|0_2\rangle + |1_2\rangle)\cdots(|0_N\rangle + |1_N\rangle) \\
=\ & \tfrac{1}{\sqrt{2^N}}(|0_1 0_2 \cdots 0_N\rangle + |0_1 0_2 \cdots 1_N\rangle + \cdots + |1_1 1_2 \cdots 1_N\rangle),
\end{aligned}
$$

where subscripts are used to denote the qubit number. It clearly shows that $H^{\otimes N}|0\rangle^{\otimes N}$ creates an equal superposition of 2^N possible logical states.

10. Show that $HXH = Z$ and $HZH = X$.

Solution:

$$
\begin{aligned}
HXH &= \tfrac{1}{\sqrt{2}}\begin{pmatrix} 1 & 1 \\ 1 & -1 \end{pmatrix}\begin{pmatrix} 0 & 1 \\ 1 & 0 \end{pmatrix}\tfrac{1}{\sqrt{2}}\begin{pmatrix} 1 & 1 \\ 1 & -1 \end{pmatrix} \\
&= \tfrac{1}{2}\begin{pmatrix} 1 & 1 \\ 1 & -1 \end{pmatrix}\begin{pmatrix} 1 & -1 \\ 1 & 1 \end{pmatrix} = \begin{pmatrix} 1 & 0 \\ 0 & -1 \end{pmatrix} = Z.
\end{aligned}
$$

Now we can prove the second identity either by using the same method or by using the first identity and the fact that $H^2 = I$ as $HZH = H(HXH)H = H^2 X H^2 = IXI = X$.

11. Write down the truth table corresponding to the following gate:

$$
\begin{aligned}
G &= \tfrac{1}{\sqrt{2}}((|00\rangle + |01\rangle)\langle 00| + (|01\rangle - |11\rangle)\langle 01| \\
&+ (|11\rangle - |00\rangle)\langle 10| + (|11\rangle + |10\rangle)\langle 11|)
\end{aligned}
$$

and also provide the matrix that represents this gate. How will you check whether G is a valid quantum gate or not?

Solution: The truth table is shown in Table 4.3 and we can obtain the matrix by replacing $|00\rangle$, $|01\rangle$, $|10\rangle$, $|11\rangle$ by their equivalent column matrices and the corresponding bras by row matrices. Then the simple matrix multiplication and addition will yield the equivalent matrix as

$$
U = \begin{pmatrix}
\frac{1}{\sqrt{2}} & 0 & -\frac{1}{\sqrt{2}} & 0 \\
\frac{1}{\sqrt{2}} & \frac{1}{\sqrt{2}} & 0 & 0 \\
0 & 0 & 0 & \frac{1}{\sqrt{2}} \\
0 & -\frac{1}{\sqrt{2}} & \frac{1}{\sqrt{2}} & \frac{1}{\sqrt{2}}
\end{pmatrix}.
$$

Input state	Output state
$|00\rangle$	$\frac{|00\rangle+|01\rangle}{\sqrt{2}}$
$|01\rangle$	$\frac{|01\rangle-|11\rangle}{\sqrt{2}}$
$|10\rangle$	$\frac{|11\rangle-|00\rangle}{\sqrt{2}}$
$|11\rangle$	$\frac{|11\rangle+|10\rangle}{\sqrt{2}}$

Table 4.3: Truth table of the gate G given in the Solved Example 11.

Now one can easily check that it is not a valid quantum operation as it is not a unitary operator. The same is shown below.

$$
UU^{\dagger} = \begin{pmatrix} \frac{1}{\sqrt{2}} & 0 & -\frac{1}{\sqrt{2}} & 0 \\ \frac{1}{\sqrt{2}} & \frac{1}{\sqrt{2}} & 0 & 0 \\ 0 & 0 & 0 & \frac{1}{\sqrt{2}} \\ 0 & -\frac{1}{\sqrt{2}} & \frac{1}{\sqrt{2}} & \frac{1}{\sqrt{2}} \end{pmatrix} \begin{pmatrix} \frac{1}{\sqrt{2}} & \frac{1}{\sqrt{2}} & 0 & 0 \\ 0 & \frac{1}{\sqrt{2}} & 0 & -\frac{1}{\sqrt{2}} \\ -\frac{1}{\sqrt{2}} & 0 & 0 & \frac{1}{\sqrt{2}} \\ 0 & 0 & \frac{1}{\sqrt{2}} & \frac{1}{\sqrt{2}} \end{pmatrix}
$$

$$
= \begin{pmatrix} 1 & \frac{1}{2} & 0 & -\frac{1}{2} \\ \frac{1}{2} & 1 & 0 & -\frac{1}{2} \\ 0 & 0 & \frac{1}{2} & \frac{1}{2} \\ -\frac{1}{2} & -\frac{1}{2} & \frac{1}{2} & \frac{3}{2} \end{pmatrix} \neq I.
$$

12. Verify whether the following representation of a CNOT gate is correct or not: $U_{\text{CNOT}} = |0\rangle|0\rangle\langle0|\langle0| + |0\rangle|1\rangle\langle0|\langle1| + |1\rangle|1\rangle\langle1|\langle0| + |1\rangle|0\rangle\langle1|\langle1|$.
    Solution: This is clearly the representation of a CNOT gate as the operation

$$
\begin{aligned}
U_{\text{CNOT}} &= |0\rangle|0\rangle\langle0|\langle0| + |0\rangle|1\rangle\langle0|\langle1| + |1\rangle|1\rangle\langle1|\langle0| + |1\rangle|0\rangle\langle1|\langle1| \\
&= |00\rangle\langle00| + |01\rangle\langle01| + |11\rangle\langle10| + |10\rangle\langle11|
\end{aligned}
$$

    maps $|00\rangle, |01\rangle, |10\rangle$ and $|11\rangle$ to $|00\rangle, |01\rangle, |11\rangle$ and $|10\rangle$ respectively.

13. Find the matrix representation of a NOT gate in the Hadamard basis.
    Solution: A NOT gate in Hadamard basis $\{|+\rangle, |-\rangle\}$ implies that it transforms $|+\rangle$ to $|-\rangle$ and $|-\rangle$ to $|+\rangle$. Thus the NOT gate is

$$
\begin{aligned}
\text{NOT} &= |+\rangle\langle-| + |-\rangle\langle+| \\
&= \frac{1}{2}\begin{pmatrix} 1 \\ 1 \end{pmatrix}\begin{pmatrix} 1 & -1 \end{pmatrix} + \frac{1}{2}\begin{pmatrix} 1 \\ -1 \end{pmatrix}\begin{pmatrix} 1 & 1 \end{pmatrix} = \begin{pmatrix} 1 & 0 \\ 0 & -1 \end{pmatrix}.
\end{aligned}
$$

14. Prove that $R_x(\pi)R_y(\frac{\pi}{2}) = H$ up to a global phase factor.

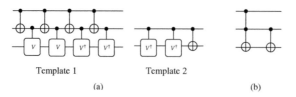

Template 1   Template 2

(a)   (b)

Figure 4.11: (a) Two quantum templates, (b) a quantum circuit to be simplified using the templates shown in (a). For detail see Solved Example 16.

Solution:

$$R_x(\pi)R_y\left(\tfrac{\pi}{2}\right)$$

$$= \begin{pmatrix} \cos\left(\tfrac{\pi}{2}\right) & -i\sin\left(\tfrac{\pi}{2}\right) \\ -i\sin\left(\tfrac{\pi}{2}\right) & \cos\left(\tfrac{\pi}{2}\right) \end{pmatrix} \begin{pmatrix} \cos\left(\tfrac{\pi}{4}\right) & -\sin\left(\tfrac{\pi}{4}\right) \\ \sin\left(\tfrac{\pi}{4}\right) & \cos\left(\tfrac{\pi}{4}\right) \end{pmatrix}$$

$$= \begin{pmatrix} 0 & -i \\ -i & 0 \end{pmatrix} \begin{pmatrix} \tfrac{1}{\sqrt{2}} & -\tfrac{1}{\sqrt{2}} \\ \tfrac{1}{\sqrt{2}} & \tfrac{1}{\sqrt{2}} \end{pmatrix}$$

$$= -i \,\tfrac{1}{\sqrt{2}} \begin{pmatrix} 1 & 1 \\ 1 & -1 \end{pmatrix}$$

$$= H.$$

In the last step we ignored a common phase factor $-i = e^{i\frac{3\pi}{2}}$.

15. Prove that $XR_Y(\theta)X = R_Y(-\theta)$.

    Solution: First we note that

$$XYX = \begin{pmatrix} 0 & 1 \\ 1 & 0 \end{pmatrix}\begin{pmatrix} 0 & -i \\ i & 0 \end{pmatrix}\begin{pmatrix} 0 & 1 \\ 1 & 0 \end{pmatrix} = \begin{pmatrix} 0 & i \\ -i & 0 \end{pmatrix} = -Y.$$

   Now we know that $R_y(\theta) = I\cos\left(\tfrac{\theta}{2}\right) - iY\sin\left(\tfrac{\theta}{2}\right)$. Therefore,

$$\begin{aligned} XR_y(\theta)X &= XIX\cos\left(\tfrac{\theta}{2}\right) - iXYX\sin\left(\tfrac{\theta}{2}\right) \\ &= I\cos\left(\tfrac{\theta}{2}\right) - i(-Y)\sin\left(\tfrac{\theta}{2}\right) \\ &= I\cos\left(-\tfrac{\theta}{2}\right) - iY\sin\left(-\tfrac{\theta}{2}\right) \\ &= R_y(-\theta). \end{aligned}$$

16. Two templates are given in Fig. 4.11a. First convert the *NCT* circuit shown in Fig. 4.11b to an equivalent *NCV* circuit and then use the given templates to optimize the *NCV* circuit. What are the circuit costs of the circuit shown in Fig. 4.11b in an *NCT* gate library and in an *NCV* gate library? Also find (a) linear and nonlinear quantum cost (b) depth in an *NCT* library and in an *NCV* library and (c) width of the circuit.

    Solution: The given circuit is shown in Fig. 4.12a. The circuit can be transformed into an *NCV* circuit by substituting the Toffoli gate

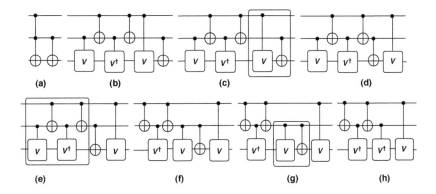

Figure 4.12: Solution of Solved Example 16.

by quantum primitives. The *NCV* circuit obtained in this way is shown in Fig 4.12b. Commutation rule is applied on the gates shown in a rectangular box in Fig 4.12c and it results in Fig 4.12d. Template 1 from Fig 4.11a is then applied in the rectangular box in Fig 4.12e and it results in Fig. 4.12f. Similarly Template 2 is applied in the rectangular box in Fig 4.12g and it results in Fig 4.12h. The circuit in Fig. 4.12h is the simplified *NCV* circuit.

Now Fig. 4.12a is the given *NCT* circuit with two gates, which cannot be further reduced using an *NCT* library so its circuit cost is 2, width is 3 as there are 3 qubits and depth is 2. Now the linear quantum cost of the circuit is the sum of quantum cost of Toffoli and CNOT, i.e., 5+1=6. To obtain the nonlinear quantum cost we have optimized the equivalent *NCV* circuit and obtain Fig. 4.12h, from which we can see that the nonlinear quantum cost is 5. Further, from the optimized *NCV* circuit shown in Fig. 4.12h we find that the depth of the circuit is 5, gate count is 5 and width is 3.

This example clarifies many ideas, for example, values of cost metrics depend on the choice of gate library, linear quantum cost is different from nonlinear quantum cost, *NCT* circuits need to be transformed to *NCV* circuits for computation of quantum cost, a template may not reduce the gate count directly as in Fig. 4.12e→Fig. 4.12f, the template matching does not reduce the gate count directly, but it helped us to reduce gate count by using the second template in Fig. 4.12g etc.

17. Obtain nonlinear quantum cost of the circuit of function 3_17 shown in Fig. 4.13. Necessary optimization of the circuit may be done using the commutation rule and matching rule (forward and backward).
    Solution: The given circuit (Fig. 4.13 also shown in Fig. 4.14a) is an *NCT* circuit. To obtain the quantum cost we need to optimize the

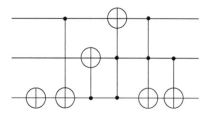

Figure 4.13: Reversible circuit for function 3_17 given in the benchmark page of D. Maslov *et al.* [72]. In Solved Example 17 our task is to find the quantum cost of this circuit.

equivalent NCV circuit. We can either substitute the Toffoli gates present in Fig. 4.14a by the equivalent NCV circuit and then optimize the NCV circuit as we have done in the previous example or we may first try to optimize the given NCT circuit and then replace the Toffoli gate present in the optimized NCT circuit by the equivalent NCV circuit and further optimize the obtained NCV circuit to obtain the quantum cost. Here we have followed the second approach. The given NCT circuit is shown in Fig. 4.14a. In Fig. 4.14b, commutation rule is applied and the arrow shows the movement of the CNOT gate. In Fig. 4.14c we show the NCT circuit obtained after commutation is applied, i.e., before substitution of quantum primitive gates (note that applied commutation rule has not reduced the gate count but will be found useful later when 2 two qubit gates will appear consecutively on the same qubit lines) and in Fig. 4.14d the quantum circuit is obtained by substituting the Toffoli gates with primitives. In Fig. 4.14e, a template matching tool is applied to the circuit (at positions indicated by underbars) to yield Fig. 4.14f where a quantum circuit with reduced gate count is obtained. In Fig. 4.14g, moving rule is applied and two movements have been done in the circuit as indicated by the arrows. As a result of application of moving rule we obtain Fig. 4.14h, where new gates are introduced (each dashed box is a new gate) to obtain a nonlinear quantum cost. Finally the quantum cost of the given circuit is found to be 7.

18. Find the quantum cost of EPR circuit and Toffoli gate.
    Solution: Quantum cost of EPR circuit is 1 as the single qubit Hadamard gate can be combined with the CNOT gate. Similarly, quantum cost of Toffoli is 5. See Fig. 4.15, which shows standard implementation of Toffoli using 7 primitive quantum gates. The Hadamard gates present at the beginning and at the end can be combined with the $C - V$ gates and that reduces the quantum cost to 5.

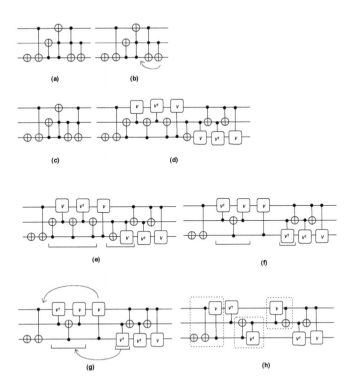

Figure 4.14: Solution of Solved Example 17. (b)-(h) show optimization of (a). From (h) we can observe that the quantum cost of the given circuit is 7.

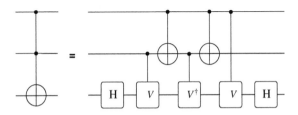

Figure 4.15: Quantum cost of Toffoli gate is 5 as the single qubit gates can be absorbed. For details see Solved Example 18.

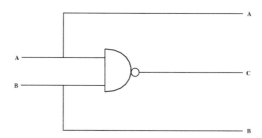

Figure 4.16: Reversible NAND gate constructed using fan-out operations increases number of qubit lines (width). For details see Solved Example 19.

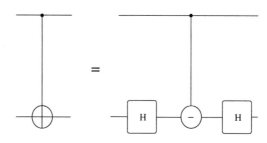

Figure 4.17: A circuit identity. For details see Solved Examples 20 and 21.

19. Can you convert an irreversible NAND gate into a reversible one just by using fan-out operations? If yes, then why don't we use this trick to convert all the existing irreversible circuits into their reversible counterparts, which will be more energy efficient?

    Solution: Yes, we can do that by making copy of each input state (using fan-out operations) as shown in Fig. 4.16. However, this process will considerably increase the width (number of qubit lines) of the circuit, which is not desirable.

20. CMINUS gate is defined as $C - Z$ gate. Now provide its matrix and prove the circuit identity shown in Fig. 4.17.

    Solution: In matrix form CMINUS $= C - Z = \begin{pmatrix} 1 & 0 & 0 & 0 \\ 0 & 1 & 0 & 0 \\ 0 & 0 & 1 & 0 \\ 0 & 0 & 0 & -1 \end{pmatrix}$.

Now in the given circuit identity

$$\text{RHS} = (I_2 \otimes H)\,\text{CMINUS}\,(I_2 \otimes H)$$

$$= \begin{pmatrix} 1 & 0 \\ 0 & 1 \end{pmatrix} \otimes \frac{1}{\sqrt{2}} \begin{pmatrix} 1 & 1 \\ 1 & -1 \end{pmatrix} \begin{pmatrix} 1 & 0 & 0 & 0 \\ 0 & 1 & 0 & 0 \\ 0 & 0 & 1 & 0 \\ 0 & 0 & 0 & -1 \end{pmatrix}$$

$$\begin{pmatrix} 1 & 0 \\ 0 & 1 \end{pmatrix} \otimes \frac{1}{\sqrt{2}} \begin{pmatrix} 1 & 1 \\ 1 & -1 \end{pmatrix}$$

$$= \frac{1}{2} \begin{pmatrix} 1 & 1 & 0 & 0 \\ 1 & -1 & 0 & 0 \\ 0 & 0 & 1 & 1 \\ 0 & 0 & 1 & -1 \end{pmatrix} \begin{pmatrix} 1 & 0 & 0 & 0 \\ 0 & 1 & 0 & 0 \\ 0 & 0 & 1 & 0 \\ 0 & 0 & 0 & -1 \end{pmatrix}$$

$$\begin{pmatrix} 1 & 1 & 0 & 0 \\ 1 & -1 & 0 & 0 \\ 0 & 0 & 1 & 1 \\ 0 & 0 & 1 & -1 \end{pmatrix}$$

$$= \frac{1}{2} \begin{pmatrix} 1 & 1 & 0 & 0 \\ 1 & -1 & 0 & 0 \\ 0 & 0 & 1 & 1 \\ 0 & 0 & 1 & -1 \end{pmatrix} \begin{pmatrix} 1 & 1 & 0 & 0 \\ 1 & -1 & 0 & 0 \\ 0 & 0 & 1 & 1 \\ 0 & 0 & -1 & 1 \end{pmatrix}$$

$$= \begin{pmatrix} 1 & 0 & 0 & 0 \\ 0 & 1 & 0 & 0 \\ 0 & 0 & 0 & 1 \\ 0 & 0 & 1 & 0 \end{pmatrix}$$

$$= \text{CNOT} = \text{LHS}.$$

21. Show that the CNOT gate can be constructed from Hadamard gates and the controlled−Z gate. Demonstrate that the construction is correct by multiplying the corresponding matrices.
    Solution: Same as the previous solved example.

22. Given that if U is an arbitrary single qubit gate (unitary operation) then

$$\text{Controlled}-U = \begin{pmatrix} I & O \\ O & U \end{pmatrix},$$

where I, O are respectively the Identity and Null operations (gate) in C^2. Use this general form of controlled gate to explicitly provide the matrix of CNOT and $C - Y$ gates.
    Solution: Here I, O and U can be represented by 2×2 matrices. For example, in the first case U is a NOT operation then $U = \begin{pmatrix} 0 & 1 \\ 1 & 0 \end{pmatrix}$, $O = \begin{pmatrix} 0 & 0 \\ 0 & 0 \end{pmatrix}$ and $I = \begin{pmatrix} 1 & 0 \\ 0 & 1 \end{pmatrix}$ and just by replacing these three

matrices in the given expression we obtain

$$\text{Controlled}-U = \text{CNOT} = \begin{pmatrix} 1 & 0 & 0 & 0 \\ 0 & 1 & 0 & 0 \\ 0 & 0 & 0 & 1 \\ 0 & 0 & 1 & 0 \end{pmatrix}.$$

Similarly, in the second case $U = Y = \begin{pmatrix} 0 & -i \\ i & 0 \end{pmatrix}$. Therefore,

$$\text{Controlled}-Y = C - Y = \begin{pmatrix} 1 & 0 & 0 & 0 \\ 0 & 1 & 0 & 0 \\ 0 & 0 & 0 & -i \\ 0 & 0 & i & 0 \end{pmatrix}.$$

4.8   Further reading

1. D. P. DiVincenzo, Quantum gates and circuits, Proc. R. Soc. Lond. A, **454** (1998) 261-276, quant-ph/9705009v1.

2. J. L. O'Brien, Optical quantum computing, Science, **318** (2007) 318 1567-1570, quant-ph/0803.1554v1. It is an excellent short review which discusses how one can do all optical quantum computing using only single photon sources, linear optical elements, and single photon detectors.

3. G. F. Viamontes, I. L. Markov and J. P. Hayes, Quantum circuit simulation, Springer, Dordrecht, Netherlands (2009).

4. QuaSi: This is an excellent simulator of quantum circuits in Java. Users can either design and simulate their own circuit or use the existing circuits for Shor's algorithm, Grover's algorithm, etc. This has an applet version which you can run online and a full downloadable version. Documentation is available in English. This can even factorize powers of primes ($25 = 5 \times 5$ or $343 = 7 \times 7 \times 7$) using Shor's algorithm. This particular feature is absent in Open Qubit. This is designed by IAKS, University of Karlsruhe. It's available for free but it needs Java 1.1 or higher version to be installed in your system. All the required links are available at http://iaks-www.ira.uka.de/home/matteck/QuaSi/aboutquasi.html.

4.9   Exercises

1. Construct AND, NOT, OR, NAND and NOR gates using circuits with just Fredkin gates. Can we consider Fredkin gate as a universal gate?

2. Provide the matrix form of the controlled$-H$ gate, where H is the Hadamard gate.

3. Prove the following useful circuit identity: $HYH = -Y$.

4. Show that up to a global phase $R_z(\theta) = P(\theta)$ and use that to show $R_z\left(\frac{\pi}{4}\right) = T$ up to a global phase.

5. What does a CCFredkin gate do?

6. Provide the matrix representation of CCZ and $CCP\left(\frac{\pi}{6}\right)$.

7. Express Deutsch, Fredkin and Toffoli gates in bra-ket notation.

8. The Hadamard, phase (S), CNOT, and $\frac{\pi}{8}$ gates form a universal gate set. However, one of these gates is unnecessary. Which one, and why? If it is unnecessary why is that kept in the set?

9. Check which of these gates are linear: (a) Hadamard (b) SWAP (c) NOR.

10. Show that AND is a nonlinear gate.

11. Justify following statements: (a) Circuit cost$\leq$ quantum cost, (b) Delay$\geq$ depth, (c) $\{$all quantum gates$\} \supset \{$all reversible gates$\}$, (d) Hadamard gate cannot be achieved classically.

12. Prove that $XR_Z(\theta)X = R_Z(-\theta)$.

13. Prove that all reversible gates with 2 inputs and 2 outputs are linear.

14. Prove the following circuit identities: (i) NOT $\equiv R_x\left(\pi\right)Ph\left(\frac{\pi}{2}\right)$, (ii) $\sqrt{\text{NOT}} \equiv R_x\left(\frac{\pi}{2}\right)Ph\left(\frac{\pi}{4}\right)$, (iii) $H \equiv R_x\left(\pi\right)R_y\left(\frac{\pi}{2}\right)Ph\left(\frac{\pi}{2}\right)$, where $Ph(\theta) = \exp\left(i\theta\right)\begin{pmatrix} 1 & 0 \\ 0 & 1 \end{pmatrix}$ is equivalent to a global phase.

15. A black box contains a classical gate which accepts n bits as inputs and produces a single bit as its output. Show that 2^n such Boolean gates are possible.

16. In binary logic how many 2-bit irreversible gates are possible? Show that the number is less than the number of possible 2-bit classical reversible gates.

17. Show that in the Bloch sphere, $R_x(\theta)$ and $R_y(\theta)$ can be visualized as rotations through an angle θ about the X axis and Y axis respectively.

18. Express the CNOT gate in the Dirac notation when the second qubit is the control qubit and the first qubit is the target qubit. Also provide its matrix form.

19. Use the circuit identity given in Fig. 4.17 to construct a template.

20. Verify that $\sqrt{\text{SWAP}} = \frac{1}{1+i} \begin{pmatrix} 1+i & 0 & 0 & 0 \\ 0 & 1 & i & 0 \\ 0 & i & 1 & 0 \\ 0 & 0 & 0 & 1+i \end{pmatrix}$. Also check that $\sqrt{\text{SWAP}}$ is a valid quantum gate.

Chapter 5

Quantum algorithms

An algorithm is a systematic procedure or a set of instructions to perform a particular computational task. Many people to associate algorithms with programming and believe that all algorithms were developed after the introduction of the modern computer (computer languages). This is a common misconception. Actually, algorithms are independent of machine and language and there are many nice algorithms that are more than a thousand years old. For example, Euclid's algorithm to find the greatest common divisor was introduced around 300 BC. Although the journey of the algorithm started before Christ, the specific word "algorithm" appeared much later. In fact, the word "algorithm" originates from the name of the Persian mathematician Abu Abdullah Muhammad bin Musa al-Khwarizmi[1], who was part of the royal court in Baghdad and who lived from about 780 to 850 AD. Most probably the word "algebra" also originated from his works. Here it would be apt to note that in our childhood when we learned addition, multiplication, etc. we essentially learned a set of algorithms as those were the systematic procedures for doing specific computational tasks.

A quantum algorithm is a physical process, which performs a computational task with the help of quantum effects. There exist a limited number of quantum algorithms, which execute the job faster than their existing classical counterparts. Among these algorithms the most popular are Grover's algorithm for unsorted database search and Shor's algorithm for factorization. It is interesting to note that David Deutsch was the first person who explicitly described a computational task that can be performed faster using quantum means than by any classical algorithm. Deutsch's original algorithm, which was a probabilistic one, has been modified by many people, and here we will present a modified Deutsch's algorithm. The present version of Deutsch's algorithm is very easy to understand and it can convincingly establish the fact that quantum algorithms can perform

---

[1]Readers can find a small biography of Al-khwarizmi at
http://www-groups.dcs.st-and.ac.uk/~history/Biographies/Al-Khwarizmi.html

certain jobs faster than their existing classical counterparts.

This chapter is dedicated to quantum algorithms. In the following section we describe Deutsch's algorithm. In the subsequent sections we discuss the Deutsch Jozsa algorithm, Grover's algorithm, Simon's algorithm and Shor's algorithm.

5.1   Deutsch's algorithm

Consider a function $f(x)$ having one bit domain and range. In a more lucid language the previous statement means both x and $f(x)$ can only have values 0 and 1. Thus we can have only four single bit functions $f(x) : \{0,1\} \to \{0,1\}$. Out of these four functions two are constant ($f(0) = f(1) = 0$ and $f(0) = f(1) = 1$) and two are balanced ($f(0) = 0$, $f(1) = 1$ and $f(0) = 1, f(1) = 0$) in the sense that output values 0 and 1 occur an equal number of times. Now assume that a black box or an oracle is given to us that can compute the value of the function $f(x)$ for a given input x. Our task is to determine whether the function computed by the oracle is constant or balanced. Alternatively, the problem may be visualized as follows. Consider a coin: if both sides of the coin have the same symbol then it represents a constant function. If the symbols are different then the coin represents a balanced function. Our task is to check whether both sides of the coin have the same symbol or not.

David Deutsch is one of the founders of quantum computing. He was born in 1953 in Hafia, Israel. Presently he works at the University of Oxford. In 1985 he introduced the idea of a universal quantum computer. In the same year he discovered the first quantum algorithm. His works provided physical meaning to the Church-Turing hypothesis. He has also contributed enormously to the field of quantum circuits and quantum error correction.
Photo credit: Lulie Tanett, photo courtesy: D. Deutsch.

To solve this problem classically we need at least two queries. To be precise, we need to know values of both $f(0)$ and $f(1)$, because knowing one of them will not give us any information about the other and we will not be able to conclude anything about whether the function is constant or balanced. But quantum mechanically it is possible to solve this problem in a single query. Let us see how it happens.

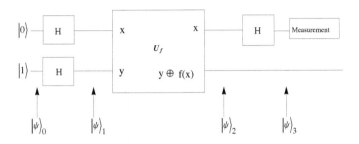

Figure 5.1: Circuit for the implementation of Deutsch's algorithm.

We start from a two qubit input state (cf. Fig. 5.1)

$$|\psi\rangle_0 = |01\rangle \tag{5.1}$$

and apply Hadamard gates on both the qubits to yield

$$|\psi\rangle_1 = \left[\frac{|0\rangle + |1\rangle}{\sqrt{2}}\right]\left[\frac{|0\rangle - |1\rangle}{\sqrt{2}}\right]. \tag{5.2}$$

Now we define a two qubit unitary transformation U_f, which transforms the state $|x, y\rangle$ into the state $|x, y \oplus f(x)\rangle$. Thus $U_f|x, y\rangle = |x, y \oplus f(x)\rangle$. It is clear that on operation of the unitary operator U_f the input state $|x, y\rangle$ remains unchanged if $f(x) = 0$ and it transforms to $|x, \bar{y}\rangle$ if $f(x) = 1$. Now if we apply U_f to $|\psi\rangle_1$ in that case $|x\rangle = \left[\frac{|0\rangle + |1\rangle}{\sqrt{2}}\right]$, $|y\rangle = \left[\frac{|0\rangle - |1\rangle}{\sqrt{2}}\right]$ and $\bar{y} = -y$. Thus we have $U_f\left(|x\rangle\left[\frac{|0\rangle - |1\rangle}{\sqrt{2}}\right]\right) = |x\rangle\left[\frac{|0\rangle - |1\rangle}{\sqrt{2}}\right]$ if $f(x) = 0$ and $U_f\left(|x\rangle\left[\frac{|0\rangle - |1\rangle}{\sqrt{2}}\right]\right) = -|x\rangle\left[\frac{|0\rangle - |1\rangle}{\sqrt{2}}\right]$ if $f(x) = 1$ or in brief, U_f : $\left(|x\rangle\left[\frac{|0\rangle - |1\rangle}{\sqrt{2}}\right]\right) \rightarrow (-1)^{f(x)}|x\rangle\left[\frac{|0\rangle - |1\rangle}{\sqrt{2}}\right]$. Now we can easily see that

$$|\psi\rangle_2 = U_f|\psi\rangle_1 = \begin{cases} \left[\frac{|0\rangle + |1\rangle}{\sqrt{2}}\right]\left[\frac{|0\rangle - |1\rangle}{\sqrt{2}}\right] & \text{if } f(0) = f(1) = 0 \\ -\left[\frac{|0\rangle + |1\rangle}{\sqrt{2}}\right]\left[\frac{|0\rangle - |1\rangle}{\sqrt{2}}\right] & \text{if } f(0) = f(1) = 1 \\ \left[\frac{|0\rangle - |1\rangle}{\sqrt{2}}\right]\left[\frac{|0\rangle - |1\rangle}{\sqrt{2}}\right] & \text{if } f(0) = 0, f(1) = 1 \\ -\left[\frac{|0\rangle - |1\rangle}{\sqrt{2}}\right]\left[\frac{|0\rangle - |1\rangle}{\sqrt{2}}\right] & \text{if } f(0) = 1, f(1) = 0, \end{cases} \tag{5.3}$$

or in brief,

$$|\psi\rangle_2 = \begin{cases} \pm\left[\dfrac{|0\rangle+|1\rangle}{\sqrt{2}}\right]\left[\dfrac{|0\rangle-|1\rangle}{\sqrt{2}}\right] & \text{if } f(0) = f(1) \Rightarrow \text{constant } f(x) \\[2ex] \pm\left[\dfrac{|0\rangle-|1\rangle}{\sqrt{2}}\right]\left[\dfrac{|0\rangle-|1\rangle}{\sqrt{2}}\right] & \text{if } f(0) \neq f(1) \Rightarrow \text{balanced } f(x). \end{cases} \tag{5.4}$$

Finally we apply a Hadamard gate on the first qubit to obtain

$$|\psi\rangle_3 = \begin{cases} \pm|0\rangle\left[\dfrac{|0\rangle-|1\rangle}{\sqrt{2}}\right] & \text{if } f(0) = f(1) \Rightarrow \text{constant } f(x) \\[2ex] \pm|1\rangle\left[\dfrac{|0\rangle-|1\rangle}{\sqrt{2}}\right] & \text{if } f(0) \neq f(1) \Rightarrow \text{balanced } f(x). \end{cases} \tag{5.5}$$

From (5.5) it is clear that a measurement of the first qubit in the computational basis will always yield 0 (1) for constant (balanced) function. Thus a single measurement on the first qubit will yield whether the function is constant or balanced, so a single query is sufficient to solve the problem. In any classical algorithm one will need at least two evolutions of $f(x)$. Consequently, this algorithm definitely establishes that for certain problems quantum algorithms can be more efficient than their classical counterparts. The gain is not too much in this simple case because we have only reduced the number of evolutions by one (from two evolutions to one evolution). In the next section we will consider a generalized version of the present algorithm to show that it is even possible to reduce the number of evolutions from 2^n to 1.

5.2 Deutsch Jozsa (DJ) algorithm

This is a generalized version of the previous algorithm. Here x can have any value from 0 to $2^n - 1$ but $f(x)$ can have values 0 and 1 only. In addition, there is an intrinsic promise or a constraint on $f(x)$ that $f(x)$ is either constant or balanced[2]. Now the task is to find out whether $f(x)$ is constant or balanced. Classically we will need at least $2^{n-1} + 1$ queries to know it with certainty. This is because even if $f(x)$ is balanced in the worst case we may get 2^{n-1} consecutive zeroes (or ones) before finally getting a 1 (0).

The algorithm is similar to the previous one but now we start with the $(n+1)$-qubit input state (see Fig. 5.2)

$$|\psi\rangle_0 = |0\rangle^{\otimes n}|1\rangle \tag{5.6}$$

and apply Hadamard transformation on all the qubits to yield

$$|\psi\rangle_1 = \sum_{x\in\{0,1\}^n} \frac{|x\rangle}{(\sqrt{2})^n}\left[\frac{|0\rangle-|1\rangle}{\sqrt{2}}\right]. \tag{5.7}$$

---

[2]Due to these intrinsic promises these kinds of algorithms are often referred to as promise algorithms.

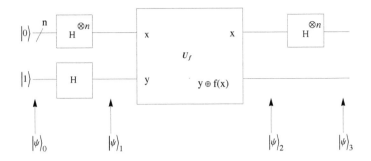

Figure 5.2: Circuit for Deutsch Jozsa algorithm.

Next the function f is evaluated by using the unitary transformation $U_f |x, y\rangle = |x, y \oplus f(x)\rangle$ and that gives us

$$|\psi\rangle_2 = \sum_x \frac{(-1)^{f(x)}|x\rangle}{\sqrt{2^n}} \left[\frac{|0\rangle - |1\rangle}{\sqrt{2}} \right]. \tag{5.8}$$

In the next step we apply Hadamard gates in parallel to the first n qubits (i.e., we apply $H^{\otimes n}$ on $|x\rangle$). Now by separately checking the cases $x = 0$ and $x = 1$ it is easy to observe that for a single qubit

$$H|x\rangle = \sum_{z=0}^{1} (-1)^{xz} \frac{|z\rangle}{\sqrt{2}}. \tag{5.9}$$

Thus the action of n Hadamard gates on the n-qubit state $|x\rangle$ can be visualized as

$$
\begin{aligned}
H^{\otimes n}|x\rangle &= H^{\otimes n}|x_1 x_2 .... x_n\rangle = H|x_1\rangle H|x_2\rangle \cdots H|x_n\rangle \\
&= \sum_{z_1=0}^{1} (-1)^{x_1 z_1} \frac{|z_1\rangle}{\sqrt{2}} \sum_{z_2=0}^{1} (-1)^{x_2 z_2} \frac{|z_2\rangle}{\sqrt{2}} \cdots \sum_{z_n=0}^{1} (-1)^{x_n z_n} \frac{|z_n\rangle}{\sqrt{2}} \\
&= \prod_{i=1}^{n} \left(\sum_{z_i=0}^{1} (-1)^{x_i z_i} \frac{|z_i\rangle}{\sqrt{2}} \right) = \frac{1}{\sqrt{2^n}} \sum_{z \in \{0,1\}^n} (-1)^{x.z} |z\rangle,
\end{aligned}
\tag{5.10}
$$

where

$$x.z = x_1 z_1 \oplus x_2 z_2 ...... \oplus x_n z_n \tag{5.11}$$

is the bit wise inner product of x and z, modulo 2. Using the above results we can write the state after the application of n Hadamard gates on the first n qubits of the state $|\psi\rangle_2$ as

$$|\psi\rangle_3 = \sum_{z \in \{0,1\}^n} \sum_{x \in \{0,1\}^n} \frac{(-1)^{x.z + f(x)}|z\rangle}{2^n} \left[\frac{|0\rangle - |1\rangle}{\sqrt{2}} \right]. \tag{5.12}$$

Now at the end of the circuit, a measurement is done on the first n qubits and we check whether the measurement yields all zeroes or not. If we obtain all zeroes then the function is constant; otherwise the function is balanced. To clarify this point we can note that the coefficient of $|0\rangle^{\otimes n}$ in $|\psi\rangle_3$ is $\sum_x \frac{(-1)^{f(x)}}{2^n}$. Therefore, if the function is constant then the coefficient is $+1$ for $f(x) = 0$ and -1 for $f(x) = 1$ respectively. In both cases the probability of getting $|0\rangle^{\otimes n}$ after the measurement is unity. On the other hand if $f(x)$ is balanced then the coefficient of $|0\rangle^{\otimes n}$ in $|\psi\rangle_3$ is 0 (because for a balanced function $f(x) = 1 \Rightarrow (-1)^{f(x)} = -1$ for half of the cases and for the other half of the cases $f(x) = 0 \Rightarrow (-1)^{f(x)} = 1$ and as a result the coefficient of $|0\rangle^{\otimes n}$ in $|\psi\rangle_3$ is $\sum_x \frac{(-1)^{f(x)}}{2^n} = 0$. Thus a single query of function $f(x)$ determines with certainty whether the function is constant or balanced. This result is impressive because classically one can determine the nature of the function with certainty only after $2^{n-1} + 1$ queries of the function. Still, it would not be reasonable to claim that the gain in this quantum algorithm is exponential. It is so because for a balanced function probability of obtaining the same response for k consecutive queries decreases exponentially with k and as a result one can use a few queries of a probabilistic classical algorithm and successfully determine the nature of the function with a very small probability of failure[3]. This algorithm strongly establishes the power of quantum algorithm but it does not get much attention because this algorithm does not have any practical application.

5.3   Grover's algorithm

Grover's algorithm is designed to efficiently search unsorted databases. To visualize the situation assume that there is a telephone directory (book) where $N = 2^n$ names and their telephone numbers have been noted alphabetically. Now you have a phone number and you want to find out the name of the person who owns that phone. A classical computer cannot do this directly since there is no order in the list of numbers. So a classical computer will go through the phone book until it finds the solution (the name). While doing the exhaustive search, the classical computer may be successful in getting the solution (name) in the first trial in the best case, and it may take N trials in the worst case. So on average it will require $\frac{N}{2}$ trials to find out the solution (name). Therefore, in O notation we need $O(N)$ computational steps to obtain the targeted result after searching the unsorted database of N items.

---

[3]Suppose we get 0 (1) in the first query of the classical algorithm, then the probability that we get the same result in the next $k - 1$ queries is $\frac{1}{2^{k-1}}$.

Think awhile: An unsorted database search is an example of a one-way problem, which is easy in one way and hard in another. In other words, it is a problem that is difficult to solve but once a solution is obtained, it can be verified quickly. For example, multiplication of two big prime numbers is very easy but once the product is given to you, it is very difficult to factorize it. This fact is the backbone of RSA cryptographic technique. Now from a large conventional telephone directory if you have a name it is simple to find out the number, but if the number is given to you then you need $O(N)$ trials to find the name. Can you exploit this apparent one-sidedness of the problem to design a cryptographic protocol? Also try to design a simple algorithm which will break your cryptographic protocol. Compare the complexity of both proposals.

Lov Kumar Grover was born in 1961 in Meerut, India. For a short period he was associated with Cornell University in Ithaca, New York. Subsequently he joined Bell Labs in New Jersey. His major contribution to quantum computing was to design the Grover's search algorithm described above, but his contribution to the study of quantum algorithms is not limited to this algorithm alone. His rigorous investigation has revealed many facets of quantum algorithms.
Photo courtesy: L. K. Grover.

A simple alternative to formulate this problem exists in terms of an oracle $f(x)$ that equals 1 for x that solves the problem of interest. So in our case x are entries in the database and $f(x)$ tests whether x is a solution or not. Since $N = 2^n$ in a quantum mechanical protocol we will at least need n qubits to describe all the N logical states. We start by preparing all the n qubits in state $|0\rangle$ (i.e., our initial state is $|\psi\rangle_0 = |0\rangle^{\otimes n}$) and then we apply n Hadamard gates to n qubits in parallel to produce equal superposition of all the possible states as

$$|\psi\rangle_1 = \frac{1}{\sqrt{N}} \sum_{x=0}^{N-1} |x\rangle. \tag{5.13}$$

Note that the states $|x\rangle$ in the above expression are n-qubit states. For example, $|0\rangle = |0_1 0_2 \cdots 0_n\rangle$ or $|x = 3\rangle = |0_1 0_2 \cdots 1_{n-1} 1_n\rangle$. This compact notation will also be used in Simon's algorithm and Shor's algorithm.

Now if we assume that out of N possible states M are solution (i.e., if there exist M values of x for which $f(x) = 1$) then we can define the solution state as

$$|1\rangle = \frac{1}{\sqrt{M}} \sum_{f(x)=1} |x\rangle \tag{5.14}$$

and the nonsolution state as

$$|0\rangle = \frac{1}{\sqrt{N-M}} \sum_{f(x)\neq 1} |x\rangle. \tag{5.15}$$

Thus we can rewrite the total state in terms of solution and nonsolution states as

$$|\psi\rangle_1 = \sqrt{\frac{N-M}{N}} |0\rangle + \sqrt{\frac{M}{N}} |1\rangle. \tag{5.16}$$

If an angle θ is defined as

$$\sin(\theta) = \sqrt{\frac{M}{N}} \tag{5.17}$$

then the state $|\psi\rangle_1$ can be written as

$$|\psi\rangle_1 = \cos(\theta)|0\rangle + \sin(\theta)|1\rangle = \begin{pmatrix} \cos(\theta) \\ \sin(\theta) \end{pmatrix}. \tag{5.18}$$

Next we define an operator as

$$U_I = I - 2|1\rangle\langle 1| = \begin{pmatrix} 1 & 0 \\ 0 & 1 \end{pmatrix} - 2 \begin{pmatrix} 0 \\ 1 \end{pmatrix} \begin{pmatrix} 0 & 1 \end{pmatrix} = \begin{pmatrix} 1 & 0 \\ 0 & -1 \end{pmatrix}. \tag{5.19}$$

This operator flips the sign of the solution states (the states that satisfy the oracle). Apparently this appears to be knowing the solution beforehand, but it is not. Actually quantum mechanics allows the oracle to be applied simultaneously on all the states, but flipping of the solution state does not solve the problem because to find out the solution we still have to search all the N states and check which of them are flipped. To improve this situation another operator is defined as

$$U_\psi = 2|\psi\rangle_{11}\langle\psi| - I = \begin{pmatrix} 2\cos^2\theta - 1 & 2\sin\theta\cos\theta \\ 2\sin\theta\cos\theta & 2\sin^2\theta - 1 \end{pmatrix}. \tag{5.20}$$

This operator inverts the coefficients of an arbitrary state with respect to their mean value. A sequential application of U_I and U_ψ results in a rotation of $2\theta = 2\sqrt{\frac{M}{N}}$ in the $\{|0\rangle, |1\rangle\}$ space[4]. This can be visualized easily because

$$G = U_\psi U_I = \begin{pmatrix} 2\cos^2\theta - 1 & -2\sin\theta\cos\theta \\ 2\sin\theta\cos\theta & 1 - 2\sin^2\theta \end{pmatrix} = \begin{pmatrix} \cos 2\theta & -\sin 2\theta \\ \sin 2\theta & \cos 2\theta \end{pmatrix} \tag{5.21}$$

gives us a simple rotation matrix in two dimensions. G is known as Grover iterator or Grover matrix. After every iteration of G, the probability of finding the solution state increases as

$$\begin{aligned} G|\psi\rangle_1 &= \begin{pmatrix} \cos 2\theta & -\sin 2\theta \\ \sin 2\theta & \cos 2\theta \end{pmatrix} \begin{pmatrix} \cos(\theta) \\ \sin(\theta) \end{pmatrix} \\ &= \begin{pmatrix} \cos(2\theta)\cos(\theta) - \sin(2\theta)\sin(\theta) \\ \sin(2\theta)\cos(\theta) + \cos(2\theta)\sin(\theta) \end{pmatrix} \\ &= \begin{pmatrix} \cos(3\theta) \\ \sin(3\theta) \end{pmatrix} \\ &= \cos(3\theta)|0\rangle + \sin(3\theta)|1\rangle. \end{aligned}$$

Finally, we obtain the solution state with certainty when the coefficient of solution state $|1\rangle$ becomes unity. If we assume that it happens after G is iteratively applied J times, then

$$\begin{aligned} \theta + 2\theta J &= \frac{\pi}{2} \\ J &= \frac{\pi}{4}\frac{1}{\theta} - \frac{1}{2} \\ &= \frac{\pi}{4}\sqrt{\frac{N}{M}} - \frac{1}{2}. \end{aligned} \tag{5.22}$$

Thus after $O\left(\sqrt{N}\right)$ iterations the state will be in the solution state with certainty, after which a single measurement can obtain the solution that satisfies the oracle.

Example 5.1: Find 1 item out of 4 items using Grover's algorithm.
Solution: Here we want G as a matrix for $N = 4 = 2^2$ and $M = 1$. Therefore, our initial equal superposition state is

$$|\psi\rangle_1 = \frac{1}{\sqrt{4}}(|00\rangle + |01\rangle + |10\rangle + |11\rangle). \tag{5.23}$$

---

[4]Since only a small fraction of the states satisfy the oracle $M \ll N$ so in our case $\theta = \sin\theta = \sqrt{\frac{M}{N}}$.

Now we can assume that out of these four states M are solution states. In general, M can be $1, 2, 3$ or 4, but it is specified that $M = 1$. In this case only one of the four possible states is solution, so we can directly rewrite the state as

$$|\psi\rangle_1 = \frac{\sqrt{3}}{\sqrt{4}} \sum_{f(x)=0} |x\rangle + \frac{1}{\sqrt{4}} |x\rangle|_{f(x)=1} = \frac{\sqrt{3}}{\sqrt{4}}|0\rangle + \frac{1}{\sqrt{4}}|1\rangle = \begin{pmatrix} \frac{\sqrt{3}}{2} \\ \frac{1}{2} \end{pmatrix}. \quad (5.24)$$

Now we may assume $\sin(\theta) = \frac{1}{2}$. Therefore $\cos(\theta) = \frac{\sqrt{3}}{2}$ and

$$|\psi\rangle_1 = \cos(\theta)|0\rangle + \sin(\theta)|1\rangle. \quad (5.25)$$

We need to be careful at this point. We cannot assume $\sin(\theta) = \theta$ because N is small and as a result $\sqrt{\frac{M}{N}}$ is not small. In this case $\theta = \frac{\pi}{6}$, and consequently Grover iterator G is

$$G = \begin{pmatrix} \cos\frac{\pi}{3} & -\sin\frac{\pi}{3} \\ \sin\frac{\pi}{3} & \cos\frac{\pi}{3} \end{pmatrix} = \begin{pmatrix} \frac{1}{2} & -\frac{\sqrt{3}}{2} \\ \frac{\sqrt{3}}{2} & \frac{1}{2} \end{pmatrix}. \quad (5.26)$$

Now when G operates on $|\psi\rangle$ then we have

$$G|\psi\rangle = \begin{pmatrix} \frac{1}{2} & -\frac{\sqrt{3}}{2} \\ \frac{\sqrt{3}}{2} & \frac{1}{2} \end{pmatrix} \begin{pmatrix} \frac{\sqrt{3}}{2} \\ \frac{1}{2} \end{pmatrix} = \begin{pmatrix} 0 \\ 1 \end{pmatrix} = |1\rangle.$$

Thus we obtain the solution state in a single step which is shorter than $\sqrt{N} = 2$, but this is obvious, because G rotates the solution state by an angle 2θ and we have started with $\theta = \frac{\pi}{3}$, therefore, after single iteration we will reach $\theta = \pi/2$ or $\sin(\theta) = 1$. As N increases we will need $O\left(\sqrt{N}\right)$ iterations. Now we can note that if we try to find the solution classically then the probability that we obtain the solution in the first, second and third query is $\frac{1}{4}, \frac{1}{4}, \frac{1}{2}$ respectively. The probability of getting the solution in the third query is $\frac{1}{2}$ as we know that there is only one solution state, so if the first three queries could not find the solution then the fourth state is the solution; we don't need to execute the fourth query. Therefore, on average we will need $\frac{1}{4} \times 1 + \frac{1}{4} \times 2 + \frac{1}{2} \times 3 = 2\frac{1}{4} = 2.25$ queries.

Problem 5.1: Boolean satisfiability problem (SAT) belongs to the class NP-complete. Can we use Grover's algorithm to reduce the complexity of SAT? If yes, then how?

Think awhile: I was teaching Grover's algorithm in a class and as a typical example I mentioned: Assume that you have a telephone directory (of N telephone numbers) and a phone number is given to you. Now you have to find out the name of the owner of the phone number. How many searches do you need? One student told me that he could do it classically in a single step. I was surprised and I asked him how. He replied, "It is simple; I'll make a call to that number and ask his name." What is your opinion about this protocol of searching?

Cartoon 5.1: What is wrong in the student's logic?

5.4 Simon's algorithm

Similar to Deutsch's algorithm and the Deutsch Jozsa algorithm, Simon's algorithm is also a promise algorithm. Here we have an oracle or a black box that computes a function

$$f : \{0,1\}^n \to \{0,1\}^n. \tag{5.27}$$

The outputs of the function are n bit strings, rather than single bit like the output of the Deutsch Jozsa algorithm. The promise about $f(x)$ is as follows

$$f(x) = f(y) \text{ iff } y = x \oplus a, \tag{5.28}$$

where the period a is an n bit string and is random over 2^n possible values[5]. The task is now to determine a with probability of success $\geq \frac{3}{4}$ and with as few calls to the oracle as possible. Note that for $n = 1$, the computational task in Deutsch's algorithm, the Deutsch Jozsa algorithm and Simon's algorithm is the same.

---

[5] $y = x \oplus a$ implies $x \oplus y = a$.

5.4.1   Classical approach

Consider a classical probabilistic algorithm whose input at the ith call of the oracle is denoted by x_i. Also assume that we have made m calls of the oracle. Thus after m queries we learn $f(x_1), f(x_2), f(x_3), \cdots, f(x_m)$. We can then pair wise compare $f(x_i)$ and $f(x_j)$ for all i, j shorter than m and $i \neq j$. If we obtain $f(x_i) = f(x_j)$ for a particular pair then the problem is solved because in that case $x_i \oplus x_j = a$. But a is random over 2^n possible values and after m oracle calls we can only check $^mC_2$ probable values of a. Thus the probability of obtaining a in $\leq m$ oracle calls is

$$\frac{^mC_2}{2^n} = \frac{\frac{m(m-1)}{2}}{2^n} \approx \frac{m^2}{2^{n+1}}. \tag{5.29}$$

This clearly shows that to obtain a probability of success (to find out a) $= \frac{1}{2}$ we need $m \approx 2^{\frac{n}{2}}$ oracle calls. Hence the algorithm has to run for exponential time to obtain a correct answer with a probability larger than a constant. Therefore, we can say that Simon's algorithm does not belong to the BPP class with respect to the classical algorithms known to date. However, it belongs to BQP. To establish that, in the following subsection we will show that Simon's problem can be solved in polynomial time (with a bounded error) by a quantum algorithm.

5.4.2   Quantum approach

Let us see how the situation changes when we approach Simon's problem with the help of a quantum algorithm.

Simon 1: The algorithm uses two quantum registers each of n qubits. The registers are initialized to basis state $|\psi\rangle_0 = |0\rangle^{\otimes n}|0\rangle^{\otimes n}$. The first n qubits of $|\psi\rangle_0$ form register 1 and the remaining n qubits form register 2. Then n Hadamard gates are applied to n qubits of the first register in parallel to produce equal superposition of all the possible states in the first register. After the Hadamard transformations the combined state is now

$$|\psi\rangle_1 = \frac{1}{\sqrt{2^n}} \sum_{x=0}^{2^n-1} |x\rangle|0\rangle. \tag{5.30}$$

Once again we note that the states $|x\rangle$ and $|0\rangle$ in (5.30) are n-qubit states. For example, $|0\rangle = |0_1 0_2 \cdots 0_n\rangle$ or $|x = 3\rangle = |0_1 0_2 \cdots 1_{n-1} 1_n\rangle$. This compact notation will be used in Shor's algorithm, too. There the size of register 1 and register 2 will not be the same.

Simon 2: The values of the first register works as input and $f(x)$ is computed through the oracle. The result is stored in the second register and the state becomes

$$|\psi\rangle_2 = \frac{1}{\sqrt{2^n}} \sum_{x=0}^{2^n-1} |x\rangle |f(x)\rangle. \qquad (5.31)$$

Simon 3: The second register is measured. Suppose the measurement yields $f(x_0)$. As a consequence of the measurement the first register collapses to only those values of x which are compatible with $f(x_0)$. According to the definition (5.27-5.28) there are only two such values of x, one is x_0 and the other one is $x_0 \oplus a$. Therefore, the quantum state after the measurement is

$$|\psi\rangle_3 = \frac{1}{\sqrt{2}}(|x_0\rangle + |x_0 \oplus a\rangle)|f(x_0)\rangle. \qquad (5.32)$$

The task of the second register is over and we can discard it from our notation and simply write the state as

$$|\psi\rangle_4 = \frac{1}{\sqrt{2}}|x_0\rangle + \frac{1}{\sqrt{2}}|x_0 \oplus a\rangle. \qquad (5.33)$$

Simon 4: Hadamard transformation is applied to the first register[6] and (5.10) is used to express the output state of the first register as

$$|\psi\rangle_5 = \frac{1}{2^{\frac{(n+1)}{2}}} \sum_{y=0}^{2^n-1} \left[(-1)^{x_0 \cdot y} + (-1)^{(x_0 \oplus a) \cdot y}\right] |y\rangle = \sum_{y=0}^{2^n-1} \alpha_y |y\rangle$$

$$(5.34)$$

where $\alpha_y = \frac{1}{2^{\frac{(n+1)}{2}}}\left[(-1)^{x_0 \cdot y} + (-1)^{(x_0 \oplus a) \cdot y}\right]$ and $|\alpha_i|^2$ is the probability of getting the first register in the state $|i\rangle$ while measured. There are two possible cases: Either $a.y = 0$ or $a.y = 1$. If $a.y = 1$, then $\left[(-1)^{x_0 \cdot y} + (-1)^{(x_0 \oplus a) \cdot y}\right] = \left[(-1)^{x_0 \cdot y} - (-1)^{x_0 \cdot y}\right] = 0$, i.e., $\alpha_y = 0$. So if a measurement is done on the register it will not yield those values of y for which $a.y = 1$. In all other cases (i.e., when $a.y = 0$) we have

$$\alpha_y = \frac{\pm 1}{2^{\frac{(n-1)}{2}}}.$$

Therefore, probability of getting a particular y which satisfies $a.y = 0$ is

$$|\alpha_y|^2 = \frac{1}{2^{(n-1)}}.$$

Simon 5: A measurement is done on the first register. This will yield a value of y at random from $\{0,1\}^n$ which satisfies $a.y = 0$. Suppose the measurement yields y_1. Therefore, we have

$$a.y_1 = 0. \qquad (5.35)$$

---

[6]Remember that x and a are n-bit strings so we have to use $H^{\otimes n}$.

Unfortunately, (5.35) is not sufficient to find a which is an n-bit string by definition.

Simon 6: Repeat the above procedure until $(n - 1)$ linearly independent $y_1, \cdots, y_{n-1}$ are obtained. Solve the equations $a.y_i = 0$, $i = 1, \cdots, (n - 1)$ to determine a.

5.4.3   Complexity analysis

Suppose we have already observed m linearly independent equations involving y_i's, i.e., we have already obtained $y_1, y_2, \cdots, y_m$ which are linearly independent. These linearly independent vectors span a vector space V of dimension C^{2^m}. This is a subspace of the total vector space (spanned by the n linearly independent vectors) of dimension C^{2^n}. Now if the next run of the algorithm yields a new equation with an associated vector y_{m+1}, then y_{m+1} will be independent of all the previous y_i's provided it lies outside the subspace V. Thus the probability that the next run of the algorithm produces a linearly independent y_{m+1} is

$$
\begin{aligned}
p_m &= \frac{\text{Total space spanned by all } n \text{ vectors} - \text{Subspace spanned by } m \text{ vectors}}{\text{Total space spanned by all } n \quad \text{vectors}} \\
&= \frac{2^n - 2^m}{2^n} = 1 - \frac{1}{2^{(n-m)}}.
\end{aligned}
$$

$$(5.36)$$

So the probability that the n consecutive run of the algorithm yields n linearly independent equations is

$$
p = \left(1 - \frac{1}{2^n}\right)\left(1 - \frac{1}{2^{n-1}}\right) \cdots \left(1 - \frac{1}{4}\right)\left(1 - \frac{1}{2}\right). \qquad (5.37)
$$

Now the question is: Can we find a lower bound to this equation? This question is important because if we can find a constant lower bound p_{min} then we will be able to tell that at least $p_{min} \times 100\%$ of the time the algorithm succeeds to find n independent solutions in n trials. Without doing rigorous mathematics we can note that when $0 \leq x_i \leq 1$ then $(1 - x_1)(1 - x_2) = 1 - x_1 - x_2 - x_1 x_2 > 1 - (x_1 + x_2)$. Therefore, from (5.37) we have

$$
p \geq \left[1 - \left(\frac{1}{2^n} + \frac{1}{2^{n-1}} + \dots + \frac{1}{2}\right)\right] \geq \frac{1}{4}.
$$

Thus the probability of failure is bounded and under quantum algorithm, Simon's problem belongs to BQP complexity class.

5.5   Shor's algorithm

Shor's algorithm is the most interesting quantum algorithm known until now. Some number theoretic tricks and physicists' well known technique of finding out the period of a function by using Fourier transform work as the backbone of Shor's algorithm. Let us start with the number theoretic tricks and show that the task of finding out a factor of a number can be reduced to finding out the period of a function, and the period can be obtained efficiently by using quantum Fourier transform (QFT).

Peter Shor was born on 14th August, 1959. At the time of this writing he is a professor in the Mathematics Department of MIT, Cambridge, MA. He obtained his Ph.D. from the same department in 1985. He is one of the pioneers in quantum computing research. He is well known for Shor's algorithm. However, his major contributions to the field also include nine qubit error correction code, CSS code and fault-tolerant computation.

Photo credit: Ayush Bhandari, photo courtesy: Peter Shor.

5.5.1 A little bit of number theory

Two numbers are called coprime to each other if the greatest common divisor (*gcd*) of them is 1. Therefore, two arbitrary integers a and b are coprime to each other if $gcd(a, b) = 1$. For example, 4 and 9 are coprime. Now, assume that we have to factorize a large odd integer[7] N. Let us choose a coprime x of N such that $x < N$. The choice of x is not very critical, since if we choose any arbitrary $x < N$, and find that x and N are not coprime (i.e., $gcd(x, N) = f \neq 1$), then f is a factor of N and we may restart our search for other factors of N by choosing a coprime of N/f instead of choosing a coprime of N. The *gcd* can be determined by one of the oldest number theoretic algorithms by Euclid. This nice little algorithm was introduced by Euclid around 300 BC.

---

[7]If N is even then we already know that 2 is a factor and we can start the factorization process from $\frac{N}{2}$.

5.5.1.1   Euclid's algorithm

If $a > b$ and $gcd(a, b) = z$; then $a - b$, $a - 2b$, $a - 3b$, $\cdots$ are exactly divisible by z. In this monotonically decreasing series gradually the terms will be shorter than the smaller number of the pair (i.e., b in this case). For the smallest value of k which satisfies $a - kb < b$, we obtain the remainder $r = a - kb$ by division of a by b. And r is also divisible by z. This point will be clearer if we provide an explicit example.

Example 5.2: Consider $a = 33$ and $b = 6$, which implies $z = 3$. Now we can easily see that $a - b = 27$, $a - 2b = 21$, $a - 3b = 15$, $a - 4b = 9$ and $r = a - 5b = 3 < b$ all are divisible by 3 (z) and the last term is simply the remainder of 33 divided by 6.

If $r = 0$ then, $z = b$ and the problem is solved. If $r \neq 0$, then our problem reduces to a similar but simpler problem involving smaller numbers as

$$z = gcd(a, b) = gcd(b, r). \tag{5.38}$$

The above procedure can be repeated and z can be expressed as gcd of pairs of smaller numbers. The last nonzero remainder obtained in this process is the desired gcd (i.e., z)[8]. In the above example, $z = gcd(33, 6) = gcd(6, 3) = 3$.

5.5.1.2   Period of the modular exponential function

By using Euclid's algorithm either we get the factors of N or we obtain a coprime x of N. Now we want to calculate $x^i mod N$. The smallest positive integer r which satisfies

$$x^r mod N = 1 \tag{5.39}$$

is called the order of $x mod N$. This implies that

$$x^r = k.N + 1 \tag{5.40}$$

for some k and therefore,

$$x^{r+1} = k.N.x + x \tag{5.41}$$

such that

$$x^{r+1} mod N = x mod N. \tag{5.42}$$

From the last equation it is easy to visualize that r is the period of modular exponential function

$$f_N(i) = x^i mod N. \tag{5.43}$$

Since $f_N(i)$ is the remainder of x^i divided by N, it cannot have more than N different values before repeating and consequently, $r \leq N$. Now the

---

[8]A   nice   discussion   and   a   pseudo   code   as   well   as   the   c code   for   implementation   of   this   algorithm   are   available   at http://en.wikipedia.org/wiki/Euclidean_algorithm#Description_of_the_algorithm

following three different cases may arise: a) r is odd, b) r is even and $x^{\frac{r}{2}} mod N = -1$ and c) r is even and $x^{\frac{r}{2}} mod N \neq -1$.

Among these three cases the first two do not provide us any information regarding factors of N. However, the third one provides us very important information regarding factors of N. To make it clear assume that $x^{\frac{r}{2}} = y$. Then from (5.39) it follows that

$$(x^r - 1) mod N = 0$$
$$\Rightarrow \left[(x^{\frac{r}{2}} + 1)(x^{\frac{r}{2}} - 1)\right] mod N = 0.$$

Thus N must have at least one common factor with $(x^{\frac{r}{2}} + 1)$ and $(x^{\frac{r}{2}} - 1)$. But it is not a trivial factor since the factor is not equal to N. This is because $x^{\frac{r}{2}} mod N \neq -1$ implies that $(x^{\frac{r}{2}} + 1) mod N \neq 0$ and thus $(x^{\frac{r}{2}} + 1)$ is not an exact multiple of N. On the other hand, $(x^{\frac{r}{2}} - 1)$ cannot be an exact multiple of N since if it is so then we will have $x^{\frac{r}{2}} mod N = 1$ and that will reduce the order to $\frac{r}{2}$. Now we can conclude that if case 3 is satisfied then at least one of the two numbers, $gcd(N, x^{\frac{r}{2}} \pm 1)$ is a nontrivial factor of N. As we have already seen that gcd can be efficiently computed by Euclid's algorithm, so the problem of factorization of a large number is essentially the same as the problem of finding out the period of a modular exponential function $f_N(i)$. But classically there does not exist any efficient algorithm to find out the period of $f_N(i)$. Before we state how the period can be efficiently determined by using quantum Fourier transform, we would like to give an example which will elaborate the factorization strategy to be adopted here.

Example 5.3: Assume that we have to find out factors of 21 and we have chosen 2 as a coprime to 21. Now our first task is to find out the series $2^i mod N$ for $i = 1, 2, 3, \cdots$. We find the series is $2, 4, 8, 16, 11, 1, 2, 4, 8, \cdots$ Thus the series repeats its values after every 6 terms and consequently, the period $r = 6$. Therefore, $gcd(21, 2^3 + 1) = gcd(21, 9) = 3$ and $gcd(21, 2^3 - 1) = gcd(21, 7) = 7$ are the possible candidates of nontrivial factors[9] of 21. In this case both of them are factors of 21.

5.5.1.3 Continued fraction representation

This representation is used to find the period in Shor's algorithm. The role of continued fraction representation in Shor's algorithm will be clear when we describe the main (quantum) part of the algorithm in Subsection 5.5.4. But before we do that we need to introduce the continued fraction representation. This is a way to express a real number R as

$$R = a_0 + \cfrac{1}{a_1 + \cfrac{1}{\ddots + \frac{1}{a_n}}}, \qquad (5.44)$$

---

[9]In this example of factorization of 21, if we choose 4 as a coprime of 21 then the required period will be $r = 3$ since $4^3 mod 21 = 64 mod 21 = 1$. Thus the period is odd and as a result the algorithm fails.

where n is a non-negative integer, a_0 is an integer, and $a_i : i \in \{1, 2, \cdots, n\}$ is a positive integer. Here we have assumed that R can be expressed as a finite continued fraction. The procedure to express R in this form is simple. We write R as sum of its integer part r and fractional part (thus we write $R = r + f$) then reciprocate the fractional part and express R as $R = r + \frac{1}{\frac{1}{f}}$.

Subsequently we express $\frac{1}{f}$ as sum of its integer part and fractional part and continue until the fractional part becomes zero and we obtain an expression of the form (5.44). In the continued fraction representation we write R as $[a_0, a_1, \cdots, a_n]$. The procedure can be made clearer through a few examples.

Example 5.4: (a) Let us consider $R = \frac{15}{4}$. Then the above procedure yields

$$
\begin{aligned}
R = \frac{15}{4} &= 3 + \frac{3}{4} \\
&= 3 + \frac{1}{\frac{4}{3}} \\
&= 3 + \frac{1}{1 + \frac{1}{3}}.
\end{aligned}
$$

Therefore, in continued fraction representation $\frac{15}{4} = [3, 1, 3]$.

(b) Consider $R = \frac{43}{30}$. Now

$$
\begin{aligned}
R &= 1 + \frac{13}{30} \\
&= 1 + \frac{1}{\frac{30}{13}} \\
&= 1 + \frac{1}{2 + \frac{4}{13}} \\
&= 1 + \frac{1}{2 + \frac{1}{\frac{13}{4}}} \\
&= 1 + \frac{1}{2 + \frac{1}{3 + \frac{1}{4}}}.
\end{aligned}
$$

Therefore, $R = [1, 2, 3, 4]$.

5.5.2   The strategy

Keeping the above discussions in mind we can now state our strategy of factorization as follows [73]:

Step 1: If N is even then return the factor 2. Otherwise, continue to the next step[10].

Step 2: Check whether $N = a^b$ for integers a and b such that $a \geq 1$ and $b \geq 2$. If $N = a^b$ then return the factor a.

Step 3: Randomly choose an integer $x \in \{2, 3, \cdots, N - 1\}$ and compute $gcd(x, N)$. If $gcd(x, N) > 1$ then return the factor $gcd(x, N)$. If $gcd(x, N) = 1$ (i.e., if x is a coprime of N) then continue to the next step.

---

[10]If N is even and we need factors other than 2 then we have to restart the algorithm with $\frac{N}{2}$ to find out the other factors.

Step 4: Find out the order r of $x \bmod N$. If r is even and $x^{\frac{r}{2}} \bmod N \neq -1$ then continue to the next step. Otherwise, restart from **Step 3** with a different x.

Step 5: Compute $gcd(x^{\frac{r}{2}} \pm 1, N)$ and check whether one of them is (or both of them are) nontrivial factor (factors) of N. If so, then return the factor (factors). Otherwise, restart from **Step 3** with a different x.

Up to **Step 3** it is a pure classical algorithm of factorization. Further, it is important to note that **Step 5** and part of **Step 4** are also computed classically. The quantum part or precisely Shor's contribution appears only in **Step 4**. Thus Shor's main contribution is to show that we can efficiently find out the order of $x \bmod N$ with the help of quantum Fourier transformation.

5.5.3 Quantum Fourier transformation

Before we define quantum Fourier transform (QFT) we need to define discrete Fourier transform (DFT), which is nothing but Fourier transform of a discrete data set. Suppose we have a discrete data set $\{x_0, x_1, \cdots, x_{N-1}\}$ then the DFT of this data set is

$$y_k = \frac{1}{\sqrt{N}} \sum_{j=0}^{N-1} x_j e^{\frac{2\pi ijk}{N}}, \tag{5.45}$$

where x_j and y_k are complex numbers and the index k varies from 0 to $N-1$. Thus the DFT maps a discrete data set into another discrete data set. The concept of DFT will be clearer if we provide a few simple examples.

Example 5.5: Suppose you have a data set $x_j = \{\sqrt{2}, 2\}$. Compute DFT of this set.

Solution: Here $x_0 = \sqrt{2}$, $x_1 = 2$ and $N = 2$. Therefore,

$$y_k = \frac{1}{\sqrt{2}} \sum_{j=0}^{1} x_j e^{\frac{2\pi ijk}{2}} = \frac{1}{\sqrt{2}} \left(\sqrt{2} + 2e^{i\pi k} \right).$$

Thus,

$$y_0 = \frac{1}{\sqrt{2}} \left(\sqrt{2} + 2 \right) = 1 + \sqrt{2},$$

and

$$y_1 = \frac{1}{\sqrt{2}} \left(\sqrt{2} + 2e^{i\pi} \right) = \frac{1}{\sqrt{2}} \left(\sqrt{2} - 2 \right) = 1 - \sqrt{2}.$$

Example 5.6: Find y_0 for $x_j = \{1 + i, \sqrt{2}, 3 - 2i, 4\}$.

Solution: Here $N = 4$ and

$$y_0 = \frac{1}{\sqrt{N}} \sum_{j=0}^{N-1} x_j = \frac{1}{2} \left((1+i) + \left(\sqrt{2} \right) + (3 - 2i) + 4 \right) = 4 + \frac{\sqrt{2} - i}{2}.$$

Thus we have learned how to find out DFT for a discrete data set (i.e., for a set of complex numbers). Now a quantum state $|\psi\rangle = \sum_{n=0}^{N-1} c_n|n\rangle$ is usually described by the state vector $\begin{pmatrix} c_0 \\ c_1 \\ \vdots \\ c_{N-1} \end{pmatrix}$, which is nothing but a set of complex numbers $\{c_i\}$. When we take DFT of this particular set of complex numbers (i.e., DFT of amplitudes of a quantum state) then we obtain QFT. Thus QFT is the classical DFT to the coefficients of a quantum state. We may visualize QFT as an operator F which maps the quantum state $|\psi\rangle = \sum_{n=0}^{N-1} c_n|n\rangle$ into another state $|\psi'\rangle = \sum_{k=0}^{N-1} b_k|k\rangle$ where $\{b_k\}$ are DFT of $\{c_n\}$. Thus we have

$$|\psi'\rangle = F|\psi\rangle = F\sum_{n=0}^{N-1} c_n|n\rangle = \sum_{k=0}^{N-1} b_k|k\rangle \tag{5.46}$$

and from (5.45)

$$b_k = \frac{1}{\sqrt{N}} \sum_{n=0}^{N-1} c_n e^{\frac{2\pi i n k}{N}}. \tag{5.47}$$

To qualify as a physically acceptable transformation F has to be unitary. The unitarity can be proved easily but before we do so we would like to note that the amplitudes of $|\psi'\rangle$ (i.e., $\{b_k\}$) are linear in original amplitudes c_n. Therefore F is a linear operator. We can easily write F in Dirac notation as

$$F = \sum_{m,k=0}^{N-1} \frac{e^{\frac{2\pi i m k}{N}}}{\sqrt{N}} |k\rangle\langle m|. \tag{5.48}$$

We can check that F is the correct transformation as

$$\begin{aligned} F|\psi\rangle &= F\sum_{n=0}^{N-1} c_n|n\rangle = \sum_{m,k,n=0}^{N-1} \frac{c_n e^{\frac{2\pi i m k}{N}}}{\sqrt{N}} |k\rangle\delta_{m,n} \\ &= \sum_{k,n=0}^{N-1} \frac{c_n e^{\frac{2\pi i n k}{N}}}{\sqrt{N}} |k\rangle = \sum_{k=0}^{N-1} b_k|k\rangle = |\psi'\rangle. \end{aligned} \tag{5.49}$$

Now using (5.48) we can write

$$F^\dagger = \sum_{m',k'=0}^{N-1} \frac{e^{-\frac{2\pi i m' k'}{N}}}{\sqrt{N}} |m'\rangle\langle k'| \tag{5.50}$$

and easily observe that

$$\begin{aligned} F^\dagger F &= \sum_{m',k',m,k=0}^{N-1} \frac{e^{\frac{2\pi i}{N}(mk-m'k')}}{N} |m'\rangle\langle m|\delta_{k,k'} \\ &= \sum_{m',m,k=0}^{N-1} \frac{e^{\frac{2\pi i}{N}(m-m')k}}{N} |m'\rangle\langle m| \\ &= \sum_{m',m=0}^{N-1} |m'\rangle\langle m|\delta_{m,m'} \\ &= \sum_{m=0}^{N-1} |m\rangle\langle m| = I. \end{aligned} \tag{5.51}$$

Similarly, we can show $FF^\dagger = I$ and hence F is a unitary operation that can be physically realized. Thus QFT can be physically implemented. It is straightforward to observe that F can also be written in matrix form as

$$F = \frac{1}{\sqrt{N}} \begin{pmatrix} e^{\frac{2\pi i}{N} 0 \times 0} = 1 & e^{\frac{2\pi i}{N} 0 \times 1} = 1 & e^{\frac{2\pi i}{N} 0 \times 2} = 1 & \cdots & 1 \\ e^{\frac{2\pi i}{N} 1 \times 0} = 1 & e^{\frac{2\pi i}{N} 1 \times 1} = \omega & e^{\frac{2\pi i}{N} 1 \times 2} = \omega^2 & \cdots & \omega^{(N-1)} \\ e^{\frac{2\pi i}{N} 2 \times 0} = 1 & e^{\frac{2\pi i}{N} 2 \times 1} = \omega^2 & e^{\frac{2\pi i}{N} 2 \times 2} = \omega^4 & \cdots & \omega^{2(N-1)} \\ \vdots & \vdots & \vdots & \ddots & \vdots \\ 1 & \omega^{(N-1)} & \omega^{2(N-1)} & \cdots & \omega^{(N-1)(N-1)} \end{pmatrix}$$

$$= \frac{1}{\sqrt{N}} \begin{pmatrix} 1 & 1 & 1 & \cdots & 1 \\ 1 & \omega & \omega^2 & \cdots & \omega^{(N-1)} \\ 1 & \omega^2 & \omega^4 & \cdots & \omega^{2(N-1)} \\ \vdots & \vdots & \vdots & \ddots & \vdots \\ 1 & \omega^{(N-1)} & \omega^{2(N-1)} & \cdots & \omega^{(N-1)(N-1)} \end{pmatrix},$$

$$(5.52)$$

where $\omega = e^{\frac{2\pi i}{N}}$. Before we proceed further we would like to provide two simple examples of QFT.

Example 5.7: Consider an arbitrary single qubit state $|\psi\rangle = \alpha|0\rangle + \beta|1\rangle$ and obtain its QFT.

Solution: Method 1: Here $N = 2$, $c_0 = \alpha$ and $c_1 = \beta$. Therefore, $b_k = \frac{1}{\sqrt{2}} \sum_{n=0}^{1} c_n e^{\pi i n k}$ and we have

$$b_0 = \frac{1}{\sqrt{2}} \sum_{n=0}^{1} c_n = \frac{\alpha+\beta}{\sqrt{2}}$$
$$b_1 = \frac{1}{\sqrt{2}} \sum_{n=0}^{1} c_n e^{\pi i n} = \frac{\alpha-\beta}{\sqrt{2}}.$$

Method 2: As $N = 2$, so $|\psi'\rangle = F|\psi\rangle = \frac{1}{\sqrt{2}} \begin{pmatrix} 1 & 1 \\ 1 & e^{\frac{2\pi i}{2}} \end{pmatrix} \begin{pmatrix} \alpha \\ \beta \end{pmatrix} =$

$\frac{1}{\sqrt{2}} \begin{pmatrix} 1 & 1 \\ 1 & -1 \end{pmatrix} \begin{pmatrix} \alpha \\ \beta \end{pmatrix} = \frac{1}{\sqrt{2}} \begin{pmatrix} \alpha + \beta \\ \alpha - \beta \end{pmatrix} = \frac{\alpha+\beta}{\sqrt{2}}|0\rangle + \frac{\alpha-\beta}{\sqrt{2}}|1\rangle.$

It is easy to identify that for a single qubit $F = \frac{1}{\sqrt{2}} \begin{pmatrix} 1 & 1 \\ 1 & -1 \end{pmatrix}$ is nothing but a Hadamard gate. In other words a single Hadamard gate implements QFT for a single qubit. But to implement QFT for multi-qubit systems we need more complex circuits. As an example, such a circuit for a three qubit case is shown in Fig. 5.3, where $R_k = p\left(\frac{2\pi}{2^k}\right) = \begin{pmatrix} 1 & 0 \\ 0 & \exp\left(\frac{2\pi i}{2^k}\right) \end{pmatrix}$ is a phase gate which is already described in Section 4.1.

Example 5.8: Consider an arbitrary two qubit state $|\psi\rangle = a|00\rangle + b|01\rangle + c|10\rangle + d|11\rangle$ and obtain its QFT.

Here $N = 4$. Therefore, $b_k = \frac{1}{2} \sum_{n=0}^{3} c_n e^{\frac{\pi i n k}{2}}$ and we have

$$b_0 = \frac{1}{2} \sum_{n=0}^{3} c_n = \frac{1}{2}(a + b + c + d)$$
$$b_1 = \frac{1}{2} \sum_{n=0}^{3} c_n e^{\pi i n/2} = \frac{1}{2}(a + be^{\pi i/2} + ce^{\pi i} + de^{3\pi i/2})$$
$$b_2 = \frac{1}{2} \sum_{n=0}^{3} c_n e^{\pi i n} = \frac{1}{2}(a + be^{\pi i} + ce^{2\pi i} + de^{3\pi i})$$
$$b_3 = \frac{1}{2} \sum_{n=0}^{3} c_n e^{3\pi i n/2} = \frac{1}{2}(a + be^{3\pi i/2} + ce^{3\pi i} + de^{9\pi i/2}).$$

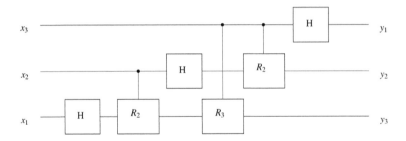

Figure 5.3: A circuit for implementation of QFT in a three qubit state.

Now the whole background is ready to finally describe the core of Shor's algorithm.

5.5.4   Main (quantum) part of the algorithm

As **Step 1-Step 3** of our factorization strategy are performed classically, so the quantum part of the algorithm starts from **Step 4** of the strategy. Thus the task is to find out the order. Let us see how it is done in Shor's algorithm. From the first three steps of the strategy we already know that the integer to be factorized is N and x is a coprime to it.

Before we describe the order finding algorithm, we would like to draw your attention towards the fact that the first few steps of this algorithm are analogous to that of Simon's algorithm. Here also we need two quantum registers. Let us call them register 1 and register 2. Register 1 is a t-qubit register such that $2 \log_2 N < t < 2 \log_2 \sqrt{2}N$ or in other words $N^2 < 2^t < 2N^2$ and register 2 is an l-qubit register where l is the number of qubits needed to store N. For example, for $N = 15$, we have $\log_2 N = 3.9$ and $\log_2 \sqrt{2}N = 4.4$. Therefore, $2 \log_2 N = 7.8 < t < 2 \log_2 \sqrt{2}N = 8.8$ and consequently $t = 8$ and as $\log_2 N = 3.9$ so $l = 4$ qubits are required to store N. Similarly for $N = 55$, $t = 12$ and $l = 6$. Now we describe the actual protocol in the following steps:

Shor 1: Create two quantum memory registers described above and initialize all the qubits of both the registers in $|0\rangle$. Therefore, our initial state is

$$|\psi\rangle_0 = |0\rangle^{\otimes t}|0\rangle^{\otimes l}.$$

Shor 2: Use t Hadamard gates to create an equal superposition of 2^t states in the first register. This will transform $|\psi\rangle_0$ to

$$|\psi\rangle_1 = \frac{1}{\sqrt{2^t}} \sum_{k=0}^{2^t-1} |k\rangle|0\rangle. \tag{5.53}$$

Remember that the state of register 1 is a t-qubit state but the state of the second register is a l-qubit state. The compact notation used here is the same as the notation used in Simon's algorithm.

Shor 3: Perform a modular exponent of the first register onto the second to yield

$$|\psi\rangle_2 = \frac{1}{\sqrt{2^t}} \sum_{k=0}^{2^t-1} |k\rangle |x^k mod N\rangle. \tag{5.54}$$

Here we would like to note that $x^k mod N < N$ implies that a l-qubit register that can store N is sufficient to store $x^k mod N$ and it is independent of the size of register 1. Consequently, even if we choose a larger register 1, the size of register 2 would remain the same.

Shor 4: Measure the state of register 2. If it yields

$$x^k mod N = y_0 \tag{5.55}$$

then the state of register 1 collapses to a superposition state of those values of k which are compatible with y_0. As $x^k mod N$ is periodic in k with an order r so the register 1 collapses to the state

$$|\psi\rangle_3 = \frac{1}{\sqrt{M}} \sum_{m=0}^{M-1} |d + mr\rangle \tag{5.56}$$

where d is the lowest possible value of k which is compatible with y_0 and M is the number of values of k satisfying (5.55). We may note that $M \approx \frac{2^t}{r} \gg 1$.

The task of the second register is over and we may now omit it from our notation. Further, we may note that if we measure $|\psi\rangle_3$ it will collapse to one of the possible states and we will not get any information about r. We could have obtained r by measuring $|\psi\rangle_3$ if we had multiple copies of $|\psi\rangle_3$. However, copying is not allowed and if we follow the steps up to this point to generate a copy of $|\psi\rangle_3$ then we may not succeed in generating a copy of $|\psi\rangle_3$, as during the second run of the protocol the second register may collapse randomly to some value $y_1 \neq y_0$ and in that case $|\psi\rangle_3$ will be superposition of those values of k which are compatible with y_1.

Shor 5: To find r perform a QFT (as defined by (5.46) and (5.47)) of $|\psi\rangle_3$, which will yield

$$
\begin{aligned}
|\psi\rangle_4 &= \frac{1}{\sqrt{M}} \sum_{m=0}^{M-1} \frac{1}{\sqrt{2^t}} \sum_{n=0}^{2^t-1} \exp\left(\frac{2\pi i(d+mr)n}{2^t}\right) |n\rangle \\
&= \frac{1}{\sqrt{2^t M}} \sum_{n=0}^{2^t-1} \exp\left(\frac{2\pi idn}{2^t}\right) \left(\sum_{m=0}^{M-1} \xi^m |n\rangle\right)
\end{aligned}
$$

where $\xi = \exp\left(2\pi i \frac{nr}{2^t}\right)$.

Shor 6: Measure the state of register 1. The probability of obtaining state $|n\rangle$ is

$$p(n) = \frac{1}{2^t M} \left| \sum_{m=0}^{M-1} \xi^m \right|^2.$$

If $\frac{nr}{2^t}$ is not an integer or very close to an integer, then the terms in the sum will add incoherently (i.e., they will destructively interfere). As $p(n) \to 0$ for such states, they are extremely unlikely to be observed. But if $\frac{nr}{2^t} \approx C$ (i.e., if $\frac{n}{2^t} \approx \frac{C}{r}$), where C is an integer, then $\xi \approx 1$ and $p(n) = \frac{1}{2^t M} \left| \sum_{m=0}^{M-1} \xi^m \right|^2 = \frac{M^2}{2^t M} = \frac{M}{2^t} \approx \frac{1}{r}$. Thus the probability of observing a state for which $\xi \approx 1$ is much more. Consequently, most of the time we will observe a state for which $\frac{n}{2^t} \approx \frac{C}{r}$.

The quantum part of the algorithm finishes here. We have obtained n as outcome of our measurement on register 1 and we know 2^t as we had prepared a t-qubit register. But still we don't know appropriate values of C and r. We just know both of them are integers. Thus our task is now to find out an integer fraction $\frac{C}{r}$ which is very close to $\frac{n}{2^t}$ and $r < N$. The same is done classically and by using the method of continued fraction.

Shor 7: Compute the convergents $\frac{C}{r_1}$ to $\frac{n}{2^t}$ for which the denominator $r_1 < N$. However, if $\frac{C}{r}$ is equal to or very close to an integer then so is $\frac{2C}{r} = \frac{C}{\frac{r}{2}}, \frac{3C}{r} = \frac{C}{\frac{r}{3}}$. Consequently, instead of r_1 the period of $x \bmod N$ may be $r = pr_1$ provided $pr_1 < N$. So once r_1 is obtained, we check small integer multiples of r_1 as possible values of r. To be precise, we compute $x^{pr_1} \bmod N$ for $p = 1, 2, 3, \cdots$ till we obtain a value of p for which $x^{pr_1} \bmod N = 1$.

Finally, once r is found, we return to **Step 5** of the strategy to find out the $gcd(x^{\frac{r}{2}} \pm 1, N)$. We will now elaborate the above described steps through a particular example.

Example 5.9: Consider $N = 15$ and $x = 7$ as its coprime. We have already mentioned that $t = 8$ and $l = 4$ in this case. So our initial state is $|\psi\rangle_0 = |0\rangle^{\otimes 8} |0\rangle^{\otimes 4}$. After the Hadamard transformations the state becomes

$$|\psi\rangle_1 = \frac{1}{\sqrt{256}} \left(|0\rangle + |1\rangle + \cdots + |255\rangle \right) |0\rangle.$$

Now we apply modular expansion and obtain

$$
\begin{aligned}
&|\psi\rangle_2 \\
=\ & \frac{1}{\sqrt{256}} \left(|0\rangle |7^0 \bmod 15\rangle + |1\rangle |7^1 \bmod 15\rangle + \cdots + |255\rangle |7^{255} \bmod 15\rangle \right) \\
=\ & \frac{1}{\sqrt{256}} \left(|0\rangle |1\rangle + |1\rangle |7\rangle + |2\rangle |4\rangle + |3\rangle |13\rangle \right. \\
&+\ \left. |4\rangle |1\rangle + |5\rangle |7\rangle + \cdots + |255\rangle |13\rangle \right).
\end{aligned}
$$

We measure register 2. Suppose we obtain it as $|7\rangle$ then register 1 collapses to

$$|\psi\rangle_3 = \frac{1}{\sqrt{64}}\left(|1\rangle + |5\rangle + \cdots + |253\rangle\right).$$

Apply QFT on $|\psi\rangle_3$ to obtain

$$|\psi\rangle_4 = \frac{1}{\sqrt{256 \times 64}} \sum_{n=0}^{255} \exp\left(\frac{2\pi i n}{256}\right)\left(\sum_{m=0}^{63} \xi^m |n\rangle\right)$$

where $\xi = \exp\left(2\pi i \frac{n \times r}{256}\right)$. Now we measure register 1. Assume that we obtain $|128\rangle$. Here $\frac{n}{2^t} = \frac{1}{2} = 0 + \frac{1}{2}$ which can be written in continued fraction form as $[0,2]$. It converges in two steps and consequently $r_1 = 2$. Now we check $x^{r_1} mod N = 7^2 mod 15 = 4 \neq 1$ so r_1 is not the period. Next we check $2r_1$ as possible order of $x mod N$ and we find that the $7^4 mod 15 = 1$. Thus the required period is 4. The probability obtaining the above-mentioned measurement outcome (i.e., $|128\rangle$) is

$$P(128) = \frac{1}{256 \times 64}\left|\sum_{m=0}^{31} \exp\left(2\pi i \frac{128 \times 4}{256}\right)^m\right|^2 = 0.0625.$$

The use of continued fraction was not clear in the last example, so let us provide another example.

Example 5.10: In the above example assume that the measurement on the first register has yielded $|n\rangle = |127\rangle$. In this case

$$P(127) = \frac{1}{256 \times 64}\left|\sum_{m=0}^{31} \exp\left(2\pi i \frac{127 \times 4}{256}\right)^m\right|^2 = 0.0253 = 2.53\%.$$

Here $\frac{n}{2^t} = \frac{127}{256} = 0 + \frac{1}{\frac{256}{127}} = 0 + \frac{1}{2 + \frac{1}{\frac{127}{2}}} = 0 + \frac{1}{2 + \frac{1}{63 + \frac{1}{2}}}$. Thus $\frac{127}{256} = [0, 2, 63, 2]$ which converges in 4 steps. Thus $r_1 = 4$. We have already checked that 4 is the order of $7 mod 15$.

Example 5.11: Consider $N = 55$ and $x = 13$ as its coprime. We have already mentioned that $t = 12$ and $l = 6$ in this case. So our initial state is $|\psi\rangle_0 = |0\rangle^{\otimes 12}|0\rangle^{\otimes 6}$. After the Hadamard transformations the state becomes

$$|\psi\rangle_1 = \frac{1}{\sqrt{4096}}\left(|0\rangle + |1\rangle + \cdots + |4095\rangle\right)|0\rangle.$$

Now we apply modular expansion and obtain

$$\begin{aligned} &|\psi\rangle_2 \\ =\ & \frac{1}{\sqrt{4096}}\left(|0\rangle|13^0 mod 55\rangle + |1\rangle|13^1 mod 55\rangle + \cdots + |4091\rangle|7^{255} mod 15\rangle\right) \\ =\ & \frac{1}{\sqrt{4096}}\left(|0\rangle|1\rangle + |1\rangle|13\rangle + |2\rangle|4\rangle + |3\rangle|52\rangle + |4\rangle|16\rangle + \cdots + |4091\rangle|2\rangle\right). \end{aligned}$$

We measure register 2. Suppose we obtain it as $|52\rangle$ then register 1 collapses to

$$|\psi\rangle_3 = \frac{1}{\sqrt{205}} \left(|3\rangle + |23\rangle + \cdots + |4083\rangle \right).$$

Apply QFT on $|\psi\rangle_3$ to obtain

$$|\psi\rangle_4 = \frac{1}{\sqrt{4096 \times 205}} \sum_{n=0}^{4095} \exp\left(\frac{2\pi i n}{4096}\right) \left(\sum_{m=0}^{204} \xi^m |n\rangle \right)$$

where $\xi = \exp\left(2\pi i \frac{n \times r}{4096}\right)$. Now we measure register 1. Assume that we obtain $|4095\rangle$, probability of which is quite low. However, $\frac{n}{2^t} = \frac{2047}{4096} = 0 + \frac{1}{\frac{4096}{2047}} = 0 + \frac{1}{2 + \frac{1}{\frac{2047}{2}}} = 0 + \frac{1}{2 + \frac{1}{1023 + \frac{1}{2}}}$. Thus $\frac{2047}{4096} = [0, 2, 1023, 2]$, which converges in four steps. Thus $r_1 = 4$. Now we systematically check that $x^{r_1} mod N = 13^4 mod 55 = 16 \neq 1$, $x^{2r_1} mod N = 13^8 mod 55 = 36 \neq 1$, $x^{3r_1} mod N = 13^{12} mod 55 = 26 \neq 1$, $x^{4r_1} mod N = 13^{16} mod 55 = 31 \neq 1$, $x^{5r_1} mod N = 13^{20} mod 55 = 1$. Thus the required period is 20.

5.6 Solution of Pell's equation and the principal ideal problem

One of the oldest problems in number theory is to find out the solutions of Pell's equation. Given a positive non square integer d, Pell's equation is

$$x^2 - dy^2 = 1 \tag{5.57}$$

and the task is to find out the positive integer solutions of (5.57)[11]. In 1768 Lagrange had shown that there are an infinite number of solutions of Pell's equation. He also showed that if (x_1, y_1) is the least positive solution of (5.57), then all positive solutions are given by (x_n, y_n) for $n = 1, 2, 3, \cdots$ where

$$x_n + y_n\sqrt{d} = (x_1 + y_1\sqrt{d})^n. \tag{5.58}$$

Therefore, the actual problem is to find the least positive solution (x_1, y_1). This cannot be done in polynomial time. But to solve the problem it is enough to compute the regulator R, which is defined as

$$R = \ln(x_1 + y_1\sqrt{d}). \tag{5.59}$$

---

[11]Fermat was the first person to study this equation extensively, and he deserved the credit, but Euler erroneously attributed this equation to Pell.

Around thousand years ago the original algorithm for solving this equation was provided by the Indian mathematicians Brahmagupta and Bhaskara II [74]. Possibly this is the second oldest number theoretic algorithm after Euclid's algorithm. For a detailed and interesting discussion see http://www-groups.dcs.st-and.ac.uk/~history/HistTopics/Pell.html

Hallgern [74] has provided a polynomial time quantum algorithm to compute the regulator and thus to solve Pell's equation in polynomial time. To do so, he has essentially used the benefit of QFT in finding out the period of a function of interest. Shor's algorithm cannot be used directly to solve this problem. However, technically Hallgern's algorithm for solving Pell's equation is analogous to Shor's algorithm. Keeping this in mind, we have not described Hallgern's work in detail in this book. Interested readers may refer to [74] for an elaborate description of the algorithm. The background provided here in connection to Shor's algorithm is sufficient for the understanding of Hallgern's work. In the same paper [74] he also provided a polynomial time algorithm for another related problem known as the principal ideal problem (PIP). It is believed that the Pell's equation and PIP are harder than factoring and therefore they may be used for encryption and Shor's algorithm will not be able to decrypt them directly. But once a scalable quantum computer is built, a classical cryptographic protocol that depends on Pell's equation or on PIP will not remain secure because of Hallgern's algorithm. This is the importance of Hallgern's algorithm.

5.7 Discussion

Let us begin the conclusion of this chapter with an interesting question: How many quantum algorithms exist? It is difficult to tell the exact number, but the number is very small. For example, in this chapter we have discussed a few algorithms. How many more are there? To be honest, there are not many useful quantum algorithms reported to date. A comprehensive list of the quantum algorithms reported to date can be found at http://math.nist.gov/quantum/zoo/. However, all the reported algorithms are not independent. For example, Deutsch's algorithm is essentially a special case of the Deutsch Jozsa (DJ) algorithm. In recent past Peter Shor tried to address this issue in an interesting paper [75] entitled, "Why haven't more quantum algorithms been found?" In his opinion there are two main reasons. Firstly, it is difficult to think and visualize a quantum algorithm while you really stay in a classical world, and secondly there does not exist a quantum computer where you can run your ideas and improve them by trial-error-correction-trial schemes. Here we would like to add another point. It is not really difficult to design a quantum algorithm but it makes some sense only if it can reduce the complexity with respect to the best known classical algorithm that can solve the same problem. This is because of the fact that all the classical algorithms can also be processed in a quantum computer but not vice versa. Still, there have been many efforts and there are some very interesting quantum algorithms. However, many questions are still open and we believe that the power of quantum algorithms has not been completely explored till now. We are optimistic to see many more interesting and powerful quantum algorithms in the near future. Before we finish the discussion it would be apt to note that most of the algorithms discussed in this chapter have been experimentally realized.

Consider the simplest possible case of DJ algorithm where f is a function from one bit to one bit. Specifically this is the case of Deutsch's algorithm. But interestingly, this is the simplest case of Simon's algorithm, too. This particular case of DJ algorithm was first demonstrated by I. L. Chuang *et al.* in 1998 [76] using bulk NMR technique where nuclear spins were used as qubits. Specifically, they used a chloroform (CH_3Cl) molecule to implement the algorithm. This was the first experimental demonstration of a quantum algorithm. Since then several physical implementations of DJ algorithm have been reported. For example, in 2003 D. Wei *et al.* demonstrated a seven-qubit DJ algorithm using liquid state NMR technique [77] and recently, Z. Wu *et al.*, have demonstrated DJ algorithm in a five-qubit NMR system [78]. As Deutsch's algorithm is a special case of DJ algorithm, we can implement Deutsch and DJ algorithms in various ways. In Section 1.3 we mentioned some of the initial NMR-based experimental implementations of Grover's algorithm and Shor's algorithm. To be precise, we mentioned that in 1999, L. M. K. Vandersypen *et al.* implemented Grover's database-search algorithm using a three-qubit quantum computer [30]. In the next year, L. M. K. Vandersypen *et al.* succeeded in implementing Shor's order finding algorithm using a 5-qubit NMR quantum computer [32] and in 2001, L. M. K. Vandersypen *et al.* implemented a 7-qubit NMR quantum computer [33], which had successfully deduced the prime factors of 15. Since then Shor's algorithm has been implemented using different techniques ([79, 80, 81, 82] and references therein). In 2009 A. Politi, J. C. F. Matthews and J. L. O'Brien implemented Shor's algorithm on a photonic chip and successfully factorized 15. In 2012, E. Lucero *et al.* [82] also factorized 15 by implementing Shor's algorithm with a Josephson phase qubit quantum processor (super-conductive implementation). In the same year experimentalists factorized a relatively larger number as in a very interesting optical experiment E. M. López *et al.* [79] successfully factorized 21 using Shor's algorithm. The most interesting feature of this experiment is that in this experimental t-qubit control register (register 1) is replaced by a single qubit that is recycled t times. Another interesting experimental work was recently reported by N. Xu *et al.* [81]. In this work they successfully factorized 143 using a liquid-crystal NMR quantum computer.

In the previous chapter we saw that quantum gates can be implemented experimentally. In this chapter we have seen that Shor's algorithm and the other quantum algorithms discussed in this chapter are experimentally realizable. However, due to the unavailability of a large quantum computer we cannot really utilize the advantages of a quantum computer. For example, to date we cannot efficiently factorize a large composite number. Thus the most important questions at this stage are: If we can build a 7-qubit quantum computer then why can we not build a 700-qubit quantum computer? What are the difficulties associated with the construction of a large quantum computer? Are there ways to circumvent those difficulties? We address these questions in the next chapter.

5.8 Solved examples

1. Determine the eigenvalues of the Grover iterator G in terms of θ.
   Solution: Eigenvalues of $G = \begin{pmatrix} \cos 2\theta & -\sin 2\theta \\ \sin 2\theta & \cos 2\theta \end{pmatrix}$ should satisfy

   the secular equation $Det \begin{pmatrix} \cos 2\theta - \lambda & -\sin 2\theta \\ \sin 2\theta & \cos 2\theta - \lambda \end{pmatrix} = 0$, which im-

   plies $(\cos 2\theta - \lambda)^2 = -\sin^2 2\theta$ or, $\lambda = \cos 2\theta \pm i \sin 2\theta$. Therefore the eigenvalues are $\cos 2\theta + i \sin 2\theta$ and $\cos 2\theta - i \sin 2\theta$.

2. Find out the order of $4 mod 15$ and use that to find out the factors of 15.
   Solution: If we compute $4^i mod 15$ for $i = 1, 2, 3, \cdots$ then we obtain the series $4, 1, 4, 1, \cdots$. Thus the series repeats its values after every 2 terms and consequently, the order $r = 2$. Therefore, $gcd(15, 4 + 1) = gcd(15, 5) = 5$ and $gcd(15, 4 - 1) = gcd(15, 3) = 3$ are the possible candidates of nontrivial factors of 15. In this case both of them are factors of 15.

3. Provide circuits for all possible unitary operation U_f in Deutsch's algorithm.
   Solution: Here U_f is a 2-qubit unitary operation that maps $|x, y\rangle$ into the state $|x, y \oplus f(x)\rangle$. So

   $$\begin{aligned} U_f &= |0\rangle|0 + f(0)\rangle\langle 00| + |0\rangle|1 + f(0)\rangle\langle 01| \\ &+ |1\rangle|0 + f(1)\rangle\langle 10| + |1\rangle|1 + f(1)\rangle\langle 11|. \end{aligned}$$

   Since the function $f(x)$ is either constant or balanced so we have only 4 possible cases:
   Case 1. If $f(0) = f(1) = 0$ then $U_{f1} = I \otimes I$ as no changes happen.
   Case 2. If $f(0) = f(1) = 1$ then $U_{f2} = |01\rangle\langle 00| + |00\rangle\langle 01| + |11\rangle\langle 10| + |10\rangle\langle 11|$. Thus the first qubit remains unchanged and the second qubit flips so $U_{f2} = I \otimes X$
   Case 3. If $f(0) = 0, f(1) = 1$ then $U_{f3} = |00\rangle\langle 00| + |01\rangle\langle 01| + |11\rangle\langle 10| + |10\rangle\langle 11|$. Clearly this is the CNOT gate.
   Case 4. If $f(0) = 1, f(1) = 0$ then $U_{f4} = |01\rangle\langle 00| + |00\rangle\langle 01| + |10\rangle\langle 10| + |11\rangle\langle 11|$. Clearly this is the 0-control CNOT gate, which flips the target qubit (second qubit) when the control qubit (first qubit) is in the state $|0\rangle$. With the conventional CNOT gate the same operation can be visualized by inserting two NOT gates in the first qubit line, one before the CNOT operation and another after

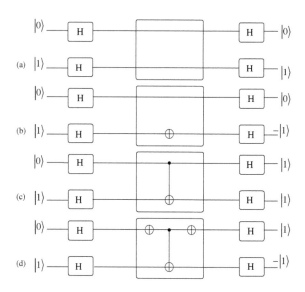

Figure 5.4: Explicit quantum circuits for Deutsch's algorithm (a)-(d) are showing the cases 1-4, respectively. The cases 1-4 are discussed in the Solved Example 3.

the CNOT operation or equivalently by inserting a NOT gate in the second qubit line after the CNOT. Thus

$$U_{f4} = (\text{NOT} \otimes I)\text{CNOT}(\text{NOT} \otimes I) = (I \otimes \text{NOT})\text{CNOT}.$$

Explicit circuits representing U_{f1} to U_{f4} are shown in Fig. 5.4. Now we may visualize Deutsch's problem as follows: One of these 4 possible circuits is inside a black box. We cannot see inside the black box and we have to tell whether the circuit inside the black box gives us constant output or balanced output.

4. Assume that in Deutsch's algorithm there is a missing gate fault (a gate is missing) in the second qubit line. Due to the fault, the Hadamard gate present in the second qubit line before (U_f) is not working. Also assume that the input state is $|\psi_0\rangle = |00\rangle$. Now run the same algorithm as described in this chapter and show that even in this situation we obtain correct results in 75% of the cases. Also show that this algorithm produces correct results with certainty in 25% of the cases.

   Solution: As we follow the same algorithm, here the input state for U_f will be $|\psi_1\rangle = \frac{1}{\sqrt{2}}(|00\rangle + |10\rangle)$. The state after the operation of

U_f will be

$$|\psi\rangle_2 = U_f |\psi\rangle_1 = \begin{cases} \left[\dfrac{|00\rangle+|10\rangle}{\sqrt{2}}\right] & \text{if } f(0) = f(1) = 0 \\[2mm] \left[\dfrac{|01\rangle+|11\rangle}{\sqrt{2}}\right] & \text{if } f(0) = f(1) = 1 \\[2mm] \left[\dfrac{|00\rangle+|11\rangle}{\sqrt{2}}\right] & \text{if } f(0) = 0,\ f(1) = 1 \\[2mm] \left[\dfrac{|01\rangle+|10\rangle}{\sqrt{2}}\right] & \text{if } f(0) = 1,\ f(1) = 0. \end{cases} \qquad (5.60)$$

Finally we apply a Hadamard gate on the first qubit to obtain

$$|\psi\rangle_3 = \begin{cases} |00\rangle \text{ if } f(0) = f(1) = 0 \\ |01\rangle \text{ if } f(0) = f(1) = 1 \\ \frac{1}{2}(|00\rangle + |01\rangle + |10\rangle - |11\rangle) \text{ if } f(0) = 0,\ f(1) = 1 \\ \frac{1}{2}(|00\rangle + |01\rangle - |10\rangle + |11\rangle) \text{ if } f(0) = 1,\ f(1) = 0. \end{cases}$$
$$(5.61)$$

Now we measure the first qubit, if we get it in state $|1\rangle$ then with certainty we know that the function is balanced as the first qubit is always in state $|0\rangle$ for the constant functions. If we obtain $|0\rangle$ then we assume the function is constant. But this conclusion may be wrong in half of those cases where the function is actually balanced. So the algorithm will yield the wrong result in $\frac{1}{2} \times \frac{1}{2} = \frac{1}{4}$ of the cases. Consequently, our conclusions will be correct in $1 - \frac{1}{4} = \frac{3}{4}$ of the cases. However, we are certain about our conclusion iff the outcome is $|1\rangle$. Thus our conclusions are correct with certainty with probability $\frac{1}{4}$.

5. Show that the unitary operator U_f defined as $U_f|x, y\rangle = |x, y \oplus f(x)\rangle$ is a self-inverse operator, where $f(x) \in \{0, 1\}$.
   Solution: $U_f U_f|x, y\rangle = U_f|x, y \oplus f(x)\rangle = |x, y \oplus f(x) \oplus f(x)\rangle = |x, y\rangle$ as $f(x) \in \{0, 1\}$ and consequently, $U_f U_f = I$, which implies that the operator U_f is a self-inverse operator.

6. Show that the operator defined in (5.20) is unitary.
   Solution: The given operator is

$$U_\psi = \begin{pmatrix} 2\cos^2\theta - 1 & 2\sin\theta\cos\theta \\ 2\sin\theta\cos\theta & 2\sin^2\theta - 1 \end{pmatrix} = \begin{pmatrix} \cos 2\theta & \sin 2\theta \\ \sin 2\theta & -\cos 2\theta \end{pmatrix}.$$

Here $U_\psi = U_\psi^\dagger$, therefore it would be sufficient to show that $U_\psi U_\psi = I$. Now

$$\begin{aligned} U_\psi U_\psi &= \begin{pmatrix} \cos 2\theta & \sin 2\theta \\ \sin 2\theta & -\cos 2\theta \end{pmatrix} \begin{pmatrix} \cos 2\theta & \sin 2\theta \\ \sin 2\theta & -\cos 2\theta \end{pmatrix} \\ &= \begin{pmatrix} \cos^2 2\theta + \sin^2 2\theta & 0 \\ 0 & \cos^2 2\theta + \sin^2 2\theta \end{pmatrix} \\ &= I. \end{aligned}$$

7. To factorize a l bit number Shor's algorithm for factorization requires time proportional to $O\left(l^2\left(\log_2 l\right)\left(\log_2 \log_2 l\right)\right)$. Assume that the quantum computer can factorize a 50 digit binary number in one hour. Now compute the time required to factorize a (a) 300 bit number, (b) 1000 bit number and (c) 2000 bit number.
   Solution: For convenience let us write $F(l) = l^2\left(\log_2 l\right)\left(\log_2 \log_2 l\right)$. Now it is given that the quantum computer can factorize a 50 digit number in one hour. Thus in one hour it can do $F(50) = 35227$ number of operations. Therefore, to factorize a l digit number it would require $\frac{F(l)}{F(50)}$ hours. Now to factorize, (a) 300 bit number it would take $\frac{F(300)}{F(50)} = 63.92$ hours which is almost 2.66 days, (b) 1000 bit number it would take $\frac{F(1000)}{F(50)} = 938.4$ hours which is almost 39 days, and (c) 2000 bit number it would take $\frac{F(2000)}{F(50)} = 4302$ hours which is almost 6 months.

5.9   Further reading

1. We have discussed only a handful of quantum algorithms. A comprehensive list of quantum algorithms can be found at http://math. nist.gov/quantum/zoo/. Algorithms are classified and a short description of all the quantum algorithms and the speedup achieved in each one are systematically noted here. Here one can also find the appropriate references for the quantum algorithms of his/her interest.

2. A. O. Pittenger, An introduction to quantum computing algorithms, Progress in computer science and applied logic, Vol. **19**, Birkhäuser, Boston, MA, United States (1999).

3. A set of good review articles on quantum algorithms has been written over the last 15 years. For example, you may find it interesting to read: (a) Quantum algorithms, M. Mosca, quant-ph/0808.0369v1. (b) On quantum algorithms, R. Cleve, A Ekert, L. Henderson, C. Macchiavello and M. Mosaca, Quantum computation and quantum information theory, World Scientific, Singapore (2000) 86-100. (c) Quantum algorithms revisited, R. Cleve, A. Ekert, C. Macchiavello and M. Mosca, Proc. R. Soc. Lond. A **454** (1998) 339-354, quant-ph/9708016v1. (d) A. Ekert and R. Jozsa, Quantum computation and Shor's factoring algorithm, Rev. Mod. Phys. **68** (1996) 733–753. Apart from Shor's algorithm, this article also reviews the basic ideas of quantum computation and quantum complexity classes.

4. A small list of quantum computing simulators and corresponding web links are available at http://zksi.iitis.pl/wiki/software:start.

Julia Wallace's web page also contains a good list of quantum computing simulators at http://www.cs.kent.ac.uk/archive/people/staff/jw74/qc/simtable.htm but this page has not been updated recently. A more comprehensive list of quantum computing simulator can be found at http://www.quantiki.org/wiki/List_of_QC_simulators. We believe readers will find most of these simulators interesting.

5.10 Exercises

1. Compute the quantum Fourier transform of the state $\frac{1}{\sqrt{30}}|0\rangle + \frac{2}{\sqrt{30}}|1\rangle + \frac{5}{\sqrt{30}}|2\rangle$.

2. Show that Grover iterator G is unitary. Is it Hermitian?

3. Compute the quantum Fourier transform of all Bell states as expressed in the computational basis.

4. Briefly answer all the questions below:
   (a) Find out the order of $7 mod 15$ and use that to find out the factors of 15.
   (b) Give a simple example of balanced function.
   (c) Give a simple example of constant function.

5. Explicitly show that the Deutsch Jozsa algorithm works for input state $|\psi\rangle_0 = |001\rangle$.

6. Express $\frac{171}{52}$ in continued fraction representation.

7. Elaborately describe all the steps required to factorize 77 using Shor's algorithm.

Chapter 6

Quantum error correction

In Chapter 4 we introduced different quantum gates and quantum circuits and in Chapter 5 we described several quantum algorithms. So far the analysis of the algorithms and the circuits has been made with the assumption that the required unitary operations can be implemented exactly without any error. This is an abstraction. In an actual physical system that implements a quantum algorithm or a quantum circuit there are several sources of errors. The situation is similar in classical computation and to circumvent this problem error correcting codes are used in classical digital computing. Since qubits are more delicate and more susceptible to errors we need to introduce quantum error correcting codes. It is very important, as without such codes it will not be possible to build a scalable quantum computer. Such quantum error correcting codes are introduced in this chapter. However, we will not limit ourselves to the discussion of error correcting codes alone. We will also discuss more general aspects of errors and error correction, like decoherence, decoherence free subspace and fault-tolerance.

6.1 Quantum error correction

In Chapter 2 we briefly discussed the possibilities of appearance of errors in a classical channel. We also noted the difficulties associated with the error correction in analog computing. However, we have not discussed the following important issues:

(a) How can classical errors be corrected?

(b) What are the differences between classical and quantum errors?

(c) Can we use classical error correcting schemes to correct quantum errors?

(d) If not, then how to correct quantum errors?

In the following sections we will discuss these issues. But before we start the technical discussion we would prefer to recall the answer to a more fundamental question: Why don't we use analog classical computers? The answer is obvious: the possibility of infinitely many states makes it impossible to correct errors. Apparently, the same logic is applicable to quantum states. This is because the qubit is a two level quantum system, which can have continuum of possible states specified by

$$|\psi\rangle = \alpha|0\rangle + \beta|1\rangle,$$

where α and β are arbitrary complex numbers which satisfy $|\alpha|^2 + |\beta|^2 = 1$. In the early days of quantum information theory, Landauer [83] and others were convinced by this logic and it was believed that quantum error correction would not be possible [84]. Later on clever techniques of quantum error correction were introduced [24], but to understand them we need to understand: (a) What is an error model and (b) How are classical errors corrected?

6.2   Basic idea of an error model

When we wish to protect our information against possible errors we first need to define the particular form of error from which we wish to protect our information. Each type of error leads to an error model. We will further clarify this point soon. We do not want a bit/qubit to change its value when it is either stored or it is moving from one place to another place. But because of error model the set of bits/qubits get changed (evolved). A particular error model describes a certain type of evolution and is often referred to as a channel. For example, we are interested in *identity channel* which implies an error free channel. The simplest example of classical error model is a *bit flip channel*. In this simplest model the state of a bit flips with probability p and remains unchanged with probability $(1-p)$. As the probability of bit flip is the same for both 0 and 1, this channel is often referred to as a symmetric bit flip channel. Such a channel is shown in Fig. 6.1. We can now easily add some more complexities to this simple error model and generate new error models. The following examples elaborate how to obtain new error models by modifying the symmetric bit flip channel described above.

Example 6.1: Consider that the probability of bit flip for 0 is p and that for 1 is q, where $p \neq q$. This asymmetric bit flip channel provides us with a new and a little more complex error model.

Example 6.2: Consider that the bit flipping does not occur independently in different bits. That is, bit flip errors are correlated. This provides another example of an error model, which is a bit more general in nature.

   The above examples of classical error models are provided to clear the concept of an error model. Once the error model is known then the task at hand is to protect the information from that particular kind of error. This

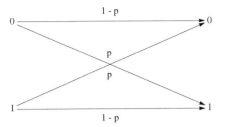

Figure 6.1: The simplest classical error model: Symmetric bit flip channel.

is done by encoding the bits in a way that is robust against that particular type of error. Usually it is done by adding a few extra bits to a logical bit that we want to protect. Thus from a single bit we obtain a string of bits. This string of bits corresponding to a particular bit value is called *codeword* as it encodes the actual bit value. Thus we have two codewords[1], one for bit value 0 and the other for bit value 1. The set of these two codewords is called *code*. Practically, a codeword adds some redundancy to the corresponding bit value and due to this redundancy even when error affects some of the bits in a codeword, the remaining bits contain sufficient information so that the actual logical bit can be recovered. As we progress in this chapter, our understanding of the meaning of code and the error model will be refined. We may now introduce a classical error correcting code that can protect information from a bit flip error model.

6.2.1   How to correct classical errors

The simplest example of a classical error correcting code is a repetition code. We replace the bit we want to protect by odd number of copies of it and then apply majority voting technique. To clear this idea let us start with an example in which we make three copies of the bit we want to protect. So we make ideally

$$0 \mapsto (000)$$
$$1 \mapsto (111).$$

Here, (000) and (111) are codewords as they encode 0 and 1 respectively and $\{(000), (111)\}$ is the code. Now an error may occur that causes one of the three bits to flip; if the first bit flips then we have

$$0 \mapsto (100)$$
$$1 \mapsto (011).$$

---

[1]We assume that we are using binary logic. If we use multivalued logic then we will have more codewords, but the essential idea will remain same.

In spite of the error, we can still decode the bit correctly by majority voting technique. If two out of three bits (i.e., majority) are $0\,(1)$ then the original bit was $0\,(1)$. This three bit repetition code will work only when at most 1 bit is flipped. If two bits are flipped in a particular case then this three bit repetition code will fail. Even if we take a four bit repetition code and two of the bits flip, then the code will not be able to lead us to any decision because the voting will end in a tie. But a five bit repetition code will work even when two bits are flipped. From this example it's clear that a $2n$ bit repetition code can detect and correct as many bit flip errors as a $2n - 1$ bit repetition code can do. However, the former needs more computational resource. This is why even number repetition codes are not used.

Let us assume that the probability of each bit to flip is p, then the probability that a particular bit does not flip is $(1 - p)$. Now two bits can flip in three different ways, e.g.,

$$(000) \longrightarrow \begin{matrix} (110) \\ (101) \\ (110) \end{matrix}$$

and

$$(111) \longrightarrow \begin{matrix} (001) \\ (010) \\ (100). \end{matrix}$$

Therefore, the total probability that two bits flip and one remains unchanged is $3pp(1 - p) = 3p^2 - 3p^3$. The probability that all three bits flip is p^3. Thus the probability that majority voting fails (i.e., the probability that more than one bit flips) is $3p^2 - 3p^3 + p^3 = 3p^2 - 2p^3$. For the success of the repetition code it is required that the probability of error after encoding must be less than the probability of error without encoding (i.e., p). Thus this repetition code protects the information against bit flip error for

$$\begin{aligned} 3p^2 - 2p^3 \quad &< \quad p \\ \text{or, } 2p^2 - 3p + 1 \quad &> \quad 0 \\ \text{or, } (p - 1)(2p - 1) \quad &> \quad 0. \end{aligned} \tag{6.1}$$

In general $p \leq 1$ but for $p = 1$ the above inequality is not satisfied. Now for $p < 1$ the first term in the inequality (6.1) is negative. Therefore, the second term has to be negative in order to have a positive product, i.e.,

$$\begin{aligned} 2p - 1 \quad &< \quad 0 \\ p \quad &< \quad \tfrac{1}{2}. \end{aligned}$$

Thus the majority voting technique is useful only when the probability of a bit flip is less than $\tfrac{1}{2}$.

6.2.2 Difference between classical error and quantum error

It is tempting to think that the majority voting technique discussed in the previous subsection can be generalized for the quantum error correction. However, in reality it has the following problems.

1. **Nocloning:** In the classical version of repetition code, we protect information by making extra copies of it. But from the nocloning theorem we know that an unknown quantum state cannot be copied with perfect fidelity.

2. **Phase flip error**: In classical information, error can cause bit flip only but more things may go wrong with a qubit. For example, there may be a phase error which can change phase of $|0\rangle$ or $|1\rangle$ or of both. Just think about the following situation

$$
\begin{aligned}
|0\rangle &\rightarrow |0\rangle \\
|1\rangle &\rightarrow -|1\rangle.
\end{aligned}
\tag{6.2}
$$

This kind of error is not possible in the classical world but this is a serious quantum error because it can transform the state $\frac{1}{\sqrt{2}}[|0\rangle + |1\rangle]$ to the orthogonal state $\frac{1}{\sqrt{2}}[|0\rangle - |1\rangle]$. The classical coding scheme cannot provide any protection against this kind of errors.

3. **Small errors:** Consider a qubit $\alpha|0\rangle + \beta|1\rangle$. Here the probabilities of obtaining 0 and 1 as outcomes of a measurement in computational basis are $p(0) = |\alpha|^2$ and $p(1) = |\beta|^2$ respectively. An error may slightly change this probability distribution to $p(0) = |\alpha|^2 + \epsilon$ and $p(1) = |\beta|^2 - \epsilon$, where $\epsilon \rightarrow 0$ is a real constant. These small changes can accumulate over time[2]. This kind of small error cannot be corrected by classical error correction technique because the classical error correcting code is designed to correct only large errors (bit flip).

4. **Measurement issue:** In a classical repetition code we need to measure the state and use the measurement outcome to recover the state. But a measurement of quantum state would destroy the state under observation and consequently, recovery of the initial quantum state would become impossible. Thus we cannot use classical repetition code for correction of quantum errors.

From the above points it is clear that simple repetition code alone will not be able to correct quantum errors. We need to look for some other technique to correct quantum errors. To start with let us try to develop an analogue of majority voting scheme.

---

[2]There are situations where we intentionally change α and β and accumulate that change for time. Just compare the situation with Grover's algorithm. A decoherence may have an effect similar to the Grover iterator. See Examples 6.5 and 6.6.

6.2.3   How to correct quantum errors

An arbitrary quantum error can be viewed as combination of bit flip and phase flip errors. So if we can design circuit (or scheme) for simultaneous correction of bit flip and phase flip errors, then it would be possible to correct any arbitrary quantum error. To provide a clear view of the quantum error correction technique we will first provide a scheme for correction of bit flip error in the following subsection. In the subsequent subsections we will show how one can correct phase flip error and combination of bit flip and phase flip errors.

6.2.3.1   Correction of quantum bit flip error: A 3-qubit code

Let us start with an arbitrary qubit $|\psi\rangle = \alpha|0\rangle + \beta|1\rangle$. We use two ancilla (extra) qubits to add redundancy. The encoding of the information is done with the help of two CNOT gates. See Fig. 6.2: after the operation of the first two CNOT gates the state would become

$$|\psi\rangle_1 = \alpha|000\rangle + \beta|111\rangle.$$

Now the bit flip error happens. Without loss of generality we may assume that the first bit flips, then the state will be transformed to

$$|\psi\rangle_2 = \alpha|100\rangle + \beta|011\rangle.$$

Now we reach the measurement stage. Here two parity checking circuit blocks are implemented. The parity checking circuit blocks are shown using two rectangular boxes in Fig. 6.2. First two CNOTs (which are shown inside the first rectangular box) check the parity of the first two qubits and the last two CNOTs (which are shown inside the second rectangular box) check the parity of the second and third qubits. To check the parity of the first two qubits we use the first two qubits as control qubits of two CNOT gates and the target qubit of both of these CNOT gates is the first ancilla (A_1) qubit. If both the control qubits are the same then the measurement on the target qubit (ancilla) yields 0 and if they are different then the measurement on ancilla yields 1. In our particular case the measurement on A_1 would yield 1 and from that we would know that either the first or second qubit is flipped. The measurement on the second ancilla (A_2) would yield zero which implies that neither the second nor the third qubit has flipped. Since our intrinsic assumption is that at most one bit has flipped, therefore, from the outcomes of these two measurements we conclude that the first qubit is flipped. From these measurements we learn where the error is. This information is called error syndrome. After knowing the error syndrome we can easily correct the error by flipping back the qubit which has been found to be flipped due to error.

We can mathematically visualize it as follows. After the introduction of error two more ancilla prepared in $|0\rangle$ are introduced and with that the

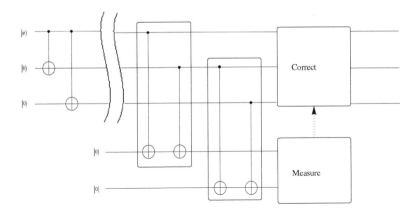

Figure 6.2: A simple scheme for correction of quantum bit flip error.

combined state of the system is

$$|\psi\rangle_3 = |\psi\rangle_2 \otimes |00\rangle = \alpha|10000\rangle + \beta|01100\rangle.$$

Now after the operation of the four CNOT gates the state becomes

$$|\psi\rangle_4 = \alpha|10010\rangle + \beta|01110\rangle = (\alpha|100\rangle + \beta|011\rangle)) \otimes |10\rangle.$$

Thus the measurement on the 4th and 5th qubit would yield 1 and 0 respectively and from that we can conclude that the first qubit is flipped. After knowing that the first qubit is flipped we can flip that back to the original state without knowing anything about α and β. This is just a quantum version of classical 3 bit repetition code. It is interesting to observe that a generalization of classical repetition code can correct quantum error, too. In the next subsection we will show that, more surprisingly, if we change basis then we can also correct phase flip. Further, we wish to note that the bit flip error can be corrected even without the measurement. Such a measurement less bit flip error correction circuit, which requires two CNOT gates and a Toffoli gate, is used in the circuit described in Fig. 6.4.

6.2.4 Correction of phase flip error: A 3-qubit code

There is no classical analogue of phase flip error. But we can show that a phase flip error can be transformed into a bit flip error. This implies that a 3-qubit bit flip error correction code can be used to correct phase flip errors, too. Let us see how a phase flip error can be transformed to a bit flip error:

Consider the Hadamard basis states

$$|+\rangle = H|0\rangle = \frac{1}{\sqrt{2}} (|0\rangle + |1\rangle)$$
$$|-\rangle = H|1\rangle = \frac{1}{\sqrt{2}} (|0\rangle - |1\rangle).$$

It is easy to visualize that $\{|+\rangle, |-\rangle\}$ forms an orthonormal basis set and a phase flip error transforms the state $|+\rangle$ to $|-\rangle$ and vice versa. Consequently, if we work in Hadamard (diagonal) basis then the phase flip errors of computational basis are equivalent to bit flip errors in diagonal basis. So to obtain the error syndrome we need to encode the state $|0\rangle$ in $|+++\rangle$ and the state $|1\rangle$ in $|---\rangle$. This can be done by applying three Hadamards at the end of three qubit lines of the encoding block used in the quantum bit flip error correction circuit. The precise encoding circuit is shown in Fig. 6.3. The first two CNOTs transform the initial state (with ancilla) $|\psi00\rangle$ to $|\psi\rangle_1 = \alpha|000\rangle + \beta|111\rangle$ as before. The three Hadamards encode the state as

$$
\begin{aligned}
|\psi\rangle_2 = H^{\otimes 3}|\psi\rangle_1 &= \frac{\alpha}{2\sqrt{2}} ((|0\rangle + |1\rangle)(|0\rangle + |1\rangle)(|0\rangle + |1\rangle)) \\
&+ \frac{\beta}{2\sqrt{2}} ((|0\rangle - |1\rangle)(|0\rangle - |1\rangle)(|0\rangle - |1\rangle)) \\
&= \alpha|+++\rangle + \beta|---\rangle.
\end{aligned}
$$

Here we can see that the Hadamard gates perform a basis transformation from the computational basis to the diagonal basis. Now if a phase flip error happens in the first qubit then the state $|\psi\rangle_2$ gets transformed to

$$|\psi\rangle_3 = \alpha|-++\rangle + \beta|+--\rangle.$$

Since the Hadamard operation is self-inverse we can use three more of them in the decoding circuit to transform this state into

$$|\psi\rangle_4 = H^{\otimes 3}|\psi\rangle_3 = \alpha|100\rangle + \beta|011\rangle.$$

Thus the phase flip error is reduced to a bit flip error and we already know how to correct it. Thus we have learned how to correct independent bit flip and phase flip errors but they may happen simultaneously. To correct an arbitrary error we need to combine the idea of bit flip error correction and phase flip error correction. When we do so then we obtain a 9-qubit code which is known as the Shor code, after its inventor. We describe the Shor code in the next subsection.

6.2.5   Shor code: A 9-qubit code that can correct an arbitrary single qubit error

We have already noted that any arbitrary quantum error can be viewed as combination of bit flip error and phase flip error and consequently, a code that can correct both the errors can essentially correct an arbitrary single

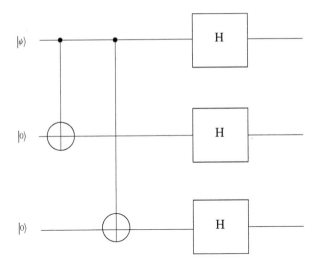

Figure 6.3: Encoding circuit for the 3-qubit phase flip code.

qubit error. This code was introduced by Peter Shor in 1995 [24]. In this coding scheme the encoding is done in two steps as shown in Fig. 6.4. The first part of this circuit encodes the qubit using the three qubit phase flip code and the second part of the circuit encodes each of these three qubits using encoding circuit used in the three qubit bit flip code (see Fig. 6.2). Clearly in this part we use three copies of the encoding circuit used for bit flip code. This method of encoding using a hierarchy of levels is known as *concatenation*. To be precise, in the first part of Shor's encoding circuit, we encode the state in a manner similar to phase-flip code where the encoding circuit maps $|0\rangle \mapsto |+++\rangle$ and $|1\rangle \mapsto |---\rangle$. Consequently, we obtain

$$|\psi\rangle_1 = \alpha|+++\rangle + \beta|---\rangle.$$

In the next step each of these three qubits are encoded using three qubit bit flip code. We have seen that in three qubit bit flip code $\alpha|0\rangle + \beta|1\rangle$ is encoded as $\alpha|000\rangle + \beta|111\rangle$. For a $|+\rangle$ state $\alpha = \beta = \frac{1}{\sqrt{2}}$, hence a $|+\rangle$ would be encoded as $\frac{1}{\sqrt{2}}|000\rangle + \frac{1}{\sqrt{2}}|111\rangle$ and similarly, a $|-\rangle$ state would be encoded as $\frac{1}{\sqrt{2}}|000\rangle - \frac{1}{\sqrt{2}}|111\rangle$. The result is a nine qubit code in which the arbitrary quantum state $\alpha|0\rangle + \beta|1\rangle$ is encoded as

$$
\begin{aligned}
|\psi\rangle_2 &= \frac{\alpha}{2\sqrt{2}}\left(|000\rangle + |111\rangle\right)\left(|000\rangle + |111\rangle\right)\left(|000\rangle + |111\rangle\right) \\
&+ \frac{\beta}{2\sqrt{2}}\left(|000\rangle - |111\rangle\right)\left(|000\rangle - |111\rangle\right)\left(|000\rangle - |111\rangle\right).
\end{aligned}
$$

Suppose we allow both bit flip and phase flip to happen on this codeword with a restriction that at most one bit flip and one phase flip may happen.

Now if a bit flip happens in the first qubit then we can find the error syndrome by checking the parity of the first two qubits and that of the second and third qubits. The first parity check will show that the first and second qubits are different so one of them is flipped. The second parity check will show that the second and third qubits are the same. Therefore, neither the second qubit nor the third qubit is flipped, and consequently the first qubit is flipped. Similarly, if the parity check between the second and the third qubit shows that one of them is flipped and the parity check between the first two qubits shows that one of them (none of them) is flipped, then we can conclude that second qubit (third qubit) is flipped. Thus two parity checks are sufficient for detection of occurrence of bit flip error in any of the first three qubits. In general, parity checks between the qubits (1,2), (2,3), (4,5), (5,6), (7,8), (8,9) will find bit flip syndrome in any of the 9 qubits. Thus in total we need to perform six parity checks. Once the bit flip syndrome is known we can correct it by flipping the corresponding qubit (say first qubit in our case). Correction can also be done without measurement as shown in Fig. 6.4.

Now if the phase flip happens on any of the first three qubits, then it will change the sign of the first block of qubits as $|000\rangle + |111\rangle$ to $|000\rangle - |111\rangle$ and vice versa. Note that the phase flip in any of the three qubits (i.e., phase of the first qubit is flipped or phase of the second qubit is flipped or phase of the third qubit is flipped) has the same effect and the error correction procedure described here will be valid for any of these three possible errors. The same is true for every block of qubits. To trace the error syndrome we first compare the sign (relative phase) of the first and second blocks of three qubits in a manner similar to the phase flip code described above. The situation is similar with the only difference being that in the previous case we have compared the sign of first and second qubits and here we need to compare the sign of first and second blocks of qubits. To be clearer we may note that $(|000\rangle + |111\rangle)(|000\rangle + |111\rangle)$ has the same sign $(+)$ in both blocks of qubits, while $(|000\rangle - |111\rangle)(|000\rangle + |111\rangle)$ has different signs. As we are considering a specific case where the phase flip has occurred on any of the first three qubits, i.e., in the first block of qubits, so our measurement will find that the signs of the first and second blocks are different but the signs of second and third blocks are the same. From these observations we can easily conclude that the phase must have flipped in the first block of three qubits. We can then recover from this by flipping the sign in the first block of three qubits back to its original value. Similarly, we can detect and correct phase flip on any of the nine qubits.

Thus Shor code corrects both bit flip and phase flip errors. It is straightforward to check that even if both bit flip and phase flip occur on the first qubit (in general on any of the nine qubits) then the above procedure can detect and correct both of them. Thus Shor code can correct combined bit-phase flip error, too. But the power of Shor code is not limited to the correction of bit flip, phase flip and combined bit-phase flip errors. It can actually correct any arbitrary single qubit error. To visualize that lucidly we may note that an arbitrary single qubit error can be described by an arbitrary single qubit operation $E = \begin{pmatrix} E_{11} & E_{12} \\ E_{21} & E_{22} \end{pmatrix}$, which maps an input state $|\psi\rangle$ to an erroneous state $E|\psi\rangle$. For example, a bit flip error is described by $X = \begin{pmatrix} 0 & 1 \\ 1 & 0 \end{pmatrix}$, phase flip error is described by $Z = \begin{pmatrix} 1 & 0 \\ 0 & -1 \end{pmatrix}$ and a combined bit flip and phase flip error is described by $ZX = iY = \begin{pmatrix} 0 & 1 \\ -1 & 0 \end{pmatrix}$. Thus these three error models are described by three Pauli matrices and error free channel is described by the Identity matrix (I). Now since any arbitrary single qubit operator can be described as a linear combination of Pauli operators, so any arbitrary single qubit error can be visualized as a linear combination of bit flip, phase flip and combined bit-phase flip errors.

To visualize the above, we assume that E can be expressed as a linear combination of I, X, Z and ZX, which physically means that we assume that any arbitrary error can be viewed as a combination of no error (I), bit flip error (X), phase flip error (Z) and combined bit flip and phase flip error ($ZX = iY$). Thus as per our assumption

$$
\begin{aligned}
E &= \begin{pmatrix} E_{11} & E_{12} \\ E_{21} & E_{22} \end{pmatrix} \\
&= a \begin{pmatrix} 1 & 0 \\ 0 & 1 \end{pmatrix} + b \begin{pmatrix} 0 & 1 \\ 1 & 0 \end{pmatrix} + c \begin{pmatrix} 1 & 0 \\ 0 & -1 \end{pmatrix} + d \begin{pmatrix} 0 & 1 \\ -1 & 0 \end{pmatrix} \\
&= \begin{pmatrix} a+c & b+d \\ b-d & a-c \end{pmatrix},
\end{aligned}
$$

which implies

$$
\begin{aligned}
E_{11} &= a+c \\
E_{12} &= b+d \\
E_{21} &= b-d \\
E_{22} &= a-c.
\end{aligned}
$$

Therefore,

$$
\begin{aligned}
a &= \tfrac{E_{11}+E_{22}}{2} \\
b &= \tfrac{E_{12}+E_{21}}{2} \\
c &= \tfrac{E_{11}-E_{22}}{2} \\
d &= \tfrac{E_{12}-E_{21}}{2}
\end{aligned}
$$

and consequently the arbitrary single qubit error

$$
E = \begin{pmatrix} E_{11} & E_{12} \\ E_{21} & E_{22} \end{pmatrix} = \tfrac{E_{11}+E_{22}}{2} I + \tfrac{E_{12}+E_{21}}{2} X + \tfrac{E_{11}-E_{22}}{2} Z + \tfrac{E_{12}-E_{21}}{2} ZX.
$$

Thus when this arbitrary error occurs on the state $|\psi\rangle$ then the state is transformed to the erroneous state

$$
\begin{aligned}
|\psi'\rangle = E|\psi\rangle \;\; = \;\; & \tfrac{E_{11}+E_{22}}{2} I|\psi\rangle + \tfrac{E_{12}+E_{21}}{2} X|\psi\rangle \\
+ \;\; & \tfrac{E_{11}-E_{22}}{2} Z|\psi\rangle + \tfrac{E_{12}-E_{21}}{2} ZX|\psi\rangle.
\end{aligned}
\tag{6.3}
$$

The unnormalized state $|\psi'\rangle$ is a superposition of four terms: $|\psi\rangle$, $X|\psi\rangle$, $Z|\psi\rangle$ and $ZX|\psi\rangle$. Now measurement of the error syndrome in Shor code collapses the superposition state $|\psi'\rangle$ into one of the four states: $|\psi\rangle$, $X|\psi\rangle$, $Z|\psi\rangle$ and $ZX|\psi\rangle$, from which the actual state $|\psi\rangle$ can be recovered by applying appropriate inversion operation. Thus Shor code can correct an arbitrary single qubit error. In general, any quantum error correction code that can correct bit flip, phase flip and combined bit-phase flip errors can correct an arbitrary single qubit error. It is important to understand the relevance of (6.3). This fundamental result shows that by correcting a discrete set of errors we can correct a larger class of continuous errors. In other words, it shows that quantum errors can be discretized which was not the case with analogue computation. Thus although we have a continuum of quantum states (like analogue computing) still we can correct errors and in principle perform quantum computing tasks even in presence of noise.

The above proof of discretization of quantum error is lucid, simple and convincing but it is not the most general one as we had intentionally excluded the environment from the discussion. Now we will briefly describe a more general proof of the above idea. This discussion is expected to clear your perception about the effect of environment which consists of everything apart from the system (qubit) under consideration. Further, you will find this concept useful in understanding the security of protocols of quantum cryptography, which will be discussed in Chapter 8. In addition, it would be helpful to introduce the concept of decoherence and decoherence free subspace. Keeping these things in mind, let us ask: What is the most generic evolution of a qubit in state $|0\rangle$ which is interacting with an environment $|E\rangle$? The most generic evolution will yield a superposition state:

$$
|0\rangle|E\rangle \mapsto \beta_1|0\rangle|E_1\rangle + \beta_2|1\rangle|E_2\rangle,
\tag{6.4}
$$

where $|\beta_1|^2 + |\beta_2|^2 = 1$. Thus with the probability $|\beta_1|^2$ the state has remained in state $|0\rangle$ and the environment has evolved to some state $|E_1\rangle$ and with probability $|\beta_2|^2$ the state has evolved to state $|1\rangle$ and the environment has evolved to some state $|E_2\rangle$. Similarly, when a qubit in state $|1\rangle$ interacts with the environment $|E\rangle$ then the most generic evolution is

$$
|1\rangle|E\rangle \mapsto \beta_3|0\rangle|E_3\rangle + \beta_4|1\rangle|E_4\rangle,
\tag{6.5}
$$

where $|\beta_3|^2 + |\beta_4|^2 = 1$.

Now we can consider a more general scenario where the qubit is in an arbitrary state $|\psi\rangle = \alpha|0\rangle + \beta|1\rangle$ and it interacts with environment $|E\rangle$. In

such a case, the state evolves as

$$|\psi\rangle|E\rangle = (\alpha|0\rangle + \beta|1\rangle)\,|E\rangle \quad \mapsto \quad \begin{aligned}&\alpha\beta_1|0\rangle|E_1\rangle + \alpha\beta_2|1\rangle|E_2\rangle \\ +\;&\beta\beta_3|0\rangle|E_3\rangle + \beta\beta_4|1\rangle|E_4\rangle.\end{aligned} \tag{6.6}$$

The above state can be decomposed as follows

$$\begin{aligned}
|\psi\rangle|E\rangle \;\mapsto\;& \alpha\beta_1|0\rangle|E_1\rangle + \alpha\beta_2|1\rangle|E_2\rangle + \beta\beta_3|0\rangle|E_3\rangle + \beta\beta_4|1\rangle|E_4\rangle \\
=\;& \tfrac{1}{2}\,(\alpha|0\rangle + \beta|1\rangle)\,(\beta_1|E_1\rangle + \beta_4|E_4\rangle) \\
+\;& \tfrac{1}{2}\,(\alpha|0\rangle - \beta|1\rangle)\,(\beta_1|E_1\rangle - \beta_4|E_4\rangle) \\
+\;& \tfrac{1}{2}\,(\alpha|1\rangle + \beta|0\rangle)\,(\beta_2|E_2\rangle + \beta_3|E_3\rangle) \\
+\;& \tfrac{1}{2}\,(\alpha|1\rangle - \beta|0\rangle)\,(\beta_2|E_2\rangle - \beta_3|E_3\rangle) \\
=\;& \tfrac{1}{2}I|\psi\rangle\,(\beta_1|E_1\rangle + \beta_4|E_4\rangle) + \tfrac{1}{2}Z|\psi\rangle\,(\beta_1|E_1\rangle - \beta_4|E_4\rangle) \\
+\;& \tfrac{1}{2}X|\psi\rangle\,(\beta_2|E_2\rangle + \beta_3|E_3\rangle) + \tfrac{1}{2}XZ|\psi\rangle\,(\beta_2|E_2\rangle - \beta_3|E_3\rangle).
\end{aligned} \tag{6.7}$$

This represents the most general evolution that can happen on a qubit and it is straightforward to observe that after this most generic evolution the state of the qubit is either unaffected, or bit flipped, or phase flipped or combined bit-phase flipped. In brief, once again we have shown that quantum error can be discretized and hence it can be corrected by Shor code or by a similar code.

Cartoon 6.1: As everything apart from the system is part of the environment, the argument used by the boy to justify his candidature seems technically correct. What is your opinion?

6.3 A little more on quantum error correction

First we would like to note that Shor's 9-qubit quantum error correction code is not the optimal code. There are more effective codes. For examples,

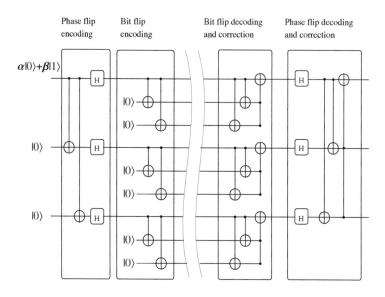

Figure 6.4: Circuit for implementation of 9-qubit Shor code. In the last two blocks we have shown that errors can be corrected without measurement. This idea can be used in the last two sections, too.

Calderbank, Shor and Steane code (CSS code) is a 7-qubit code. The minimum number of qubits required for successful implementation of a quantum error correcting code that can correct arbitrary single qubit error is 5. Such a code was given by Laflamme, Miguel, Paz and Zurek [25]. In the quantum error correction schemes described above we have assumed that the encoding and decoding of quantum state are done perfectly. The errors only appear in the channel at some time after encoding and before decoding. Physically this may be viewed as if we have noise-less quantum computers in which we can perform encoding and decoding without error and only after encoding the qubit enters a noisy channel and at the end of the channel decoding is done with the help of another perfect quantum computer. However, the actual quantum computer is made up of some quantum gates and these gates may be noisy. Here is a new problem! But we have a solution, too. The solution is to use the theory of fault-tolerant quantum computation which allows us to remove the assumption of perfect encoding and decoding. In the next section we will describe this idea. Another relevant idea which needs to be discussed here is the threshold theorem for quantum computation. In Subsection 6.3.2 we will briefly describe threshold theorem. Before we state the threshold theorem let us discuss the basic ideas of fault-tolerant quantum computation.

6.3.1 Fault-tolerant quantum computation

We have already learned that the quantum information can be protected by a quantum error correcting code. But the quantum gates that are used to decode and encode can themselves produce errors. Further, if we wish to use a normal quantum circuit for computation then it will not work on the encoded information. So we have to decode the state before each gate and encode it again after operation of the gate. But this approach does not appear interesting for two reasons. Firstly, we need to do a large number of encoding and decoding operations and secondly in between the encoding and the decoding operations, the gate may introduce an error with probability p. So a good idea could be to design a circuit, which can directly act on the encoded quantum states without leading to propagation of unacceptable (uncorrectable) error. As the error produced by such a circuit is correctable, the circuit is referred to as fault-tolerant circuit. When we implement a fault-tolerant circuit then each gate U of the original circuit is replaced by a fault-tolerant implementation U_{FT} of that gate. Usually the encoded gate U_{FT} is a small circuit built by a few unencoded gates. Now we need to apply the gates (U_{FT}) of a circuit on the encoded states in such a way that if one of the unencoded gates that are used to build the encoded gate is faulty (a gate which produces error) then the information is not lost and finally the produced error can be corrected. The idea can be elaborated through a few specific examples.

Example 6.3: Assume that we wish to implement a $CNOT_{FT}$. As the gate has to work on the encoded state so we need to define the encoding first. Let us use the code we have used for bit flip checking. Thus our code is $\{(000), (111)\}$ which maps $|0\rangle \mapsto |000\rangle$ and $|1\rangle \mapsto |111\rangle$. This particular choice of codeword implies that a conventional CNOT gate which is described as

$$CNOT : \begin{cases} |0\rangle|0\rangle \mapsto |0\rangle|0\rangle \\ |0\rangle|1\rangle \mapsto |0\rangle|1\rangle \\ |1\rangle|0\rangle \mapsto |1\rangle|1\rangle \\ |1\rangle|1\rangle \mapsto |1\rangle|0\rangle \end{cases}$$

should be transformed to the corresponding encoded gate $CNOT_{enc}$ described as

$$CNOT_{enc} : \begin{cases} |000\rangle|000\rangle \mapsto |000\rangle|000\rangle \\ |000\rangle|111\rangle \mapsto |000\rangle|111\rangle \\ |111\rangle|000\rangle \mapsto |111\rangle|111\rangle \\ |111\rangle|111\rangle \mapsto |111\rangle|000\rangle \end{cases} . \qquad (6.8)$$

It is easy to observe that this encoded gate can directly work on our encoded state but all implementations of this encoding (6.8) are not essentially fault-tolerant. To clarify this idea, we have shown two implementations of $CNOT_{enc}$ in Fig. 6.5. The circuit shown in the left is not a fault-tolerant implementation because if an error happens at the control qubit of the first unencoded CNOT gate then the target qubits of all the CNOT gates will be

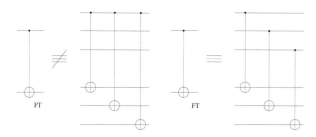

Figure 6.5: Both the circuits implement encoding described in Eqn. (6.8) but the circuit on the left is not fault-tolerant as an error occurring in the beginning of the first qubit line will propagate to all the target qubits. The circuit on the right is free of this problem and represents a fault-tolerant representation of the CNOT gate.

changed and thus the error will propagate to the entire second block. This is an uncorrectable error. Now it is easy to observe that the circuit shown in the right side of Fig. 6.5 is free from such an error and consequently it represents the fault-tolerant gate $CNOT_{FT}$.

Example 6.4: Consider a Hadamard gate. Construction of fault-tolerant Hadamard gate H_{FT} is relatively simpler as $H_{FT} \equiv H^{\otimes 3}$, where each Hadamard gate is implemented separately via single qubit rotations. Any error caused by one of the unencoded Hadamard gates will not affect the other two and consequently, the error will not propagate.

Now as we have H_{FT} and $CNOT_{FT}$ we can easily provide fault-tolerant versions of bit flip and phase flip error correction circuits. Here as an example we have provided a fault-tolerant EPR circuit in Fig. 6.6. Note that we have done error correction after each step, this enables the circuit to tolerate one error in each step. Unless this is done the scheme will fail. We can easily visualize it if we assume that one error happens in the first qubit line at H_{FT} and another one happens in the second qubit line at $CNOT_{FT}$. In this situation the circuit shown in Fig. 6.6 will work fault-tolerantly but if we drop the error correction between H_{FT} and $CNOT_{FT}$ then the scheme will fail.

> **Is there any difference between fault and error?**
> Fault is a location in a circuit where a gate or storage error occurs while error is associated with a qubit in a block that deviates from the ideal state. Fault happens at fixed points in the circuit but error may happen when the qubit is in fly.

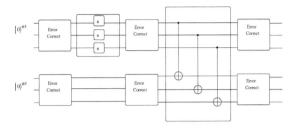

Figure 6.6: A fault-tolerant implementation of EPR circuit. The first box that contains three Hadamards represents H_{FT} and the second box containing three CNOT gates represents CNOT_{FT}.

6.3.2 Threshold theorem for quantum computation

Relevance of fault-tolerant quantum computation is associated with the threshold theorem which states that a quantum computer with noise can quickly and accurately simulate an ideal quantum computer, provided the level of noise is below a certain threshold. To be precise, we need the threshold of accuracy of each quantum gate at such a level that a fault-tolerant realization of the gate can surpass the error. If it can be achieved for all the gates of a universal gate library, then we will be able to make a scalable quantum computer even in presence of decoherence (noise). Thus the important task in hand is to determine the threshold (either theoretically or semi-empirically) for a particular implementation and to experimentally design fault-tolerant quantum gates required to build the quantum computer. We cannot avoid decoherence in a realistic situation. Consequently, the future possibility of experimental realization of a scalable quantum computer would depend on our ability to determine the threshold and to implement fault-tolerant gates that can surpass the threshold determined by us.

6.4   Decoherence and decoherence free subspace

Decoherence is an external influence that destroys the quantum superposition. To be precise, when a closed quantum system interacts with the environment then pure states can get entangled with the external world, leading to mixed states. This is called decoherence. In information theory, decoherence is referred to as noise. In Subsection 3.2.3.1 we provided a lucid definition of decoherence as the loss of coherence (i.e., the change in relative phase between the amplitude of states). Thus we already know the basic definition of the decoherence but we need to understand it in greater detail. To do so, let us start with the isolated single qubit state

$|\psi\rangle = \alpha|0\rangle + \beta|1\rangle = |\alpha|e^{i\phi_\alpha}|0\rangle + |\beta|e^{i\phi_\beta}|1\rangle$, whose density matrix is

$$\rho = |\psi\rangle\langle\psi| = \begin{pmatrix} |\alpha|^2 & \alpha\beta^* \\ \alpha^*\beta & |\beta|^2 \end{pmatrix} = \begin{pmatrix} |\alpha|^2 & |\alpha||\beta|e^{i(\phi_\alpha-\phi_\beta)} \\ |\alpha||\beta|e^{-i(\phi_\alpha-\phi_\beta)} & |\beta|^2 \end{pmatrix}. \quad (6.9)$$

We have already mentioned that decoherence means loss of coherence. This implies decay of the off-diagonal elements of the density matrix (6.9). In reality this is a time dependent process and in general, density matrix changes over time as

$$\rho_{\text{decoherence}} = \begin{pmatrix} f(t)|\alpha|^2 & e^{-\frac{t}{\tau}}|\alpha||\beta|e^{i(\phi_\alpha-\phi_\beta)} \\ e^{-\frac{t}{\tau}}|\alpha||\beta|e^{-i(\phi_\alpha-\phi_\beta)} & 1-f(t)|\alpha|^2 \end{pmatrix}, \quad (6.10)$$

where τ, which is called the decoherence time[3], is the characteristic time scale of the decoherence process. If we choose $f(t)=1$ for our convenience, then we obtain[4]

$$\rho_{\text{decoherence}} = \begin{pmatrix} |\alpha|^2 & e^{-\frac{t}{\tau}}|\alpha||\beta|e^{i(\phi_\alpha-\phi_\beta)} \\ e^{-\frac{t}{\tau}}|\alpha||\beta|e^{-i(\phi_\alpha-\phi_\beta)} & |\beta|^2 \end{pmatrix}. \quad (6.11)$$

In (6.11) off-diagonal terms effectively vanish after time $t > \tau$ and we obtain

$$\rho_{\text{decoherence}} = \begin{pmatrix} |\alpha|^2 & 0 \\ 0 & |\beta|^2 \end{pmatrix}. \quad (6.12)$$

The following example specifically shows that the coupling of a qubit with its environment may lead to the damping out of the off-diagonal terms of density matrix.

Example 6.5: Consider a coupling model such that the qubit of the system remains unchanged but the environment which is initially at $|0\rangle$ transforms to $|1\rangle$ ($|2\rangle$), with probability p if the qubit is in $|0\rangle$ ($|1\rangle$). Thus the effect of this coupling can be summarized as

$$|00\rangle \rightarrow |0\rangle \otimes \left(\sqrt{1-p}|0\rangle + \sqrt{p}|1\rangle\right)$$
$$|10\rangle \rightarrow |1\rangle \otimes \left(\sqrt{1-p}|0\rangle + \sqrt{p}|2\rangle\right).$$

This is usually called the phase damping channel. The significance of the nomenclature will be clear at the end of this example. Now it is straightforward to visualize that the effect of this coupling on an arbitrary qubit $|\psi\rangle = \alpha|0\rangle + \beta|1\rangle$ is

$$|\psi\rangle|E\rangle = (\alpha|0\rangle + \beta|1\rangle)|0\rangle \rightarrow \alpha|0\rangle \otimes \left(\sqrt{1-p}|0\rangle + \sqrt{p}|1\rangle\right)$$
$$+ \beta|1\rangle \otimes \left(\sqrt{1-p}|0\rangle + \sqrt{p}|2\rangle\right).$$

---

[3]In context of NMR decoherence time is called transverse relaxation time T_2, which corresponds to the decoherence of the transverse nuclear spin magnetization.

[4]In Example 6.6 we will describe a specific case where diagonal terms of density matrix will also evolve with time.

The coupled state is entangled as expected and the corresponding density matrix is

$$
\begin{aligned}
\rho \; &= \; |\alpha|^2 \left((1-p)|00\rangle\langle 00| + \sqrt{p(1-p)}\,(|00\rangle\langle 01| + |01\rangle\langle 00|) + p|01\rangle\langle 01| \right) \\
&+ \; \alpha\beta^* \left((1-p)|00\rangle\langle 10| + \sqrt{p(1-p)}\,(|00\rangle\langle 12| + |01\rangle\langle 10|) + p|01\rangle\langle 12| \right) \\
&+ \; \beta\alpha^* \left((1-p)|10\rangle\langle 00| + \sqrt{p(1-p)}\,(|10\rangle\langle 01| + |12\rangle\langle 00|) + p|12\rangle\langle 01| \right) \\
&+ \; |\beta|^2 \left((1-p)|10\rangle\langle 10| + \sqrt{p(1-p)}\,(|10\rangle\langle 12| + |12\rangle\langle 10|) + p|12\rangle\langle 12| \right).
\end{aligned}
$$

Now if we index the original qubit as A and environment as E for convenience then we can easily write the reduced density matrix of the qubit as

$$
\begin{aligned}
\rho^A = Tr_E(\rho) \; &= \; |\alpha|^2 \left((1-p)|0\rangle\langle 0| + p|0\rangle\langle 0| \right) + \alpha\beta^*(1-p)|0\rangle\langle 1| \\
&+ \; \beta\alpha^*(1-p)|1\rangle\langle 0| + |\beta|^2 \left((1-p)|1\rangle\langle 1| + p|1\rangle\langle 1| \right) \\
&= \; \begin{pmatrix} |\alpha|^2 & (1-p)\alpha\beta^* \\ (1-p)\beta\alpha^* & |\beta|^2 \end{pmatrix}.
\end{aligned}
$$

If we compare this with the density matrix of the qubit before coupling which is given in (6.9), then we can easily observe that the sole effect of the coupling with the environment is the introduction of the $(1-p)$ factor with the off-diagonal terms. As $(1-p) < 1$, it is essentially damping the off-diagonal terms. Now after n iterations of this coupling interaction we will have

$$
\rho_n^A = \begin{pmatrix} |\alpha|^2 & (1-p)^n \alpha\beta^* \\ (1-p)^n \beta\alpha^* & |\beta|^2 \end{pmatrix}.
$$

As $(1-p) < 1$ so $(1-p)^n \to 0$ when $n \to \infty$ and consequently, the off-diagonal terms present in the density matrix vanish after a large number of iterations of this coupling interaction and we obtain (6.12). Thus the interaction with the environment described in this example is decoherence. Alternatively, we can visualize it as follows. Assume that the time required for each iteration is Δt then it is logically expected that $p \propto \Delta t$ or, $p = \Gamma \Delta t$ where Γ is the proportionality constant. Now if we allow the system to evolve for time t then $n = \frac{t}{\Delta t}$ and after time t the factor associated with the off-diagonal elements of the reduced density matrix is $(1-p)^n = (1 - \Gamma \Delta t)^{\frac{t}{\Delta t}}$. Finally, let us recall the well-known result that if $x \ll 1$ then $\exp(-x) = 1 - x$ and consider $\Delta t \to 0$ then $(1-p) = (1 - \Gamma \Delta t) \approx \exp(-\Gamma \Delta t)$ and $(1-p)^n \approx \exp(-n\Gamma\Delta t) = \exp\left(-\frac{t}{\Delta t} \times \Gamma \Delta t\right) = \exp(-\Gamma t)$. Clearly this leads to

$$
\rho^A(t) = \begin{pmatrix} |\alpha|^2 & \exp(-\Gamma t)\alpha\beta^* \\ \exp(-\Gamma t)\beta\alpha^* & |\beta|^2 \end{pmatrix}.
$$

Note that the effect of phase damping channel is limited to the off-diagonal terms only.

Let us now consider another interesting example which is known as amplitude damping channel.

Example 6.6: Consider that the effect of interaction with the environment is as follows:

$$|00\rangle \rightarrow |00\rangle$$
$$|10\rangle \rightarrow \left(\sqrt{1-p}|10\rangle + \sqrt{p}|01\rangle\right).$$

If we consider $|0\rangle$ and $|1\rangle$ as the ground state and the first excited state respectively then this interaction with environment causes spontaneous decay ($|1\rangle_A \rightarrow |0\rangle_A$) with probability p. The effect of this coupling on an arbitrary qubit $|\psi\rangle = \alpha|0\rangle + \beta|1\rangle$ is

$$|\psi\rangle|E\rangle = (\alpha|0\rangle + \beta|1\rangle)\,|0\rangle \rightarrow \alpha|00\rangle + \beta\left(\sqrt{1-p}|10\rangle + \sqrt{p}|01\rangle\right)$$

and the corresponding density matrix is

$$
\begin{aligned}
\rho =\ & |\alpha|^2|00\rangle\langle00| + \alpha\beta^*\left(\sqrt{1-p}|00\rangle\langle10| + \sqrt{p}|00\rangle\langle01|\right) \\
& + \beta\alpha^*\left(\sqrt{1-p}|10\rangle\langle00| + \sqrt{p}|01\rangle\langle00|\right) \\
& + |\beta|^2\left((1-p)|10\rangle\langle10| + \sqrt{p(1-p)}\left(|10\rangle\langle01| + |01\rangle\langle10|\right) + p|01\rangle\langle01|\right).
\end{aligned}
$$

Therefore, the reduced density matrix of the system after the interaction is

$$
\begin{aligned}
\rho^A = Tr_E(\rho) =\ & |\alpha|^2|0\rangle\langle0| + \alpha\beta^*\sqrt{(1-p)}|0\rangle\langle1| \\
& + \beta\alpha^*\sqrt{(1-p)}|1\rangle\langle0| + |\beta|^2\left((1-p)|1\rangle\langle1| + p|0\rangle\langle0|\right) \\
=\ & \begin{pmatrix} |\alpha|^2 + p|\beta|^2 & \sqrt{(1-p)}\alpha\beta^* \\ \sqrt{(1-p)}\beta\alpha^* & (1-p)|\beta|^2 \end{pmatrix} \\
=\ & \begin{pmatrix} 1 - (1-p)|\beta|^2 & \sqrt{(1-p)}\alpha\beta^* \\ \sqrt{(1-p)}\beta\alpha^* & (1-p)|\beta|^2 \end{pmatrix}.
\end{aligned}
$$

Note that the diagonal terms (i.e., amplitudes) also evolve with time. Now we may follow the last part of the previous example and obtain

$$
\rho^A(t) = \begin{pmatrix} 1 - \exp(-\Gamma t)|\beta|^2 & \exp(-\frac{\Gamma}{2}t)\alpha\beta^* \\ \exp(-\frac{\Gamma}{2}t)\beta\alpha^* & \exp(-\Gamma t)|\beta|^2 \end{pmatrix}.
$$

Thus the diagonal terms in the density matrix will evolve faster than the off-diagonal terms. Technically, it can be stated that the transverse relaxation time $T_2 = 2\Gamma$ is twice the longitudinal relaxation time $T_1 = \Gamma$. After T_2 we will have $\rho^A(t > T_2) = |0\rangle\langle0|$ which clearly shows that it is an amplitude damping channel.

Decoherence may be viewed as a result of the coupling between two physical systems (i.e., the qubit and the environment) which were initially isolated. The coupling leads to the loss of information from the system into the environment. It is important to understand that the decoherence is different from dissipation. Dissipation implies loss of energy to the environment, while decoherence implies the loss of coherence, i.e., super-positions, and may not involve any loss of energy. Loss of superposition is essentially loss of quantum character. Consequently, we need to do all the computational tasks before the state decoheres. Thus decoherence time τ alone is not important. The important factor is the ratio $\frac{\tau}{t_g}$, where t_g is the time required for operation of a quantum gate. Thus $N_{op} = \frac{\tau}{t_g}$ provides an estimate of number of quantum operations that can be done before the state decoheres (i.e., before the required superposition is lost). For example, in ion trap experimentally achieved values are $\tau \approx 10$s and $t_g = 10^{-6}$s. Therefore, $N_{op} = 10^7$. A similar number of operations N_{op} is expected in NMR too as the theoretical estimates are $\tau \approx 10^4$s and $t_g = 10^{-3}$s. Here it would be apt to note that if we can do error correction then we may be able to perform the computation for a period longer than the decoherence time. However, the error rates are usually proportional to $\frac{1}{N_{op}}$ and it is believed that if $\frac{1}{N_{op}} \approx 10^{-4}$ then the error correction is possible. This implies that for a useful realization of qubit system we must have $N_{op} \geq 10^4$.

6.4.1 Decoherence free subspace

Decoherence is the largest obstacle in the construction of a scalable quantum computer, but it is interesting that there may exist subspaces of the system's Hilbert space, where the system is decoupled from the environment and thus its evolution is completely unitary. Such a subspace is called decoherence free subspace (DFS). In DFS, quantum information is isolated from the environment and consequently, environment cannot cause loss of coherence. In other words DFS is a space where you can hide your qubits from the environment. The study of DFS is extremely important for practical implementation of a quantum computer, which cannot be isolated from its environment. Here it would be apt to recall that the relative phase, that is, a phase that only multiplies a single term in a superposition, affects the measurement results, whereas the global phase has no effect on the measurement outcomes. Consequently, if the environment introduces the relative phase then it is a problem but if the environment introduces only the global phase then the effect of the environment is irrelevant.

We already know that a set of states (state vectors) makes a state space. So we may say that a set of decoherence free (DF) states (the states for which the probability for a decoherence event to take place is zero) makes DFS. The construction of DFS is a tricky business. This is done by introducing logical qubits, which are denoted by $|0_L\rangle$ and $|1_L\rangle$. The logical qubits are different from physical qubits and they are chosen in such a way that the effect of the environment on any arbitrary superposition of them is restricted to the introduction of the global phase only. The state space spanned by such logical qubits is DFS. We can understand it better from the following examples.

Example 6.7: Consider that the logical qubits are encoded into the Bell states by the following mapping:

$$|0\rangle \mapsto |0_L\rangle = |\phi^-\rangle$$
$$|1\rangle \mapsto |1_L\rangle = |\phi^+\rangle$$.

Also consider a dephasing-noise environment described by

$$|0\rangle \rightarrow e^{i\langle\phi_0\rangle}|0\rangle$$
$$|1\rangle \rightarrow e^{i\langle\phi_1\rangle}|1\rangle$$,

where decoherence has introduced an independent random average phase shift $\langle\phi_i\rangle$, $(i = 0, 1)$ to each basis state. Now show that the subspace spanned by $|\phi^\pm\rangle$ is decoherence free as far as the given dephasing noise-environment is concerned.

Solution: The effect of this environment on the encoded states is

$$
\begin{aligned}
|\phi^\pm\rangle \quad &\rightarrow \quad \frac{e^{i\langle\phi_0\rangle}|0\rangle e^{i\langle\phi_1\rangle}|1\rangle \pm e^{i\langle\phi_1\rangle}|1\rangle e^{i\langle\phi_0\rangle}|0\rangle}{\sqrt{2}} \\
&= \quad e^{i(\langle\phi_0\rangle+\langle\phi_1\rangle)}\frac{(|01\rangle \pm |10\rangle)}{\sqrt{2}} \\
&= \quad e^{i(\langle\phi_0\rangle+\langle\phi_1\rangle)}|\phi^\pm\rangle.
\end{aligned}
$$

As the encoded states are still orthogonal and as the effect of the environment is limited to the introduction of a fixed (same) global phase in the encoded states and their linear combinations, this environment will not have any detectable effect on the subspace spanned by $|\phi^\pm\rangle$. Thus the subspace spanned by $|\phi^\pm\rangle$ is decoherence free.

Here are a few more simple examples of such DF states:

Example 6.8: Consider a system made up of many identical atoms. There exists a large DFS which is robust against spontaneous emission. To be precise, the large number of ground states of this system are DF states as they are protected against spontaneous decay. In addition excited states can also be DF states because of the symmetry of the system-environment coupling [85].

Example 6.9: Consider an antisymmetric state of two atoms, one of which is excited and coupled symmetrically to the environment. Here both the atoms are allowed to emit a photon with finite probability but their efforts destructively interfere and as a result the state remains in the initial state for a long time [85].

In this chapter, we have seen that quantum errors can be discretized and corrected, fault-tolerant quantum computation is possible with the help of fault-tolerant quantum gates and DFS exist. These observations make us optimistic about building a quantum computer and we may now ask: What are the basic requirements for physical implementation of a quantum computer? In 2000 D. P. DiVincenzo [31] first tried to address this question with a systematic approach. He listed five basic criteria for the physical implementation of a quantum computer and two additional criteria for implementation of quantum communication. These seven criteria are now known as DiVincenzo criteria. In the following section we briefly describe DiVincenzo criteria and technological issues related to the simultaneous realization of these criteria.

6.5 DiVincenzo criteria

First we discuss five DiVincenzo criteria that are expected to be satisfied by any proposal of physical implementation of a quantum computer. In the following we have elaborated the meaning of each of the criteria. However, the statements of the criteria used here are the same as those in the original work of D. P. DiVincenzo [31].

Criterion 1: A scalable physical system with well characterized qubits.
This is easy to visualize as any implementation of a quantum computer would require a large number of qubits. These qubits are required to be well defined (characterized) in the sense that their physical parameters are precisely known. For example, for a well defined qubit the interaction of the qubit with the other qubits and external fields is known. We have already discussed various possibilities of realization of qubits. To be precise, we have already noted that the spin states of nucleus or electron, the energy states of atom or quantum dot, the polarization states of photon, etc. can be used for physical realization of qubits. The implemented qubits are of two types: stationary and flying. As the names suggest stationary qubits are material qubits that are available at one place and are expected to be used for implementation of a quantum computer. For example, we may think of a quantum dot or an NMR qubit system as stationary qubit. On the other hand, for quantum communication we need qubits to propagate from one place to the other. Such qubits which propagate from one location to the other are called flying qubits. Usually flying qubits are photons. The problem associated with the qubits is that they are not stable. So we need to do as many operations as possible within the time period for which the qubits are stable.

Criterion 2: The ability to initialize the state of the qubits to a simple fiducial state, such as $|000\rangle$.
In all the quantum algorithms described in Chapter 5, we have seen that we need to prepare an initial state. For example, equal superposition state $\sum_{x=0}^{2^n-1} |x\rangle$ is required as the initial state in most of the quantum algorithms. To obtain $\sum_{x=0}^{2^n-1} |x\rangle$ we apply $H^{\otimes n}$ on $|0\rangle^{\otimes n}$. Thus to implement the quantum algorithms we must be able to initially prepare the state $|0\rangle^{\otimes n}$. Further, for the error correction we need to prepare several ancilla qubits in $|0\rangle$. These two lucid

and convincing examples show that the ability to initialize the implemented qubits into the required state is an essential requirement for successful implementation of a quantum computer.

Criterion 3: Long relevant decoherence times, much longer than the gate operation time.

In the previous section we elaborately discussed this point and have showed that $N_{op} = \frac{\tau}{t_g}$ is required to be high for a successful implementation of the quantum computer. Here it would be apt to note that t_g is associated with clock speed of the computer. If a particular implementation of a qubit can interact strongly with the environment then t_g will be small. For example, nuclear spin interacts weakly (compared to the electron spin) with the environment so the NMR based quantum computer will be slow. Further, when we use a quantum particle as qubit, then it can have many decoherence times associated with its different degrees of freedom. However, all of these decoherence times are not relevant for the specific implementation of the qubit. This is why "relevant decoherence times" is categorically mentioned in this criterion.

Criterion 4: A universal set of quantum gates.

To perform an arbitrary quantum computing task we must be able to implement any unitary operations required for the computational task. Existence of universal quantum gate library allows us to do that. In Subsection 2.3 we described several universal quantum gate libraries. Theoretically we can do all computational tasks using the gates from a universal gate library but there are many implementational issues associated with these gates. For a short discussion one may refer to the original paper of DiVincenzo [31].

Criterion 5: A qubit-specific measurement capability.

Finally, at the end of computation we need to measure the qubits to know the outcome of the computation. But we just need to measure a specific set of qubits without disturbing other qubits of our quantum computer. For example, in error correction when we measure the ancilla we must not disturb the original qubit. Often the measurement operation is not perfect. In such cases ancilla qubits are introduced and CNOT gates are used to copy the values of the output qubits (in a particular basis) to the ancilla qubits. Subsequently, the ancilla qubits and the original qubits are measured to obtain a reliable value of the outcome of the computation.

In addition to the above 5 criteria for physical implementation of a quantum computer, the following two additional criteria for implementation of quantum communication were also introduced by DiVincenzo.

Criterion 6: The ability to interconvert stationary and flying qubits.
This is essential for distributed quantum computing and also for implementation of long distance quantum communication tasks that require quantum repeaters.

Criterion 7: The ability to faithfully transmit flying qubits between specified locations.
This is essential for all quantum communication tasks including quantum teleportation, dense coding and quantum cryptography. There exist quantum communication tasks that do not require Criterion 6. However, all quantum communication tasks require Criterion 7. The relevance of this criterion will be clearer when we describe quantum communication protocols in Chapter 7 and Chapter 8.

Here it would be apt to note that DiVincenzo does not specifically mention anything about requirement of (a) the implementation of DFS, (b) fault-tolerant computation and (c) implementation of quantum error correction. But these are also required for physical implementation of a quantum computer. Thus we learn about what is required to build a scalable quantum computer. At present we have several proposals but each of them has some technological limitations. For example, in Section 1.3 we mentioned about the success in building small quantum computers using NMR. However, NMR based quantum computers are not scalable, they are not expected to go beyond 50 qubits. At present decoherence is the biggest challenge in building a scalable quantum computer and it is difficult to predict a specific time frame by which a useful quantum computer can be built. Whenever it is built the techniques that we have learned in this chapter will be very useful for its construction.

It is beyond the scope of this textbook to describe the experimental techniques that are presently used to protect qubits and quantum gates from the effect of error, fault and environment. However, we can summarize some of the recent experimental achievements. This would give us a perception of where we stand. Interested readers may consult the papers cited here and the references therein for further details.

Experimental status is different for different implementations. However, in general it is relatively easier to protect single qubits from the decoherence. It is comparatively much more difficult to maintain the coherence between more than one qubit during the operation of a multi-qubit gate, especially in hybrid systems, where different kinds of qubit decohere at different rates. Recently, T. v. d. Sar *et al.* [86] have demonstrated decoherence-protected solid state quantum gates that work at room temperature. Another recent interesting result is reported by J. M. Chow *et al.* [87]. Using superconducting single-frequency single-junction transmon qubits they have constructed a universal gate set, which approaches fault-tolerant thresholds. In another interesting work J. Zhang, R. Laflamme and D. Suter [88] demonstrated fault-tolerant implementation of Identity,

NOT and Hadamard gates for the logical qubits, which are encoded using 5-qubit perfect quantum error correcting code. Further, P. Schindler *et al.* recently demonstrated repeated quantum error correction (up to three consecutive cycles) for phase-flip errors on qubits encoded with trapped ions [89]. In summary, it is possible to do quantum error correction and implement a universal set of fault-tolerant quantum gates.

Personally, I am optimistic that a quantum computer will be built in my lifetime. With this optimism we will finish our discussion on quantum computation and move to the discussion of quantum communication protocols. The technological issues associated with the implementation of quantum communication protocols are relatively less. Specially quantum cryptographic protocols can be reasonably implemented even today.

6.6   Solved examples

1. Identify the nature of error/errors that has/have occurred in (a) $|000\rangle \mapsto |100\rangle$, (b) $\alpha|000\rangle + \beta|111\rangle \mapsto \alpha|100\rangle - \beta|011\rangle$, (c) $\alpha|000\rangle + \beta|111\rangle \mapsto \alpha|110\rangle + \beta|001\rangle$, (d) $\alpha|000\rangle + \beta|111\rangle \mapsto -\alpha|000\rangle - \beta|111\rangle$.
   Solution: Identified errors are as follows: (a) a bit flip error on the first qubit, (b) a bit flip error on the first qubit and a phase flip error, (c) both first and second qubits flip, (d) just the global phase is changed so no detectable error has occurred.

2. Assume that $\{000, 111\}$ is used as a code. Now tell us, which of the above errors are not correctable?
   Solution: The errors that have occurred in (c) are not correctable as more than one bit has flipped so majority voting technique will fail.

3. Assume that bit-flip errors act identically on every qubit of a quantum register. Now show that (a) $\{|0_L\rangle, |1_L\rangle\} = \{|\psi^+\rangle, |\phi^+\rangle\}$ and (b) $\{|0_L\rangle, |1_L\rangle\} = \{|\psi^-\rangle, |\phi^-\rangle\}$ form DFSs in this case.
   Solution: (a) Given that we encode the logical qubits by two physical qubits as follows:

$$
\begin{aligned}
|0_L\rangle &\equiv |\psi^+\rangle = \tfrac{1}{2}(|00\rangle + |11\rangle) \\
|1_L\rangle &\equiv |\phi^+\rangle = \tfrac{1}{2}(|01\rangle + |10\rangle).
\end{aligned}
$$

In the given error model the bit flip errors act identically on every qubit of the quantum register. Consequently, the effect of the bit flip error on the logical qubits is $|0_L\rangle \rightarrow |0_L\rangle$ and $|1_L\rangle \rightarrow |1_L\rangle$. Therefore, a general state $|\psi_L\rangle = \alpha|0_L\rangle + \beta|1_L\rangle$ remains unchanged. Thus $\{|0_L\rangle, |1_L\rangle\} = \{|\psi^+\rangle, |\phi^+\rangle\}$ forms the required DFS. Similarly in (b) we can find that the effect of the bit flip error on the logical qubits is $|0_L\rangle \rightarrow -|0_L\rangle$ and $|1_L\rangle \rightarrow -|1_L\rangle$. Therefore, a general state $|\psi_L\rangle = \alpha|0_L\rangle + \beta|1_L\rangle$ acquires only a meaningless global phase. Consequently, $\{|0_L\rangle, |1_L\rangle\} = \{|\psi^-\rangle, |\phi^-\rangle\}$ also forms a DFS.

4. On a quantum channel the probability that a bit flips is p. If no error correction scheme is implemented, then obtain an analytic expression for the fidelity of the output state.

   Solution: If we consider the input state as an arbitrary qubit $|\psi_1\rangle = \alpha|0\rangle + \beta|1\rangle$ then with probability p it flips to $|\psi_2\rangle = X|\psi_1\rangle = \alpha|1\rangle + \beta|0\rangle$ and with probability $(1 - p)$ it remains unchanged. Thus, in absence of the error correction the final state is an ensemble $\{1 - p, |\psi_1\rangle, p, |\psi_2\rangle\}$ so the density matrix is

$$\rho = (1 - p)|\psi_1\rangle\langle\psi_1| + p|\psi_2\rangle\langle\psi_2|$$

   and fidelity is

$$
\begin{aligned}
F(|\psi_1\rangle, \rho) &= \sqrt{\langle\psi_1|\rho|\psi_1\rangle} \\
&= \sqrt{(1 - p) + p|\langle\psi_1|\psi_2\rangle|^2} \\
&= \sqrt{(1 - p) + p|\langle\psi_1|X|\psi_1\rangle|^2} \\
&= \sqrt{1 - p(1 - 4|Re[\alpha\beta^*]|^2}.
\end{aligned}
$$

5. In the previous problem find the minimum possible value of fidelity and provide an example of quantum state for which fidelity will be minimum.

   Solution: Here $p = 0$ and $p = 1$ are the trivial cases. In all other cases both p (as $0 < p < 1$) and $|\langle\psi_1|\psi_2\rangle|^2$ are positive so $F(|\psi_1\rangle, \rho)$ is minimum when $p|\langle\psi_1|X|\psi_1\rangle|^2 = 0$, i.e., when $\langle\psi_1|X|\psi_1\rangle = Re[\alpha\beta^*] = 0$. In such cases minimum value of fidelity is $F_{min}(|\psi_1\rangle, \rho) = \sqrt{1 - p}$. It is easy to observe that the condition $\langle\psi_1|X|\psi_1\rangle = Re[\alpha\beta^*] = 0$ will be satisfied if $\alpha = ic\beta$, where c is a real constant. Now $|\alpha|^2 + |\beta|^2 = 1$ implies $c^2|\beta|^2 + |\beta|^2 = 1$ or, $c = \pm\sqrt{\frac{1}{|\beta|^2} - 1}$. For every allowed value of β we will obtain a c which will give us two states for which the output of a bit flip channel has minimum fidelity. For example, let $\beta = 0.8$, then $c = \pm\frac{3}{4}$ and $\alpha = \pm 0.6i$. Now if we choose $\alpha = 0.6i$ then $|\psi_1\rangle = 0.6i|0\rangle + 0.8|1\rangle$ and $|\psi_2\rangle = 0.6i|1\rangle + 0.8|0\rangle$. Clearly $\langle\psi_1|\psi_2\rangle = 0$.

6. In certain bit flip quantum channel probability q that nothing happens to the qubit is $\frac{8}{9}$. Obtain the minimum fidelity of the channel.

   Solution: Here the probability that nothing happens is $\frac{8}{9}$. Therefore, the probability of bit flip is $p = \frac{1}{9}$ and the minimum fidelity is

$$F_{min}(|\psi_1\rangle, \rho) = \sqrt{1 - p} = \sqrt{\frac{8}{9}} = \frac{2\sqrt{2}}{3}.$$

7. Show that a quantum bit flip channel can be described as a quantum operation $\phi(\rho)$ that maps an input state ρ to

$$\rho' = \phi(\rho) = (1 - p)\rho + pX\rho X,$$

where p is the probability that a bit flip error happens.

Solution: As before consider the input state as an arbitrary qubit $|\psi_1\rangle = \alpha|0\rangle + \beta|1\rangle$. Corresponding density matrix is $\rho = |\psi_1\rangle\langle\psi_1|$. Now with probability p the input state flips to $|\psi_2\rangle = X|\psi_1\rangle = \alpha|1\rangle + \beta|0\rangle$ and with probability $(1-p)$ it remains unchanged. Thus, in absence of the error correction the final state is an ensemble $\{1 - p, |\psi_1\rangle, p, X|\psi_1\rangle\}$ so the output density matrix is

$$\begin{aligned}\rho' = \phi(\rho) &= (1-p)\,|\psi_1\rangle\langle\psi_1| + pX|\psi_1\rangle\langle\psi_1|X^\dagger\\ &= (1-p)\rho + pX\rho X.\end{aligned}$$

In the last step we have used the fact that $X^\dagger = X$. Here we have considered the input state as a pure state but even if we consider it as a mixed state ρ the above expression of bit flip channel will remain the same as under a unitary operator U density matrix ρ evolves as $U\rho U^\dagger$. In a bit flip channel $U = X$ and with probability p the input state ρ evolves to $X\rho X^\dagger = X\rho X$ and with probability $(1-p)$ it remains unchanged. So the mixed input state ρ evolves to $\rho' = \phi(\rho) = (1-p)\rho + pX\rho X$.

8. Suppose you have used 3 bit code to correct possible bit flip error in the input state $\alpha|0\rangle + \beta|1\rangle$, with what fidelity is the output state generated?

Solution: Note that first we transform an input state $|\psi_1\rangle = \alpha|0\rangle + \beta|1\rangle \mapsto \alpha|000\rangle + \beta|111\rangle$. Now with probability p each bit can flip so with probability $3p(1-p)^2$ single bit flip error happens. In this case the final state is properly corrected and finally we obtain $\alpha|0\rangle + \beta|1\rangle$. Two of the bits flip with probability $3p^2(1-p)$ and with probability p^3 all the bits flip. In both of these cases majority voting technique will fail and we will obtain $|\psi_2\rangle = X|\psi_1\rangle = \alpha|1\rangle + \beta|0\rangle$ as the corrected state. In remaining cases, i.e., with probability $1 - 3p(1-p)^2 - 3p^2(1-p) - p^3 = (1-p)^3$ no bit flip error occurs, so the output state is $\alpha|0\rangle + \beta|1\rangle$. Consequently, after error correction we obtain a statistical ensemble $\{1 - 3p^2 + 2p^3, |\psi_1\rangle, 3p^2 - 2p^3, |\psi_2\rangle\}$. Density matrix of this mixed state is

$$\rho = \left(1 - 3p^2 + 2p^3\right)|\psi_1\rangle\langle\psi_1| + \left(3p^2 - 2p^3\right)|\psi_2\rangle\langle\psi_2|.$$

Since the input state is a pure state, we can compute the fidelity using (3.94) as

$$\begin{aligned}F\left(|\psi_1\rangle, \rho\right) &= \sqrt{\langle\psi_1|\rho|\psi_1\rangle}\\ &= \sqrt{(1 - 3p^2 + 2p^3) + (3p^2 - 2p^3)\,|\langle\psi_1|\psi_2\rangle|^2}\\ &= \sqrt{(1 - 3p^2 + 2p^3) + (3p^2 - 2p^3)\,|\langle\psi_1|X|\psi_2\rangle|^2}\\ &= \sqrt{1 - (3p^2 - 2p^3)(1 - 4|Re[\alpha\beta^*]|^2}.\end{aligned}$$

Now it is easy to observe the error correction is useful iff fidelity of error corrected state is more than that without error correction.

Thus it is straightforward to observe that error correction is useful iff $(3p^2 - 2p^3) < p \Rightarrow (2p - 1)(1 - p) < 0 \Rightarrow p < \frac{1}{2}$ as before.

9. Consider that the effect of interaction of a qubit with the environment is as follows:

$$|00\rangle \to |00\rangle$$
$$|10\rangle \to |1\rangle \otimes \left(\sqrt{1 - p}|0\rangle + \sqrt{p}|1\rangle\right).$$

Establish that two iterations of this interaction is equivalent to the phase damping channel described in Example 6.5.

Solution: The effect of this coupling on an arbitrary qubit $|\psi\rangle = \alpha|0\rangle + \beta|1\rangle$ is

$$|\psi\rangle|E\rangle = (\alpha|0\rangle + \beta|1\rangle)\,|0\rangle \to \alpha|00\rangle + \beta\left(\sqrt{1 - p}|10\rangle + \sqrt{p}|11\rangle\right),$$

and the corresponding density matrix is

$$
\begin{aligned}
\rho \;=\;\; & |\alpha|^2|00\rangle\langle00| + \alpha\beta^*\left(\sqrt{1 - p}|00\rangle\langle10| + \sqrt{p}|00\rangle\langle11|\right) \\
+\;\; & \beta\alpha^*\left(\sqrt{1 - p}|10\rangle\langle00| + \sqrt{p}|11\rangle\langle00|\right) + |\beta|^2\left((1 - p)|10\rangle\langle10| \right.\\
+\;\; & \left. \sqrt{p(1 - p)}\left(|10\rangle\langle11| + |11\rangle\langle10|\right) + p|11\rangle\langle11|\right).
\end{aligned}
$$

Therefore, the reduced density matrix of the system after the interaction is

$$
\begin{aligned}
\rho^A = Tr_E\left(\rho\right) \;=\;\; & |\alpha|^2|0\rangle\langle0| + \alpha\beta^*\sqrt{(1 - p)}|0\rangle\langle1| \\
+\;\; & \beta\alpha^*\sqrt{(1 - p)}|1\rangle\langle0| + |\beta|^2\left((1 - p)|1\rangle\langle1| + p|1\rangle\langle1|\right) \\
\;=\;\; & \begin{pmatrix} |\alpha|^2 & \sqrt{(1 - p)}\alpha\beta^* \\ \sqrt{(1 - p)}\beta\alpha^* & |\beta|^2 \end{pmatrix}.
\end{aligned}
$$

Thus after two iterations the effect of this coupling is the same as that of the phase damping channel described in Example 6.5 and consequently this is a special case of phase damping. However, transverse relaxation time T_2 for this particular channel is twice the transverse relaxation time of the channel described in Example 6.5.

10. What do you conclude if the measurement outcome in the 3-qubit bit flip error correction code described in Subsection 6.2.3.1 is (a) 00, (b) 01, and (c) 11?

Solution: (a) If the measurement outcome is 00 then none of the qubits has flipped.

(b) If the measurement outcome is 01 then the third qubit has flipped.

(c) If the measurement outcome is 11 then the second qubit has flipped.

6.7   Further reading

1. W. H. Zurek, Decoherence and the transition from quantum to classical—revisited, quant-ph/0306072.

2. A special theme issue on decoherence is recently published by Phil. Trans. R. Soc. A. **370** (2012) 4425-4609. The issue is published in September 2012. We recommend the entire issue for further reading. However, the following review articles of this special issue deserve special attention: (a) R. Raussendorf, Key ideas in quantum error correction, Phil. Trans. R. Soc. A. **370** (2012) 4541-4565. This article nicely reviews the ideas related to quantum error correction and fault-tolerant quantum computation. (b) A. Hagar, Decoherence: the view from the history and philosophy of science, Phil. Trans. R. Soc. A. **370** (2012) 4594-4609. This review is focused to condensed matter implementations, but it is an interesting review, especially Section 2 of the article where the author discusses a brief history of decoherence.

3. P. L. Knight, A. Beige and W. J. Munro, Hiding from environment: Decoherence-free subspaces in quantum information processing, ftp://ftp.cordis.europa.eu/pub/ist/docs/fet/qip2-eu-22.pdf.

4. M. A. Schlosshauer, Decoherence and the quantum to classical transition, Springer, Berlin, Germany (2007).

5. Chapter 6 of G. Benenti, G. Casati and G. Srini, Principles of quantum computation and information, Vol II, Basic tools and special topics, World Scientific, Singapore, (2007).

6. J. Preskill, Battling decoherence: The fault-tolerant quantum computer, Physics Today, June 1999, 24-30. Also available at http://www.theory.caltech.edu/~preskill/pubs/preskill-1999-phys-today.pdf.

7. Recent progress in photonic quantum simulation is reviewed in A. Aspuru-Guzik and P. Walther, Photonic quantum simulators, Nature Phys. **8** (2012) 285–291.

8. D. Gottesman, Theory of fault-tolerant quantum computation, Phys. Rev. A **57** (1998) 127–137.

9. D. Gottesman, An introduction to quantum error correction and fault-tolerant quantum computation, quant-ph/0904.2557.

6.8   Exercises

1. Explain the meaning of the following statement: In quantum error correction we essentially add redundancy and the redundancy is not

repetition.

2. Assume that bit flip errors act identically on every qubit of a quantum register. Now try to find out a DFS in this case.

3. Explicitly design a fault-tolerant SWAP gate (SWAP_{FT}).

4. Show that a quantum phase flip channel can be described as a quantum operation $\phi(\rho)$ that maps an input state ρ to

$$\rho' = \phi(\rho) = (1-p)\rho + pZ\rho Z$$

where p is the probability that a phase flip error happens.

5. Consider that the logical qubits are encoded into the Bell states by the following mapping:

$$|0\rangle \mapsto |0_L\rangle = |\psi^-\rangle$$
$$|1\rangle \mapsto |1_L\rangle = |\psi^+\rangle.$$

Also consider a dephasing-noise environment described by

$$|0\rangle \to |0\rangle$$
$$|1\rangle \to e^{i\langle\phi_1\rangle}|1\rangle.$$

Now show that the subspace spanned by $|\psi^\pm\rangle$ is not decoherence free as far as the given dephasing noise-environment is concerned.

6. On a quantum channel the probability that a phase flip error happens is p. If no error correction scheme is implemented, then obtain an analytic expression for the fidelity of the output state.

7. Show that the minimum possible value of fidelity in the previous problem is $F_{min}(|\psi_1\rangle, \rho) = \sqrt{1-p}$. Also show that the same is achieved for input states $|+\rangle$ or $|-\rangle$. [Hints: $F(|\psi_1\rangle, \rho)$ is minimum when $p|\langle\psi_1|Z|\psi_1\rangle|^2 = 0$, i.e., when $\langle\psi_1|Z|\psi_1\rangle = |\alpha|^2 - |\beta|^2 = 0 \Rightarrow |\alpha| = |\beta|$.]

8. In a phase flip quantum channel probability p that a phase flip error occurs is 0.19. Obtain the minimum fidelity of the channel.

9. Suppose you have used 3-bit code to correct possible phase flip error in the input state $\alpha|0\rangle + \beta|1\rangle$, with what fidelity is the output state generated?

Chapter 7

Quantum teleportation and superdense coding

The beauty of the quantum circuit lies in the fact that it can perform certain tasks which are impossible in the classical world. For example, consider teleportation and superdense coding. Teleportation is a quantum task in which an unknown quantum state is transmitted from a sender (Alice) to a spatially separated receiver (Bob) via an entangled quantum channel and with the help of some classical communications. Dense coding is another communication scheme in which if Alice and Bob share a prior entanglement then one of them can send n bits of classical information to the other by communicating m-qubits $(m < n)$. Before we go into the technical details of these two schemes let us try to lucidly visualize the uniqueness of teleportation. Suppose Alice and Bob are spatially separated and Alice wants to send an unknown quantum state $|\psi\rangle = \alpha|0\rangle + \beta|1\rangle$ to Bob. For some reason she is not allowed to send the state directly through a quantum channel. She cannot measure it and send the information to Bob because the moment she will try to measure the state, it will collapse either to $|0\rangle$ or to $|1\rangle$. But the state can be sent by using a simple quantum circuit (pre-shared entanglement) and some classical communications. In the process the unknown quantum state gets destroyed at Alice's end and reappears at Bob's end. The state never exists in the communication channel. This process is called quantum teleportation. The uniqueness of quantum teleportation will be clearer if we note that the Fax is not an example of teleportation because the original copy remains with the sender. Again a letter posted in a letter box is not an example of teleportation because the letter exists (travels) in its original form in the communication channel. In brief, there is no classical analogue of quantum teleportation. The same is true for dense coding. In this chapter we will describe these two unique quantum communication schemes and discuss their significance and applications.

7.1 Different types of teleportation schemes

In a teleportation scheme, Alice and Bob are spatially separated and they share an entangled state. This entangled state constitutes the quantum channel between Alice and Bob. Teleportation can be achieved by using different types of entangled states as a quantum channel. The original scheme was proposed by Bennett, Brassard, Crépeau, Jozsa, Peres and Wootters in 1993 [21]. Since then a large number of teleportation schemes and their applications have been reported. These teleportation schemes can be primarily classified into two broad classes: (a) perfect teleportation schemes and (b) probabilistic teleportation schemes. The original proposal of teleportation was meant for perfect teleportation of an unknown qubit $\alpha|0\rangle + \beta|1\rangle$. By perfect teleportation we mean that the success rate of teleportation is unity. This requires a maximally entangled quantum channel. But immediately after the pioneering work of Bennett *et al.* it was realized that teleportation is possible even when the quantum channel is non-maximally entangled. In that case the success rate of Bob will not be unity and the teleportation scheme would be called probabilistic. For example, in the next section we will show that if we use $|\psi\rangle_0 = \frac{|00\rangle + |11\rangle}{\sqrt{2}}$ as quantum channel then we obtain perfect teleportation as this state is maximally entangled. However, if we use $|\psi'\rangle_0 = a|00\rangle + b|11\rangle$ where, $|a|^2 + |b|^2 = 1$ and $|a| \neq \frac{1}{\sqrt{2}}$ as quantum channel then we obtain a scheme for probabilistic teleportation. Such a scheme will be described in Section 7.3.

Teleportation schemes are not limited to two-party teleportation (i.e., teleportation between Alice and Bob). It can be easily generalized to many party quantum teleportation schemes. Such multi-party teleportation schemes have led to many interesting applications. Here we will not be able to discuss all of them, so we will restrict ourselves to the most interesting multi-party quantum teleportation scheme. Specifically, we will only discuss a scheme for controlled teleportation (CT) or quantum information splitting (QIS). In this scheme, Alice shares prior entanglement with Bob and at least one Charlie (supervisor). Nocloning theorem ensures that Alice cannot teleport copies of the unknown qubit to both Charlie and Bob because that would imply cloning. Consequently, if Alice succeeds in teleporting the unknown qubit to Charlie and Bob then only one of them will be able to construct a copy of the unknown qubit, and that would require help of the other. Simply put, Bob (Charlie) will be able to construct the state with the help of Charlie (Bob). Thus the quantum information is split between Charlie and Bob and this is why such a scheme is referred to as QIS protocol. Usually we assume that Bob reconstructs the unknown quantum state with the help of Charlie. Keeping this in mind, Charlie is often referred to as supervisor. He is supervisor in the sense that he can control the channel between Alice and Bob. Bob can properly construct the state sent by Alice if Charlie cooperates (i.e., if Charlie sends the value of his measurements to Bob). Since Charlie can control the teleportation protocol such a protocol is referred to as CT protocol, too. It is easy to recognize from the above discussion that CT schemes are equivalent to QIS schemes. Further, since they are at least three party schemes, to implement them we need to start with at least a tripartite entangled state. Recently CT/QIS schemes have been reported using GHZ state, GHZ-like state, W state, etc. In Section 7.4 we will provide an explicit example of QIS using GHZ state.

Here we would like to note that in 1999, Hillary, Buzek and Bertaiume [90] proposed a protocol for quantum secret sharing (QSS) which is a variant of QIS with the added feature of secure quantum communication. Thus QSS may be viewed as an application of QIS. We will describe a QSS protocol in Section 8.11. If we consider that the secret to be shared is a quantum state, then a QSS scheme reduces to a scheme for quantum state sharing (QSTS), which is again a variant of QIS. We can visualize the situation as if a secret (which is a quantum state) is teleported to Charlie and Bob and one of them (Bob) can obtain the secret (the quantum state) if and only if the other one (Charlie) collaborates.

It is easy to realize that in our context, $CT = QIS = QSTS \subset QSS$. Consequently, controlled teleportation protocols find specific applications in quantum secret communication schemes. In a CT protocol the measurements of Alice and Charlie are communicated to Bob via classical communication channels, and Bob uses these information to choose the unitary operation to be applied by him. Consequently, any CT protocol may be transformed to a usual teleportation scheme either by keeping Charlie's bit with Alice or by communicating the bit to Bob. But that would only make the quantum channel more complex.

The above mentioned variants of quantum teleportation schemes can be modified in various ways. One important variant of teleportation is remote state preparation where the sender (Alice) knows the state to be teleported. Thus it may be viewed as teleportation of a known state. This requires a lesser amount of classical communication compared to a perfect teleportation scheme. Remote state preparation schemes can also be generalized to probabilistic remote state preparation and controlled remote state preparation schemes.

To provide a clear perception of each of these different teleportation schemes (e.g., perfect teleportation, probabilistic teleportation, controlled teleportation, remote state preparation, etc.) we will provide some specific examples in the following sections. To begin with, let us describe a simple scheme for perfect teleportation.

The most acknowledged personality in today's quantum computing community is Charles H. Bennett. He was born in 1943. He obtained his Ph.D. from Harvard University, Cambridge, MA, in 1970 and joined IBM Research in 1972. At present he works with IBM Research. In the last four decades he has made many pioneering contributions to quantum communication and quantum computing. In 1973, he showed that it is feasible to exploit Landauer's principle to construct a general purpose reversible computer [6]. In 1984, in collaboration with Gilles Brassard, he proposed an unconditionally secure protocol of quantum key distribution [18]. The protocol is now known as the BB84 protocol. In 1992 he proposed a similar but independent protocol of quantum key distribution, which is now known as the B92 protocol [91]. In 1992 Bennett and Wiesner introduced the idea of dense coding [20] and in 1993 Bennett, Brassard, Crépeau, Jozsa, Peres and Wootters introduced the idea of quantum teleportation [21]. These two effects do not have any classical analogue. His contributions in the field of quantitative theory of entanglement and also in the field of secure transmission of classical and quantum information through noisy channels, are remarkable. Almost three decades ago he introduced the idea of quantum cryptography and recently it was used at the 2010 Soccer World Cup to protect important information.

Photo courtesy: C. H. Bennett.

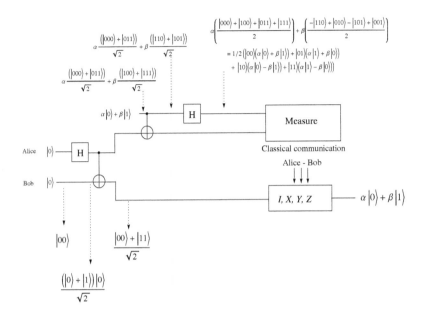

Figure 7.1: Simple teleportation circuit.

7.2 A simple scheme for perfect teleportation

A simple teleportation scheme is described in Fig. 7.1. Initially Alice and Bob share a Bell state

$$|\psi\rangle_0 = \frac{|00\rangle + |11\rangle}{\sqrt{2}}, \tag{7.1}$$

where Alice has access to the first qubit of the pair and the second qubit is with Bob. This initial Bell state can be prepared by a simple EPR circuit consisting of a Hadamard gate followed by a CNOT gate (as shown in Fig. 4.5 and as used in the left most block of Fig. 7.1). Now Alice wants to send the unknown quantum state $|\psi\rangle = \alpha|0\rangle + \beta|1\rangle$ to Bob. To do so, she uses the EPR circuit in reverse order, i.e., a CNOT followed by a Hadamard. The circuit entangles the unknown state with Alice's half of the EPR pair which is already entangled with Bob's qubit. The action of the teleportation circuit described in Fig. 7.1 can be clearly understood in the following steps:

With the unknown state the initial state of the system is

$$|\psi\rangle_1 = (\alpha|0\rangle + \beta|1\rangle) \otimes \frac{|00\rangle + |11\rangle}{\sqrt{2}} = \left(\alpha|0\rangle \frac{|00\rangle + |11\rangle}{\sqrt{2}} + \beta|1\rangle \frac{|00\rangle + |11\rangle}{\sqrt{2}}\right).$$

Alice measures	State of Bob's qubit	Bob's operation	Bob's state after the operation
00	$\alpha\|0\rangle + \beta\|1\rangle$	I	$\alpha\|0\rangle + \beta\|1\rangle$
01	$\alpha\|1\rangle + \beta\|0\rangle$	X	$\alpha\|0\rangle + \beta\|1\rangle$
10	$\alpha\|0\rangle - \beta\|1\rangle$	Z	$\alpha\|0\rangle + \beta\|1\rangle$
11	$\alpha\|1\rangle - \beta\|0\rangle$	$ZX = iY$	$\alpha\|0\rangle + \beta\|1\rangle$

Table 7.1: Table of Alice's measurement outcomes and corresponding unitary operations (gates) to be used by Bob to reproduce the unknown one-qubit quantum state $(\alpha|0\rangle + \beta|1\rangle)$.

After the operation of the CNOT gate (using the first qubit of $|\psi\rangle_1$ as the control qubit and the second qubit as the target qubit) the state becomes

$$|\psi\rangle_2 = \left(\alpha|0\rangle \frac{|00\rangle + |11\rangle}{\sqrt{2}} + \beta|1\rangle \frac{|10\rangle + |01\rangle}{\sqrt{2}} \right).$$

As Alice sends the first qubit of the state through the Hadamard gate the state of the system transforms to

$$
\begin{aligned}
|\psi\rangle_3 &= \left(\alpha \frac{(|0\rangle + |1\rangle)}{\sqrt{2}} \frac{(|00\rangle + |11\rangle)}{\sqrt{2}} + \beta \frac{(|0\rangle - |1\rangle)}{\sqrt{2}} \frac{(|10\rangle + |01\rangle)}{\sqrt{2}} \right) \\
&= \tfrac{1}{2} \left(|00\rangle(\alpha|0\rangle + \beta|1\rangle) + |01\rangle(\alpha|1\rangle + \beta|0\rangle) \right. \\
&+ \left. |10\rangle(\alpha|0\rangle - \beta|1\rangle) + |11\rangle(\alpha|1\rangle - \beta|0\rangle) \right).
\end{aligned}
\tag{7.2}
$$

From (7.2) it is clear that if Alice measures the first two qubits in computational basis and uses a classical channel to send the outcome of her measurement (which of the four possible outcomes she has obtained) to Bob, then Bob will be able to reconstruct the unknown state by applying appropriate Pauli gates on his qubit. This is so because Alice's measurement completely determines the state of Bob's qubit and that allows Bob to choose an appropriate quantum gate which can reproduce the unknown quantum state $|\psi\rangle$ at his end. Bob's appropriate choices of quantum gates (unitary operations) are listed in Table 7.1.

   Thus the unknown quantum state is teleported with the help of a shared Bell state and two classical bits. Before we discuss other variants of teleportation protocols it is important that we clear two conceptual issues. The first is that teleportation implies neither instantaneous communication nor superluminal (faster than light) communication, as without the classical communication from Alice to Bob no information can be transferred from Alice to Bob. As the classical communication between Alice to Bob cannot be superluminal, so the teleportation cannot be superluminal. The second issue is that it does not violate the nocloning principle, as the creation of $(\alpha|0\rangle + \beta|1\rangle)$ at Bob's end necessarily implies destruction of the unknown one qubit state $(\alpha|0\rangle + \beta|1\rangle)$ at Alice's end. Since there can be only one copy of $(\alpha|0\rangle + \beta|1\rangle)$, in a multiparty teleportation protocol or in QIS only one of the receivers will be able to reconstruct the unknown state $(\alpha|0\rangle + \beta|1\rangle)$ and he will require help from the other receivers. This issue will be clarified when we discuss QIS.

7.3 Probabilistic teleportation

In the previous section we described a scheme of perfect teleportation using the maximally entangled state. Here we describe a scheme of probabilistic teleportation. To do so, we assume that Alice and Bob initially share a non-maximally entangled Bell type state $|\psi'\rangle_0 = a|00\rangle + b|11\rangle$, where $|a|^2 + |b|^2 = 1$ and $|a| \neq \frac{1}{\sqrt{2}}$. Now we may follow the previous scheme of perfect teleportation to visualize the difference. Here with the unknown state the initial state of the system is

$$
\begin{aligned}
|\psi'\rangle_1 &= (\alpha|0\rangle + \beta|1\rangle) \otimes (a|00\rangle + b|11\rangle) \\
&= (\alpha|0\rangle (a|00\rangle + b|11\rangle) + \beta|1\rangle (a|00\rangle + b|11\rangle)).
\end{aligned}
$$

After the operation of the CNOT gate (using the first qubit of $|\psi'\rangle_1$ as the control qubit and the second qubit as the target qubit) the state becomes

$$
|\psi'\rangle_2 = (\alpha|0\rangle (a|00\rangle + b|11\rangle) + \beta|1\rangle (a|10\rangle + b|01\rangle)).
$$

As Alice sends the first qubit of the state through the Hadamard gate the state of the system transforms to

$$
\begin{aligned}
|\psi'\rangle_3 &= \left(\alpha \frac{(|0\rangle + |1\rangle)}{\sqrt{2}} (a|00\rangle + b|11\rangle) + \beta \frac{(|0\rangle - |1\rangle)}{\sqrt{2}} (a|10\rangle + b|01\rangle)\right) \\
&= \frac{1}{\sqrt{2}} (|00\rangle(\alpha a|0\rangle + \beta b|1\rangle) + |01\rangle(\alpha b|1\rangle + \beta a|0\rangle) \\
&+ |10\rangle(\alpha a|0\rangle - \beta b|1\rangle) + |11\rangle(\alpha b|1\rangle - \beta a|0\rangle)).
\end{aligned}
$$

Alice measures her qubits (first two qubits of $|\psi'\rangle_3$) in the computational basis and sends the measurement outcome to Bob.

Up to this point the protocol is similar to the previous protocol. But now Bob will not be able to obtain the unknown quantum state just by following the previous protocol of perfect teleportation. For example, if Alice informs that her measurement outcome is $|00\rangle$ and Bob applies I then his state will become $\frac{\alpha a|0\rangle + \beta b|1\rangle}{\sqrt{|\alpha a|^2 + |\beta b|^2}} \neq \alpha|0\rangle + \beta|1\rangle$. In fact, Bob cannot construct a single qubit unitary operation to map $\frac{\alpha a|0\rangle + \beta b|1\rangle}{\sqrt{|\alpha a|^2 + |\beta b|^2}}$ to $\alpha|0\rangle + \beta|1\rangle$ without the knowledge of α and β. Therefore, Bob has to change his strategy as follows.

Bob prepares an ancilla qubit in $|0\rangle_{Auxi}$ and applies the following unitary operation on his qubits (i.e., on the combined system of his existing qubit and ancilla):

$$
U = \begin{pmatrix}
\frac{b}{a} & \sqrt{1 - \frac{b^2}{a^2}} & 0 & 0 \\
0 & 0 & 0 & -1 \\
0 & 0 & 1 & 0 \\
\sqrt{1 - \frac{b^2}{a^2}} & -\frac{b}{a} & 0 & 0
\end{pmatrix}.
\tag{7.3}
$$

As a and b are known, construction of U is allowed. Now if Alice's measurement outcome is $|00\rangle$ then Bob applies U on his product state

$$
|\psi'\rangle_4 = \left(\frac{\alpha a|0\rangle + \beta b|1\rangle}{\sqrt{|\alpha a|^2 + |\beta b|^2}} \right) |0\rangle_{Auxi} = \frac{1}{\sqrt{|\alpha a|^2 + |\beta b|^2}} \begin{pmatrix} \alpha a \\ 0 \\ \beta b \\ 0 \end{pmatrix}
$$

and obtains

$$
\begin{aligned}
U|\psi'\rangle_4 &= \frac{1}{\sqrt{|\alpha a|^2 + |\beta b|^2}} \begin{pmatrix} \alpha b \\ 0 \\ \beta b \\ \sqrt{1 - \frac{b^2}{a^2}} \alpha a \end{pmatrix} \\
&= \frac{1}{\sqrt{|\alpha a|^2 + |\beta b|^2}} \left(b(\alpha|0\rangle + \beta|1\rangle)|0\rangle + \sqrt{1 - \frac{b^2}{a^2}} \alpha a|1\rangle|1\rangle \right).
\end{aligned}
$$

Now Bob measures the last qubit (ancilla) in the computational basis. If his measurement yields $|0\rangle$ then he obtains unknown state with unit fidelity but if his measurement on ancilla yields $|1\rangle$ then the teleportation fails. Similarly, we can check the other three possibilities (i.e., when Alice's measurement outcomes are 01 or 10 or 11). Table 7.2 shows the relations among Alice's measurement outcomes, Bob's measurement outcomes, unitary operations to be applied by Bob and Bob's conclusions.

Thus this scheme can teleport an unknown quantum state with unit fidelity, but the success rate of the scheme is not unity. This is why it is called probabilistic teleportation.

Alternatively, we can express $|\psi'\rangle_1$ as

$$
\begin{aligned}
|\psi'\rangle_1 &= \frac{|\psi^+\rangle + |\psi^-\rangle}{\sqrt{2}} \alpha a|0\rangle + \frac{|\phi^+\rangle + |\phi^-\rangle}{\sqrt{2}} \alpha b|1\rangle \\
&+ \frac{|\phi^+\rangle - |\phi^-\rangle}{\sqrt{2}} \beta a|0\rangle + \frac{|\psi^+\rangle - |\psi^-\rangle}{\sqrt{2}} \beta b|1\rangle \\
&= \frac{1}{\sqrt{2}} \left(|\psi^+\rangle(\alpha a|0\rangle + \beta b|1\rangle) + |\psi^-\rangle(\alpha a|0\rangle - \beta b|1\rangle) \right. \\
&+ \left. |\phi^+\rangle(\alpha b|1\rangle + \beta a|0\rangle) + |\phi^-\rangle(\alpha b|1\rangle - \beta a|0\rangle) \right).
\end{aligned}
$$

Alice may measure $|\psi'\rangle_1$ in Bell basis and convey her measurement outcome to Bob, who can follow the same strategy (i.e., preparation of ancilla qubit, application of unitary operator described by (7.3), measurement of the last

Alice's outcome	Bob's outcome	Bob's state	Operation applied by Bob	Final state of Bob
00	0	$\alpha|0\rangle + \beta|1\rangle$	I	$\alpha|0\rangle + \beta|1\rangle$
01	0	$\alpha|1\rangle + \beta|0\rangle$	X	$\alpha|0\rangle + \beta|1\rangle$
10	0	$\alpha|0\rangle - \beta|1\rangle$	Z	$\alpha|0\rangle + \beta|1\rangle$
11	0	$\alpha|1\rangle - \beta|0\rangle$	iY	$\alpha|0\rangle + \beta|1\rangle$
00	1	$|1\rangle$	Teleportation fails	
01	1	$|1\rangle$	Teleportation fails	
10	1	$|1\rangle$	Teleportation fails	
11	1	$|1\rangle$	Teleportation fails	

Table 7.2: Probabilistic teleportation: Relation among Alice's measurement outcomes, Bob's measurement outcomes, unitary operations used by Bob and Bob's conclusions/final state.

qubit and application of suitable single qubit operation) to reach at the same conclusion as described in Table 7.2.

Cartoon 7.1: Do you think it will be possible in the future?

7.4 Controlled teleportation or quantum information splitting

We have already mentioned that controlled teleportation (CT) implies quantum information splitting (QIS). Here we provide a simple example of CT that can be considered as QIS also. In the simplest possible example of CT, Alice, Bob and Charlie share a GHZ state $\frac{1}{\sqrt{2}}(|000\rangle + |111\rangle)_{ABC}$. Here the subscripts A, B, C stand for Alice, Bob and Charlie, respectively and when we write $\frac{1}{\sqrt{2}}(|000\rangle + |111\rangle)_{ABC}$ it implies that the first, second and third qubits are with Alice, Bob and Charlie respectively. Subscripts A', and A_i, where $i \in \{1, 2, 3, \cdots, n\}$ are also used to denote qubits that are in possession of Alice. Similarly qubits of Bob and Charlie are also indexed. This notation is used in this chapter and in the next chapter, too. Now assume that Alice has received an unknown state $(\alpha|0\rangle + \beta|1\rangle)_{A'}$. Thus the combined state of the system becomes

$$|\psi\rangle_1 = (\alpha|0\rangle + \beta|1\rangle)_{A'} \otimes \frac{1}{\sqrt{2}}(|000\rangle + |111\rangle)_{ABC},$$

which can be decomposed as follows:

$$
\begin{aligned}
|\psi\rangle_1 &= (\alpha|0\rangle + \beta|1\rangle)_{A'} \otimes \tfrac{1}{\sqrt{2}}(|000\rangle + |111\rangle)_{ABC} \\
&= \tfrac{1}{\sqrt{2}}(\alpha|0000\rangle + \alpha|0111\rangle + \beta|1000\rangle + \beta|1111\rangle)_{A'ABC} \\
&= \left(\tfrac{|\psi^+\rangle + |\psi^-\rangle}{2}\right)_{A'A} \alpha|00\rangle_{BC} + \left(\tfrac{|\phi^+\rangle + |\phi^-\rangle}{2}\right)_{A'A} \alpha|11\rangle_{BC} \\
&+ \left(\tfrac{|\phi^+\rangle - |\phi^-\rangle}{2}\right)_{A'A} \beta|00\rangle_{BC} + \left(\tfrac{|\psi^+\rangle - |\psi^-\rangle}{2}\right)_{A'A} \beta|11\rangle_{BC} \\
&= \tfrac{1}{2}(|\psi^+\rangle_{A'A}(\alpha|00\rangle + \beta|11\rangle)_{BC} + |\psi^-\rangle_{A'A}(\alpha|00\rangle - \beta|11\rangle)_{BC} \\
&+ |\phi^+\rangle_{A'A}(\alpha|11\rangle + \beta|00\rangle)_{BC} + |\phi^-\rangle_{A'A}(\alpha|11\rangle - \beta|00\rangle)_{BC}) \\
&= \tfrac{1}{2\sqrt{2}}(|\psi^+\rangle_{A'A}(\alpha|0\rangle_B(|+\rangle + |-\rangle)_C + \beta|1\rangle_B(|+\rangle - |-\rangle)_C) \\
&+ |\psi^-\rangle_{A'A}(\alpha|0\rangle_B(|+\rangle + |-\rangle)_C - \beta|1\rangle_B(|+\rangle - |-\rangle)_C) \\
&+ |\phi^+\rangle_{A'A}(\alpha|1\rangle_B(|+\rangle - |-\rangle)_C + \beta|0\rangle_B(|+\rangle + |-\rangle)_C) \\
&+ |\phi^-\rangle_{A'A}(\alpha|1\rangle_B(|+\rangle - |-\rangle)_C - \beta|0\rangle_B(|+\rangle + |-\rangle)_C)) \\
&= \tfrac{1}{2\sqrt{2}}(|\psi^+\rangle_{A'A}((\alpha|0\rangle + \beta|1\rangle)_B|+\rangle_C + (\alpha|0\rangle - \beta|1\rangle)_B|-\rangle_C) \\
&+ |\psi^-\rangle_{A'A}((\alpha|0\rangle - \beta|1\rangle)_B|+\rangle_C + (\alpha|0\rangle + \beta|1\rangle)_B|-\rangle_C) \\
&+ |\phi^+\rangle_{A'A}((\alpha|1\rangle + \beta|0\rangle)_B|+\rangle_C + (\beta|0\rangle - \alpha|1\rangle)_B|-\rangle_C) \\
&+ |\phi^-\rangle_{A'A}((\alpha|1\rangle - \beta|0\rangle)_B|+\rangle_C - (\beta|0\rangle + \alpha|1\rangle)_B|-\rangle_C)).
\end{aligned}
$$

$$(7.4)$$

From (7.4) it is clear that if Alice measures her qubits in Bell basis and Charlie measures his qubit in diagonal basis and both of them convey the result to Bob, then Bob can use appropriate Pauli gates to obtain the unknown quantum state. For example, if Alice and Charlie's measurement outcomes are $|\psi^+\rangle$ and $|+\rangle$ respectively, then as a result of their measurements Bob's qubit is collapsed to the state $(\alpha|0\rangle + \beta|1\rangle)_B$. So Bob does not need to do anything, or in other words he applies I gate. Similarly, when Alice and Charlie's measurement outcomes are $|\psi^+\rangle$ and $|-\rangle$ respectively, then as a result of their measurements Bob's qubit is collapsed to the state $(\alpha|0\rangle - \beta|1\rangle)_B$. So Bob has to apply Z gate. The explicit relation between all possible combinations of measurement outcomes of Alice and Charlie and the corresponding unitary operations of Bob are provided in Table 7.3. Here it is important to understand that the status of Charlie and Bob are the same in this protocol. If instead of Charlie, Bob measures his qubit in

Alice's measurement outcome	Charlie's measurement outcome	Bob's unitary operation
$|\psi^+\rangle$	$|+\rangle$	I
$|\psi^+\rangle$	$|-\rangle$	Z
$|\psi^-\rangle$	$|+\rangle$	Z
$|\psi^-\rangle$	$|-\rangle$	I
$|\phi^+\rangle$	$|+\rangle$	X
$|\phi^+\rangle$	$|-\rangle$	iY
$|\phi^-\rangle$	$|+\rangle$	iY
$|\phi^-\rangle$	$|-\rangle$	X

Table 7.3: Relation between measurement outcomes of Alice and Charlie and the unitary operations of Bob. In all the cases the final state of Bob is $(\alpha|0\rangle + \beta|1\rangle)_B$. Thus the unknown state is perfectly teleported to Bob with the help (control) of Charlie.

diagonal basis and conveys his measurement result to Charlie then Charlie will be able to reconstruct the unknown state. There are some modified QIS schemes where there are many Bobs but the status of all of them is not the same. For example, we may think of a situation where there are four Bobs. Out of them Bob_1 can construct the unknown state with the help of Alice and Bob_2 only but each of the other Bobs requires the help from the remaining three Bobs and Alice to reconstruct the unknown state. Such schemes are called hierarchical QIS (HQIS) because there is a hierarchy among the Bobs. A scheme of HQIS was first introduced by Wang *et al.* in 2010 [92]. Recently Shukla and Pathak [93] have generalized the Wang *et al.* scheme. In the following subsection we will briefly present a scheme of HQIS, along the line of the recently presented Shukla and Pathak scheme [93].

7.4.1 Hierarchical quantum information splitting

Let us assume that Alice has chosen 4-qubit $|\Omega\rangle$ state as channel and kept the first qubit with her and has sent the second, third and fourth qubits to Bob, Charlie and Diana, respectively. In that case

$$
\begin{aligned}
|\psi_c\rangle = |\Omega\rangle_{ABCD} &= \tfrac{1}{2}[|0000\rangle + |0110\rangle + |1001\rangle - |1111\rangle]_{ABCD} \\
&= \tfrac{1}{\sqrt{2}}[|0\rangle_A|\psi_0\rangle_{BCD} + |1\rangle_A|\psi_1\rangle_{BCD}],
\end{aligned} \tag{7.5}
$$

where $|\psi_0\rangle_{BCD} = \frac{1}{\sqrt{2}}[|000\rangle + |110\rangle]$ and $|\psi_1\rangle_{BCD} = \frac{1}{\sqrt{2}}[|001\rangle - |111\rangle]$.

Alice wishes to teleport (share) among her agents a general one qubit

Outcome of Alice's measurement	State of all agents after measurement of Alice
$\|\psi^+\rangle$	$\|\Psi^+\rangle = \frac{\|\psi_0\rangle + \lambda\|\psi_1\rangle}{\sqrt{1+\|\lambda\|^2}}$
$\|\psi^-\rangle$	$\|\Psi^-\rangle = \frac{\|\psi_0\rangle - \lambda\|\psi_1\rangle}{\sqrt{1+\|\lambda\|^2}}$
$\|\phi^+\rangle$	$\|\Phi^+\rangle = \frac{\|\psi_1\rangle + \lambda\|\psi_0\rangle}{\sqrt{1+\|\lambda\|^2}}$
$\|\phi^-\rangle$	$\|\Phi^-\rangle = \frac{\|\psi_1\rangle - \lambda\|\psi_0\rangle}{\sqrt{1+\|\lambda\|^2}}$

Table 7.4: Relation between the outcomes of Alice's measurement and the combined states of all agents after the measurement of Alice.

state

$$|\psi_s\rangle = \frac{1}{\sqrt{1+|\lambda|^2}}(|0\rangle + \lambda|1\rangle), \tag{7.6}$$

which represents an arbitrary qubit. So the combined state is

$$
\begin{aligned}
|\psi_s\rangle \otimes |\psi_c\rangle &= \frac{1}{\sqrt{1+|\lambda|^2}}(|0\rangle + \lambda|1\rangle) \otimes \frac{1}{\sqrt{2}}[|0\rangle|\psi_0\rangle + |1\rangle|\psi_1\rangle] \\
&= \frac{1}{\sqrt{2(1+|\lambda|^2)}}(|00\rangle|\psi_0\rangle + |01\rangle|\psi_1\rangle + \lambda|10\rangle|\psi_0\rangle + \lambda|11\rangle|\psi_1\rangle) \\
&= \frac{1}{2\sqrt{(1+|\lambda|^2)}}[|\psi^+\rangle(|\psi_0\rangle + \lambda|\psi_1\rangle) + |\psi^-\rangle(|\psi_0\rangle - \lambda|\psi_1\rangle) \\
&\quad + |\phi^+\rangle(|\psi_1\rangle + \lambda|\psi_0\rangle) + |\phi^-\rangle(|\psi_1\rangle - \lambda|\psi_0\rangle)].
\end{aligned} \tag{7.7}
$$

Now Alice does a Bell measurement on the first 2 qubits that are in her possession. As a consequence of the measurement of Alice, the combined state of her 3 agents collapses to $|\Psi^+\rangle$ or, $|\Psi^-\rangle$ or, $|\Phi^+\rangle$ or, $|\Phi^-\rangle$ where $|\Psi^\pm\rangle = \frac{|\psi_0\rangle \pm \lambda|\psi_1\rangle}{\sqrt{1+|\lambda|^2}}$ and $|\Phi^\pm\rangle = \frac{|\psi_1\rangle \pm \lambda|\psi_0\rangle}{\sqrt{1+|\lambda|^2}}$. To be precise, the relation between the outcome of Bell measurement of Alice and the combined state of the agents after the measurement of Alice are given in Table 7.4.

If Alice's measurement outcome is $|\psi^\pm\rangle$ then the combined state of the agents is

$$
\begin{aligned}
|\Psi^\pm\rangle_{BCD} &= \frac{1}{\sqrt{1+|\lambda|^2}}[|\psi_0\rangle_{BCD} \pm \lambda|\psi_1\rangle_{BCD}] \\
&= \frac{1}{\sqrt{2(1+|\lambda|^2)}}[|000\rangle + |110\rangle \pm \lambda(|001\rangle - |111\rangle)]_{BCD}.
\end{aligned} \tag{7.8}
$$

Similarly, if Alice obtains $|\phi^\pm\rangle$ then the combined state of the agents is

$$
\begin{aligned}
|\Phi^\pm\rangle_{BCD} &= \frac{1}{\sqrt{1+|\lambda|^2}}[|\psi_1\rangle_{BCD} \pm \lambda|\psi_0\rangle_{BCD}] \\
&= \frac{1}{\sqrt{2(1+|\lambda|^2)}}[|001\rangle - |111\rangle \pm \lambda(|000\rangle + |110\rangle)]_{BCD}.
\end{aligned} \tag{7.9}
$$

Now if the agents decide that Diana will recover the secrets, then we can

Alice's measurement outcome	Measurement outcome of Bob and Charlie	Diana's operation
$\lvert\psi^+\rangle$	$\lvert 00\rangle_{BC}$	I
$\lvert\psi^+\rangle$	$\lvert 11\rangle_{BC}$	Z
$\lvert\psi^-\rangle$	$\lvert 00\rangle_{BC}$	Z
$\lvert\psi^-\rangle$	$\lvert 11\rangle_{BC}$	I
$\lvert\phi^+\rangle$	$\lvert 00\rangle_{BC}$	X
$\lvert\phi^+\rangle$	$\lvert 11\rangle_{BC}$	XZ
$\lvert\phi^-\rangle$	$\lvert 00\rangle_{BC}$	XZ
$\lvert\phi^-\rangle$	$\lvert 11\rangle_{BC}$	X

Table 7.5: Relation between unitary operators required by Diana to recover the unknown state and the measurement outcomes of Alice, Bob and Charlie. Communication from Alice and either Charlie or Bob is sufficient for Diana as the outcomes of measurement are always the same for Charlie and Bob.

decompose (7.8) and (7.9) as

$$\lvert\Psi^\pm\rangle_{BCD} = \frac{1}{\sqrt{2(1+\lvert\lambda\rvert^2)}}[\lvert 00\rangle_{BC}(\lvert 0_D\rangle \pm \lambda\lvert 1_D\rangle) + \lvert 11\rangle_{BC}(\lvert 0_D\rangle \mp \lambda\lvert 1_D\rangle)] \tag{7.10}$$

and

$$\lvert\Phi^\pm\rangle_{BCD} = \frac{1}{\sqrt{2(1+\lvert\lambda\rvert^2)}}[\lvert 00\rangle_{BC}(\lvert 1_D\rangle \pm \lambda\lvert 0_D\rangle) - \lvert 11\rangle_{BC}(\lvert 1_D\rangle \mp \lambda\lvert 0_D\rangle)]. \tag{7.11}$$

From (7.10) and (7.11) it is clear that if Bob and Charlie measure their qubits in computational basis and only one of them sends the result to Diana, then Diana will be able to reconstruct the state sent by Alice using appropriate unitary operators as shown in Table 7.5. For example, if Alice's outcome is $\lvert\psi^+\rangle$ and that of Charlie is $\lvert 0\rangle$ then the state of Diana is collapsed to $\frac{1}{\sqrt{1+\lvert\lambda\rvert^2}}(\lvert 0\rangle + \lambda\lvert 1\rangle)$, so Diana needs to apply I. Thus Diana needs the help of Alice and either Charlie or Bob to reconstruct the unknown state sent by Alice. Now we would like to ask: what happens if the agents decide that Bob will reconstruct the state sent by Alice? From (7.8) and (7.9) it is clear that Charlie and Diana cannot measure their qubits in computational basis as that would collapse the state of Bob to $\lvert 0\rangle$ or $\lvert 1\rangle$. Further, we note that we can also decompose (7.8) and (7.9) as

$$
\begin{aligned}
\lvert\Psi^\pm\rangle_{BCD} =\ & \frac{1}{2\sqrt{(1+\lvert\lambda\rvert^2)}}[(\lvert 0_B\rangle \mp \lambda\lvert 1_B\rangle)\lvert\psi^+\rangle_{CD} + (\lvert 0_B\rangle \pm \lambda\lvert 1_B\rangle)\lvert\psi^-\rangle_{CD} \\
& + (\lvert 1_B\rangle \pm \lambda\lvert 0_B\rangle)\lvert\phi^+\rangle_{CD} - (\lvert 1_B\rangle \mp \lambda\lvert 0_B\rangle)\lvert\phi^-\rangle_{CD}]
\end{aligned}
\tag{7.12}
$$

and

$$|\Phi^{\pm}\rangle_{BCD} = \frac{1}{2\sqrt{(1+|\lambda|^2)}}[(|0_B\rangle \pm \lambda|1_B\rangle)|\phi^+\rangle_{CD} + (|0_B\rangle \mp \lambda|1_B\rangle)|\phi^-\rangle_{CD}$$
$$- (|1_B\rangle \mp \lambda|0_B\rangle)|\psi^+\rangle_{CD} + (|1_B\rangle \pm \lambda|0_B\rangle)|\psi^-\rangle_{CD}].$$

$$(7.13)$$

From (7.12) and (7.13) it is clear that Bob can reconstruct the state sent by Alice if Charlie and Diana cooperate. Here Charlie's and Diana's cooperation requires a two-qubit measurement. This can be fulfilled by either one of them communicating a qubit to the other over an authenticated quantum channel, or both performing a joint measurement. Finally, after knowing the outcome of Charlie and Diana's joint measurement, Bob can apply appropriate unitary operators on his qubit (as shown in Table 7.6) to reconstruct the unknown state. Clearly, to reconstruct the state Bob requires more information than Diana. This implies that as an agent Diana is more powerful than Bob is. Thus there is a hierarchy among the agents. This captures the essential notion of HQIS.

A special case of the above may be visualized as follows: Alice sends both C and D qubits (i.e., third and fourth qubits) to Charlie and qubit B (second qubit) to Bob. In that case Charlie will never require any assistance from Bob to reconstruct the unknown state. However, Bob will not be able to reconstruct the state without the help of Charlie.

7.5   Modified teleportation schemes

The above described ideas of quantum teleportation can be modified in various ways. Over the last two decades several ideas to perform specific communication tasks using quantum teleportation have been proposed. A few of those exciting and interesting ideas are described below.

7.5.1   Remote state preparation

So far we have discussed schemes for teleportation of an unknown qubit. Now we will describe a situation where the sender (Alice) knows the state to be teleported. However, the receiver (Bob) is completely unaware of it. Thus it may be viewed as teleportation of a known state. As in this scheme a state known to Alice is prepared at Bob's end, the scheme is called remote state preparation. Remote state preparation is possible with different kinds of quantum channels. To begin with we assume that Alice and Bob share an entangled state $|\phi^-\rangle_{AB} = \left(\frac{|01\rangle - |10\rangle}{\sqrt{2}}\right)_{AB}$. Now Alice wishes to transmit an arbitrary qubit represented as (3.108). Thus the qubit to be transmitted is $|\psi\rangle = \cos\left(\frac{\theta}{2}\right)|0\rangle + \sin\left(\frac{\theta}{2}\right)\exp(i\phi)|1\rangle$. Here, Alice knows the values of θ and ϕ and this distinguishes present protocol from the conventional protocol of teleportation.

Alice measurement outcome	Outcome of joint measurement of Charlie and Diana	Bob's operation
$\lvert\psi^+\rangle$	$\lvert\psi^+\rangle_{CD}$	Z
$\lvert\psi^+\rangle$	$\lvert\psi^-\rangle_{CD}$	I
$\lvert\psi^+\rangle$	$\lvert\phi^+\rangle_{CD}$	X
$\lvert\psi^+\rangle$	$\lvert\phi^-\rangle_{CD}$	XZ
$\lvert\psi^-\rangle$	$\lvert\psi^+\rangle_{CD}$	I
$\lvert\psi^-\rangle$	$\lvert\psi^-\rangle_{CD}$	Z
$\lvert\psi^-\rangle$	$\lvert\phi^+\rangle_{CD}$	XZ
$\lvert\psi^-\rangle$	$\lvert\phi^-\rangle_{CD}$	X
$\lvert\phi^+\rangle$	$\lvert\phi^+\rangle_{CD}$	I
$\lvert\phi^+\rangle$	$\lvert\phi^-\rangle_{CD}$	Z
$\lvert\phi^+\rangle$	$\lvert\psi^+\rangle_{CD}$	XZ
$\lvert\phi^+\rangle$	$\lvert\psi^-\rangle_{CD}$	X
$\lvert\phi^-\rangle$	$\lvert\phi^+\rangle_{CD}$	Z
$\lvert\phi^-\rangle$	$\lvert\phi^-\rangle_{CD}$	I
$\lvert\phi^-\rangle$	$\lvert\psi^+\rangle_{CD}$	X
$\lvert\phi^-\rangle$	$\lvert\psi^-\rangle_{CD}$	XZ

Table 7.6: Relation between unitary operators required by Bob to recover the unknown state and the measurement outcomes of Alice, Charlie and Diana. Joint measurement of the qubits of Charlie and Diana is required.

Now we introduce a new basis set $\{|q_1\rangle, |q_2\rangle\}$, where

$$
\begin{aligned}
|q_1\rangle &= \cos\left(\tfrac{\theta}{2}\right)|0\rangle + \sin\left(\tfrac{\theta}{2}\right)\exp(i\phi)|1\rangle = |\psi\rangle \\
|q_2\rangle &= -\sin\left(\tfrac{\theta}{2}\right)\exp(-i\phi)|0\rangle + \cos\left(\tfrac{\theta}{2}\right)|1\rangle = U_{RSP}|\psi\rangle,
\end{aligned}
\tag{7.14}
$$

where $U_{RSP} = XZP(-2\phi) = -iYP(-2\phi)$ is a unitary operator, which represents a rotation in the Bloch sphere. We can easily visualize that as

$$
\begin{aligned}
U_{RSP}|q_1\rangle = XZP(-2\phi)\begin{pmatrix} \cos\left(\tfrac{\theta}{2}\right) \\ \sin\left(\tfrac{\theta}{2}\right)e^{i\phi} \end{pmatrix} &= XZ\begin{pmatrix} \cos\left(\tfrac{\theta}{2}\right) \\ \sin\left(\tfrac{\theta}{2}\right)e^{-i\phi} \end{pmatrix} \\
&= X\begin{pmatrix} \cos\left(\tfrac{\theta}{2}\right) \\ -\sin\left(\tfrac{\theta}{2}\right)e^{-i\phi} \end{pmatrix} \\
&= \begin{pmatrix} -\sin\left(\tfrac{\theta}{2}\right)e^{-i\phi} \\ \cos\left(\tfrac{\theta}{2}\right) \end{pmatrix} \\
&= |q_2\rangle.
\end{aligned}
$$

Now using $\{|q_1\rangle, |q_2\rangle\}$ basis set we can write

$$
\begin{aligned}
|0\rangle &= \cos\left(\tfrac{\theta}{2}\right)|q_1\rangle - \sin\left(\tfrac{\theta}{2}\right)\exp(i\phi)|q_2\rangle \\
|1\rangle &= \sin\left(\tfrac{\theta}{2}\right)\exp(-i\phi)|q_1\rangle + \cos\left(\tfrac{\theta}{2}\right)|q_2\rangle.
\end{aligned}
\tag{7.15}
$$

Using (7.15) we can express the entangled state $|\phi^-\rangle_{AB} = \left(\frac{|01\rangle - |10\rangle}{\sqrt{2}}\right)_{AB}$ shared by Alice and Bob as follows:

$$
\begin{aligned}
|\phi^-\rangle_{AB} &= \tfrac{1}{\sqrt{2}}\left(|0\rangle_A|1\rangle_B - |1\rangle_A|0\rangle_B\right) \\
&= \tfrac{1}{\sqrt{2}}\left(\left(\cos\left(\tfrac{\theta}{2}\right)|q_1\rangle - \sin\left(\tfrac{\theta}{2}\right)\exp(i\phi)|q_2\rangle\right)_A \right. \\
&\quad \left(\sin\left(\tfrac{\theta}{2}\right)\exp(-i\phi)|q_1\rangle + \cos\left(\tfrac{\theta}{2}\right)|q_2\rangle\right)_B \\
&\quad - \left(\sin\left(\tfrac{\theta}{2}\right)\exp(-i\phi)|q_1\rangle + \cos\left(\tfrac{\theta}{2}\right)|q_2\rangle\right)_A \\
&\quad \left.\left(\cos\left(\tfrac{\theta}{2}\right)|q_1\rangle - \sin\left(\tfrac{\theta}{2}\right)\exp(i\phi)|q_2\rangle\right)\right)_B \\
&= \tfrac{1}{\sqrt{2}}\left(\left(\cos^2\tfrac{\theta}{2} + \sin^2\tfrac{\theta}{2}\right)|q_1q_2\rangle_{AB} - \left(\cos^2\tfrac{\theta}{2} + \sin^2\tfrac{\theta}{2}\right)|q_2q_1\rangle_{AB}\right) \\
&= \tfrac{1}{\sqrt{2}}\left(|q_1q_2\rangle_{AB} - |q_2q_1\rangle_{AB}\right).
\end{aligned}
\tag{7.16}
$$

Now Alice measures her qubit in $\{|q_1\rangle, |q_2\rangle\}$ basis set and communicates the result to Bob. From (7.16) it is clear that if Alice's measurement outcome is $|q_2\rangle$ then Bob does not need to do anything to reconstruct the qubit unknown to him and if the outcome of Alice's measurement is $|q_1\rangle$ then Bob applies $U_{RSP}^{-1} = -P(2\phi)ZX$ on his qubit and reconstructs the qubit unknown to him. Thus if Alice and Bob share a prior entanglement, then the teleportation of a known qubit requires only one projective measurement and one bit of classical communication. Therefore remote state preparation requires fewer resources than the conventional teleportation. This interesting scheme was introduced by A. K. Pati in 2000 [94].

7.5.1.1 Modified remote state preparation schemes

It is now straightforward to generalize the above scheme to probabilistic remote state preparation and joint remote state preparation. For example, if instead of the maximally entangled state $|\phi^-\rangle_{AB}$, Alice and Bob share $|\psi'\rangle_0 = a|01\rangle - b|10\rangle$, where $|a|^2 + |b|^2 = 1$ and $|a| \neq \frac{1}{\sqrt{2}}$, then we will obtain a scheme for probabilistic teleportation. Similarly, a scheme for joint remote state preparation can be constructed by using a GHZ state. Such a scheme is described in [95] as follows. Alice, Bob and Charlie share a GHZ state. Alice and Bob jointly want to transmit a qubit $|\psi\rangle = \cos\left(\frac{\theta}{2}\right)|0\rangle + \sin\left(\frac{\theta}{2}\right)\exp(i\phi)|1\rangle$ to Charlie. But incidentally, Alice knows θ and Bob knows ϕ. In such a case Alice measures her qubit using $\{|p_1\rangle = \cos\left(\frac{\theta}{2}\right)|0\rangle + \sin\left(\frac{\theta}{2}\right)|1\rangle, |p_2\rangle = \sin\left(\frac{\theta}{2}\right)|0\rangle - \cos\left(\frac{\theta}{2}\right)|1\rangle\}$ basis set and Bob measures his qubit using $\{|r_1\rangle = \frac{|0\rangle + \exp(i\phi)|1\rangle}{\sqrt{2}}, |r_2\rangle = \frac{\exp(-i\phi)|0\rangle - |1\rangle}{\sqrt{2}}\}$ basis set. If Alice and Bob communicate their measurement outcomes to Charlie, then Charlie can apply appropriate unitary operations to reconstruct the qubit, which is jointly transmitted by Alice and Bob. Thus we have a simple scheme for joint remote state preparation. An elaborate discussion of this scheme as well as its extension to probabilistic joint remote state preparation (a situation that arises when Alice, Bob and Charlie initially share $a|000\rangle + b|111\rangle$, where $|a|^2 + |b|^2 = 1$ and $|a| \neq \frac{1}{\sqrt{2}}$) can be found in [95]. We will not further elaborate on it. In Section 8.11 we will briefly introduce QSS and show that QSS can be viewed as an application of QIS.

7.6 Superdense coding

Superdense coding or dense coding schemes are essentially designed to send more information through a channel of less capacity. In the simplest example of dense coding scheme Alice can send Bob two bits of classical information by passing only a single qubit. They initially share a Bell state. Without loss of generality we may assume that they initially share $|\psi^+\rangle = \frac{|00\rangle + |11\rangle}{\sqrt{2}}$. The first qubit of the entangled state is with Alice and the second qubit is with Bob. Now Alice wants to send two bits of classical information to Bob. So Alice has to send one among four alternatives. To be precise, she has to send either 00 or 01 or 10 or 11. To do so, she encodes the 2-bit classical information in the qubit which is with her and sends the qubit to Bob. To encode four different alternatives she needs at least four different single qubit unitary operations. We may call them U_{00}, U_{01}, U_{10} and U_{11} respectively. Thus to encode $00, 01, 10$ and 11 Alice applies U_{00}, U_{01}, U_{10} and U_{11} respectively. The initial state and this mapping between classical information to be encoded and the unitary operation to be used for encoding is known to Bob. These four different encoding operations will yield four output states. If these output states are mutually orthogonal then Bob will be able to discriminate them with certainty (after he receives the qubit of Alice and measures both the qubits of his possession in appropriate basis) and thus he will be able to decode the classical information encoded by Alice. Therefore, Alice can send two bits

Classical bits that Alice wishes to send	Quantum gate (unitary operations) applied by Alice	Final state of Bob
00	I	$\frac{\lvert 00\rangle + \lvert 11\rangle}{\sqrt{2}}$
01	X	$\frac{\lvert 10\rangle + \lvert 01\rangle}{\sqrt{2}}$
10	iY	$\frac{\lvert 01\rangle - \lvert 10\rangle}{\sqrt{2}}$
11	Z	$\frac{\lvert 00\rangle - \lvert 11\rangle}{\sqrt{2}}$

Table 7.7: Dense coding using Bell states.

of classical information by sending one qubit only. This is why it is called dense coding.

Now let us come back to the simplest example that we are discussing. To encode $00, 01, 10$ and 11 Alice applies $U_{00} = I, U_{01} = X, U_{10} = ZX = iY$ and $U_{11} = Z$ respectively to the qubit in her possession (i.e., on the first qubit of $\lvert \psi^+ \rangle$). Thus if she sends 00 then she does nothing (i.e., she applies I) and consequently, the final state is the same as the initial state $\lvert \psi^+ \rangle$, if she sends 01 then she applies X gate on the first qubit and that transforms the initial state $\lvert \psi^+ \rangle$ to $\lvert \phi^+ \rangle = \frac{\lvert 10\rangle + \lvert 01\rangle}{\sqrt{2}}$. Similarly, when she encodes 10 then she applies $ZX = iY$ on the first qubit, which transforms the initial state into $\lvert \phi^- \rangle = \frac{\lvert 01\rangle - \lvert 10\rangle}{\sqrt{2}}$ and when she encodes 11 then she applies Z gate on the first qubit and that transforms the initial state into $\lvert \psi^- \rangle = \frac{\lvert 00\rangle - \lvert 11\rangle}{\sqrt{2}}$. After this encoding operation Alice sends her qubit to Bob and now Bob has the entire final state in his possession. Alice's encoding operations and corresponding final states of Bob are shown in Table 7.7. It is easy to observe that final states of Bob are the four Bell states that are mutually orthogonal. Consequently, if Bob measures them in Bell basis then he will be able to decode the information encoded by Alice since he knows the initial state and the mapping between the encoding operations and the information encoded. In this scheme Alice has sent only one qubit to Bob but she has communicated two bits of classical information. Thus she has sent more information by communicating less. The coding is dense in this sense, and this is why it is referred to as dense coding. Dense coding is purely quantum phenomenon and does not have any classical analogue. This requires that Alice and Bob initially share an entangled state. It is not limited to Bell states only. In the next section we will provide a few examples of dense coding using other entangled states and we will also include some exercises on this topic at the end of this chapter. These examples and exercises will convince you about the general nature of dense coding and about its applications in different areas of quantum communication.

7.6.1 Some more examples of dense coding

As we mentioned in the last section, dense coding is not limited to Bell state. Here we elaborate the general nature of dense coding. Assume that we have a basis set $Q = \{Q_0, Q_1, \cdots, Q_{2^n-1}\}$ of n-partite orthonormal state vectors. Further we assume that there exists a set of unitary operations $U = \{U_0^i, U_1^i, \cdots, U_{2^n-1}^i : U_j^i Q_i = Q_j\}$ such that the unitary operations can transform a particular element Q_i of set Q into all the other elements of set Q. Now Alice can encode n-bit message by using an encoding scheme in which $\{U_0^i, U_1^i, \cdots, U_{2^n-1}^i\}$ are used to encode $\{0_1 0_2 \cdots 0_n, 0_1 0_2 \cdots 1_n, \cdots, 1_1 1_2 \cdots 1_n\}$ respectively. If these encoded messages are sent to Bob then Bob can unambiguously decode the message since the states received by him are mutually orthogonal.

Dense coding is a special case of the above idea. To be precise, dense coding is possible if and only if U_j^i described above are m-qubit operators, where $m < n$. In an efficient dense coding protocol the operators are chosen in such a way that $m = \frac{n}{2}$ for even n and $m = \frac{n+1}{2}$ for odd n. Now for dense coding, Alice and Bob share an entangled state Q_i in such a way that m qubits of Q_i are with Alice and the remaining $n - m$ qubits are with Bob, who knows the initial state Q_i prepared by Alice and the mapping between encoded information and the unitary operation used for encoding. Alice may encode an n-bit message by operating any of 2^n unitary m-qubit operators available with her (say she applies U_j^i) and send her qubits to Bob. Now since the orthonormal states are distinguishable, Bob can measure his qubits (in Q basis) and find the state Q_j. Since he already knows that the initial state was Q_i, now he knows that Alice has sent him a bit string indexed as j. Thus n-bit information are sent with $m < n$ qubits. This is the essence of dense coding. This does not involve any security measure. In the next chapter we will show how dense coding can be used for secure communication but before that we wish to provide some more examples of dense coding schemes.

We have already seen that dense coding is possible with two qubit Bell states. Now we wish to investigate what happens with 3-qubit GHZ states. The dense coding operations for GHZ states are shown in Tables 7.8-7.9. Here we can easily understand that unitary operations are applied on two qubits, therefore initially the three qubit entanglement is shared in such a way that Alice has two qubits and Bob has one qubit. Now Alice applies 2-qubit unitary operation to encode three bits of classical information (as she can encode among eight alternatives, she can send $\log_2 8 = 3$ bits of classical information) and sends her two qubits to Bob who does a measurement on all the three qubits together in GHZ basis and decodes the classical information. This is a clear example of dense coding, as three bits of classical information is sent by sending only two qubits. However, it is not as efficient as the dense coding using Bell state. This is so because when we use Bell state then we can send two bits of classical information by sending

Unitary operators applied on qubits 1 and 2	Orthogonal states created on application of unitary operations on initial GHZ state $\frac{1}{\sqrt{2}}(\lvert 000\rangle + \lvert 111\rangle)$
$U_0 = I \otimes I$	$\frac{1}{\sqrt{2}}(\lvert 000\rangle + \lvert 111\rangle)$
$U_1 = Z \otimes I$	$\frac{1}{\sqrt{2}}(\lvert 000\rangle - \lvert 111\rangle)$
$U_2 = X \otimes I$	$\frac{1}{\sqrt{2}}(\lvert 100\rangle + \lvert 011\rangle)$
$U_3 = iY \otimes I$	$\frac{1}{\sqrt{2}}(-\lvert 100\rangle + \lvert 011\rangle)$
$U_4 = I \otimes X$	$\frac{1}{\sqrt{2}}(\lvert 010\rangle + \lvert 101\rangle)$
$U_5 = Z \otimes X$	$\frac{1}{\sqrt{2}}(\lvert 010\rangle - \lvert 101\rangle)$
$U_6 = X \otimes X$	$\frac{1}{\sqrt{2}}(\lvert 110\rangle + \lvert 001\rangle)$
$U_7 = iY \otimes X$	$\frac{1}{\sqrt{2}}(-\lvert 110\rangle + \lvert 001\rangle)$

Table 7.8: Dense coding of GHZ states scheme 1.

1 qubit. Consequently if we had used two qubits of communication then we could have sent four bits of classical information which is more than the classical information we sent by dense coding using GHZ state. Whenever we have an odd number of qubits in the entangled states the dense coding scheme will be less efficient than the Bell state based dense coding scheme. But it does not mean that all the entangled states having even number of qubits can be used for maximal dense coding. Before we elaborate on this point, we wish to note that in Tables 7.8 and 7.9 we have provided two different schemes for dense coding of GHZ states. There exist a few more alternative ways for the same. It is not our purpose to note all of them here but we just wish to show that dense coding using the same entangled state can be done using different sets of unitary operators. Now to convince you that all even qubit entangled states are not suitable for maximal dense coding we may note that maximal dense coding is not possible in 4-qubit W states. We can at most encode three bits of classical information by applying 2-qubit unitary operations on 4-qubit W states. The same is shown in Table 7.10.

We will not give further examples of dense coding here. Once you understand the general nature of the dense coding discussed above you will be able to generate your own examples. In the next chapter we will discuss applications of dense coding in the context of quantum cryptography.

Before we move to the discussion of quantum cryptography we would like to note that both teleportation and dense coding are experimentally realized. To be precise, in 1997 in Vienna, D. Bouwmeester *et al.* demonstrated quantum teleportation using simple optics and entangled photons generated by type II parametric down-conversion process [28]. Since then considerable progress has been made on the experimental front. For example, in 2006 Q. Zhang *et al.* successfully demonstrated teleportation of two

Unitary operators applied on qubits 1 and 2	Orthogonal states created on application of unitary operations on initial GHZ state $\frac{1}{\sqrt{2}}(|000\rangle + |111\rangle)$
$U_0 = I \otimes I$	$\frac{1}{\sqrt{2}}(|000\rangle + |111\rangle)$
$U_1 = Z \otimes I$	$\frac{1}{\sqrt{2}}(|000\rangle - |111\rangle)$
$U_2 = X \otimes I$	$\frac{1}{\sqrt{2}}(|100\rangle + |011\rangle)$
$U_3 = iY \otimes I$	$\frac{1}{\sqrt{2}}(-|100\rangle + |011\rangle)$
$U_4 = I \otimes iY$	$\frac{1}{\sqrt{2}}(-|010\rangle + |101\rangle)$
$U_5 = Z \otimes iY$	$\frac{1}{\sqrt{2}}(-|010\rangle - |101\rangle)$
$U_6 = X \otimes iY$	$\frac{1}{\sqrt{2}}(-|110\rangle + |001\rangle)$
$U_7 = iY \otimes iY$	$\frac{1}{\sqrt{2}}(|110\rangle + |001\rangle)$

Table 7.9: Dense coding of GHZ states scheme 2.

Unitary operators applied on qubits 1 and 2	Orthogonal states created on application of unitary operations on initial GHZ state $|W\rangle_0 = \frac{1}{2}(|0001\rangle + |0010\rangle + |0100\rangle + |1000\rangle)$
$U_0 = X \otimes I$	$\frac{1}{2}(|1001\rangle + |1010\rangle + |1100\rangle + |0000\rangle)$
$U_1 = iY \otimes Z$	$\frac{1}{2}(-|1001\rangle - |1010\rangle + |1100\rangle + |0000\rangle)$
$U_2 = Z \otimes X$	$\frac{1}{2}(|0101\rangle + |0110\rangle + |0000\rangle - |1100\rangle)$
$U_3 = I \otimes iY$	$\frac{1}{2}(-|0101\rangle - |0110\rangle + |0000\rangle - |1100\rangle)$
$U_4 = I \otimes Z$	$\frac{1}{2}(|0001\rangle + |0010\rangle - |0100\rangle + |1000\rangle)$
$U_5 = Z \otimes I$	$\frac{1}{2}(|0001\rangle + |0010\rangle + |0100\rangle - |1000\rangle)$
$U_6 = iY \otimes iY$	$\frac{1}{2}(|1101\rangle + |1110\rangle - |1000\rangle - |0100\rangle)$
$U_7 = X \otimes X$	$\frac{1}{2}(|1101\rangle + |1110\rangle + |1000\rangle + |0100\rangle)$

Table 7.10: Dense coding of 4-qubit W states.

qubit composite system [96] and very recently free space quantum teleportation and entanglement distribution over 100 kilometers was reported by Juan Yin *et al.* [38]. It is interesting to note that the scope of teleportation is not limited to quantum states alone. Even quantum operations (gates) can be teleported. Teleportation of quantum gates is an important factor for successful implementation of distributed quantum computing. Experimental teleportation of a CNOT gate is shown by Y.-F. Hunag *et al.* [97]. Similarly, the first experimental demonstration of dense coding was reported in 1996 by K. Mattle, H. Weinfurter, P. G. Kwiat, and A. Zeilinger [98]. They demonstrated optically that one of the three messages i.e., $\log_2 3 = 1.58$ bits[1] of information can be communicated by transmitting one qubit (i.e., by sending one of the two entangled qubits which are already shared between receiver and sender). Subsequently in 1999 Fang *et al.* [99] showed dense coding using NMR technique. Since then many groups have demonstrated dense coding. It is not our purpose to review those works. We just wish to note that the purely quantum phenomena of teleportation and dense coding are experimentally achievable.

7.7 Solved examples

1. In a teleportation scheme Alice sends an arbitrary qubit $|\psi\rangle = \sin\theta|0\rangle + \cos\theta|1\rangle$ to Bob but Bob receives a qubit $|\phi\rangle = \sin\eta|0\rangle + \cos\eta|1\rangle$. Compute the fidelity of the teleportation scheme. Comment on the maximum and minimum possible values of the fidelity.
   Solution: Since both $|\psi\rangle$ and $|\phi\rangle$ are pure state so the required fidelity is given by Equation (3.95), i.e.,

   $$\begin{aligned} F\left(|\psi\rangle, |\phi\rangle\right) &= |\langle\psi|\phi\rangle| \\ &= |\sin(\theta)\sin(\eta) + \cos(\theta)\cos(\eta)| \\ &= |\cos(\theta - \eta)|. \end{aligned}$$

   Clearly the maximum value of $F\left(|\psi\rangle, |\phi\rangle\right) = 1$ when $\theta - \eta = 0, \pi$. When $\theta - \eta = 0$ then $|\psi\rangle = |\phi\rangle$ and when $\theta - \eta = \pi$ then $|\psi\rangle = -|\phi\rangle \equiv |\phi\rangle$. Thus the fidelity is maximum when Bob receives a perfect copy of the state sent by Alice. The minimum value of fidelity is 0 and clearly the same is achieved when $|\psi\rangle$ and $|\phi\rangle$ are mutually orthogonal.

2. Consider the teleportation protocol described in Section 7.2 and show that Bob cannot obtain any information on the qubit to be teleported before Alice discloses the results of her measurement.
   Solution: After Alice performs the measurement on her qubit, Bob's state collapses to one of the four possible states (e.g., $\alpha|0\rangle + \beta|1\rangle, \alpha|0\rangle - \beta|1\rangle, \alpha|1\rangle + \beta|0\rangle, \alpha|1\rangle - \beta|0\rangle$) with equal probabilities $p = \frac{1}{4}$. Now the

---

[1] It is also called 1 trit.

density matrix of Bob's state is

$$
\begin{aligned}
\rho_B &= \tfrac{1}{4}\left(\alpha|0\rangle + \beta|1\rangle\right)\left(\langle 0|\alpha^* + \langle 1|\beta^*\right) \\
&+ \tfrac{1}{4}\left(\alpha|0\rangle - \beta|1\rangle\right)\left(\langle 0|\alpha^* - \langle 1|\beta^*\right) \\
&+ \tfrac{1}{4}\left(\alpha|1\rangle + \beta|0\rangle\right)\left(\langle 1|\alpha^* + \langle 0|\beta^*\right) \\
&+ \tfrac{1}{4}\left(\alpha|1\rangle - \beta|0\rangle\right)\left(\langle 1|\alpha^* - \langle 0|\beta^*\right) \\
&= \tfrac{1}{4}\begin{pmatrix} |\alpha|^2 & \alpha\beta^* \\ \beta\alpha^* & |\beta|^2 \end{pmatrix} + \tfrac{1}{4}\begin{pmatrix} |\alpha|^2 & -\alpha\beta^* \\ -\beta\alpha^* & |\beta|^2 \end{pmatrix} \\
&+ \tfrac{1}{4}\begin{pmatrix} |\beta|^2 & \beta\alpha^* \\ \alpha\beta^* & |\alpha|^2 \end{pmatrix} + \tfrac{1}{4}\begin{pmatrix} |\beta|^2 & -\beta\alpha^* \\ -\alpha\beta^* & |\alpha|^2 \end{pmatrix} \\
&= \tfrac{1}{2}\begin{pmatrix} 1 & 0 \\ 0 & 1 \end{pmatrix}.
\end{aligned}
$$

The density matrix is already diagonal and we can easily conclude that the eigenvalues of ρ are $\lambda_1 = \lambda_2 = \tfrac{1}{2}$ and consequently the amount of ignorance for Bob is $S(\rho) = \sum_{i=1}^{2} -\lambda_i \log_2 \lambda_i = -\tfrac{1}{2}\log_2\tfrac{1}{2} + -\tfrac{1}{2}\log_2\tfrac{1}{2} = 1$ qubit. Thus Bob knows nothing about the teleported qubit.

3. Show that the following set of 16 unitary operations can be used for maximal dense coding in 4-qubit cluster state ($|C\rangle_0 = \tfrac{1}{2}(|0000\rangle + |0011\rangle + |1100\rangle - |1111\rangle)$)

$$
\begin{aligned}
G_2 = \{&I \otimes I,\, I \otimes X,\, I \otimes iY,\, I \otimes Z,\, X \otimes I,\, X \otimes X,\, X \otimes iY,\, X \otimes Z, \\
&iY \otimes I,\, iY \otimes X,\, iY \otimes iY,\, iY \otimes Z,\, Z \otimes I,\, Z \otimes X,\, Z \otimes iY,\, Z \otimes Z\}.
\end{aligned}
$$

Solution: See Table 7.11.

4. Assume that a map of a hidden treasure is given to you. To obtain the treasure you have to guide a small robot to the location of the treasure. There are ten crossings on the path to the treasure. At each crossing the robot can go to left/right/forward/backward. Using the map, you have to guide the robot to choose the appropriate directions at each crossing. You are connected with the robot with a communication device. The restriction on the allowed communication is that at each crossing you can only send a bit or a qubit to the robot. Show that the probability of winning is very small if you play with a classical communication device. However, you will always win if you play with a quantum communication device.
Solution: As there are four possible directions at each crossing so the uncertainty at each crossing is $\log_2 4 = 2$ bits. If we play with a classical device then at each crossing we can communicate only 1 bit. So the probability of choosing the correct path is $\tfrac{1}{2}$. The events are independent and so the probability of choosing the correct path in all ten crossings is only $\left(\tfrac{1}{2}\right)^{10} = 9.7 \times 10^{-4}$ which is very low. However, if we play with quantum communication devices then we

Unitary Operations	$\lvert C\rangle_0 = \frac{1}{2}(\lvert 0000\rangle + \lvert 0011\rangle + \lvert 1100\rangle - \lvert 1111\rangle)$
$U_1 = I \otimes I$	$\frac{1}{2}(\lvert 0000\rangle + \lvert 0011\rangle + \lvert 1100\rangle - \lvert 1111\rangle)$
$U_2 = I \otimes Z$	$\frac{1}{2}(\lvert 0000\rangle - \lvert 0011\rangle + \lvert 1100\rangle + \lvert 1111\rangle)$
$U_3 = Z \otimes I$	$\frac{1}{2}(\lvert 0000\rangle + \lvert 0011\rangle - \lvert 1100\rangle + \lvert 1111\rangle)$
$U_4 = Z \otimes Z$	$\frac{1}{2}(\lvert 0000\rangle - \lvert 0011\rangle - \lvert 1100\rangle - \lvert 1111\rangle)$
$U_5 = I \otimes X$	$\frac{1}{2}(\lvert 0001\rangle + \lvert 0010\rangle - \lvert 1101\rangle + \lvert 1110\rangle)$
$U_6 = I \otimes iY$	$\frac{1}{2}(-\lvert 0001\rangle + \lvert 0010\rangle + \lvert 1101\rangle + \lvert 1110\rangle)$
$U_7 = Z \otimes X$	$\frac{1}{2}(\lvert 0001\rangle + \lvert 0010\rangle + \lvert 1101\rangle - \lvert 1110\rangle)$
$U_8 = Z \otimes iY$	$\frac{1}{2}(-\lvert 0001\rangle + \lvert 0010\rangle - \lvert 1101\rangle - \lvert 1110\rangle)$
$U_9 = X \otimes I$	$\frac{1}{2}(\lvert 0100\rangle - \lvert 0111\rangle + \lvert 1000\rangle + \lvert 1011\rangle)$
$U_{10} = X \otimes Z$	$\frac{1}{2}(\lvert 0100\rangle + \lvert 0111\rangle + \lvert 1000\rangle - \lvert 1011\rangle)$
$U_{11} = iY \otimes I$	$\frac{1}{2}(\lvert 0100\rangle - \lvert 0111\rangle - \lvert 1000\rangle - \lvert 1011\rangle)$
$U_{12} = iY \otimes Z$	$\frac{1}{2}(\lvert 0100\rangle + \lvert 0111\rangle - \lvert 1000\rangle + \lvert 1011\rangle)$
$U_{13} = X \otimes X$	$\frac{1}{2}(-\lvert 0101\rangle + \lvert 0110\rangle + \lvert 1001\rangle + \lvert 1010\rangle)$
$U_{14} = X \otimes iY$	$\frac{1}{2}(\lvert 0101\rangle + \lvert 0110\rangle - \lvert 1001\rangle + \lvert 1010\rangle)$
$U_{15} = iY \otimes X$	$\frac{1}{2}(-\lvert 0101\rangle + \lvert 0110\rangle - \lvert 1001\rangle - \lvert 1010\rangle)$
$U_{16} = iY \otimes iY$	$\frac{1}{2}(\lvert 0101\rangle + \lvert 0110\rangle + \lvert 1001\rangle - \lvert 1010\rangle)$

Table 7.11: Dense coding using 4-qubit Cluster states. The unitary operators operate on the qubits 1 and 3. To convince yourself, you need to check that all the states of the second column are mutually orthogonal.

can adopt the following strategy. Before the start of the game, we can prepare ten copies of $\lvert \psi^+ \rangle = \frac{\lvert 00\rangle + \lvert 11\rangle}{\sqrt{2}}$ and send the second qubit of each Bell state to the robot. The first qubits are kept with us. At the first crossing we shall use the first Bell state and in the second crossing we shall use the second Bell state and so on. At the ith crossing we encode the direction information applying unitary operators $U_{forward} = I, U_{backward} = X, U_{left} = Z, U_{right} = iY$ on the first qubit of the ith Bell state and send that qubit to the robot. The robot can do a Bell measurement on the qubit sent by us and the partner qubit of that which is already with the robot. The Bell measurement will decode the information sent by us. As a result the robot will always choose the correct direction. Essentially we are exploiting dense coding. Clearly the participants with quantum devices will always win in this type of game. However, the probability of winning is very small for classical participants.

5. Consider the HQIS scheme described in Subsection 7.4.1 and show that Bob has no information about the unknown quantum state sent by Alice unless Charlie and Diana send the outcome of their joint measurement.

Solution: Without loss of generality we may assume that Alice's measurement outcome is $\lvert \psi^+ \rangle$. After the measurement of Alice the combined states of Bob, Charlie and Diana is

$$
\begin{aligned}
|\Psi^{+}\rangle_{BCD} \;=\;& \tfrac{1}{2}\big((Z|\psi_s\rangle)_B|\psi^{+}\rangle_{CD} + (I|\psi_s\rangle)_B|\psi^{-}\rangle_{CD} \\
+\;& (X|\psi_s\rangle)_B|\phi^{+}\rangle_{CD} - (iY|\psi_s\rangle)_B|\phi^{-}\rangle_{CD}\big).
\end{aligned} \tag{7.17}
$$

Without access to knowledge of the state with Charlie and Diana, Bob's state is given by the reduced density operator:

$$
\frac{1}{4}\left(|\psi_s\rangle\langle\psi_s| + Z|\psi_s\rangle\langle\psi_s|Z + X|\psi_s\rangle\langle\psi_s|X + iY|\psi_s\rangle\langle\psi_s|iY\right) = \frac{I}{2}, \tag{7.18}
$$

which implies (as in Solved Example 2) that Bob gains no information without the cooperation of Charlie and Diana. Analogous observations hold for the other three cases (i.e., when Alice's measurement outcomes are $|\psi^{-}\rangle$ or $|\phi^{+}\rangle$ or $|\phi^{-}\rangle$, too.

7.8   Further reading

1. A scheme for teleportation of a CNOT gate and its significance in distributed quantum computing is lucidly described in E. Desurvire, Classical and quantum information theory: An introduction for the telecom scientist, Cambridge University Press, New York (2009) 372-376.

2. D. Bouwmeester *et al.*, Experimental quantum teleportation, Nature **390** (1997) 575-579.

3. C. H. Bennett *et al.*, Remote preparation of quantum states, IEEE Trans. Inf. Theor. **50** (2005) 56-74.

4. A. M. Steane and W. v. Dam, Physicists triumph at guess my number, Physics Today, February 2000, 35-39. Also available at http://fy.chalmers.se/~tfkhj/PhysicistsTriumph.pdf.

5. S. L. Braunstein and H. J. Kimble, Teleportation of continuous quantum variables, Phys. Rev. Lett. **80** (1998) 869–872.

6. Readers may find it interesting to see a short video at YouTube: http://www.youtube.com/watch?v=_qmSdC7aQpY.

7. G. Brassard, Teleportation as a quantum computation, Physica D, **120** (1998) 43-47, quant-ph/9605035.

8. N. Aharon and L. Vaidman, Quantum advantages in classically defined tasks, Phys. Rev. Lett. **77** (2008) 052310.

7.9 Exercises

1. In Section 7.2 we have shown that perfect teleportation is possible using $|\psi^+\rangle = \frac{|00\rangle+|11\rangle}{\sqrt{2}}$. Now show that the same is also possible using the other three Bell states.

2. Show that the operation

$$
U = \begin{pmatrix}
\frac{b}{a} & \sqrt{1-\frac{b^2}{a^2}} & 0 & 0 \\
0 & 0 & 0 & -1 \\
0 & 0 & 1 & 0 \\
\sqrt{1-\frac{b^2}{a^2}} & -\frac{b}{a} & 0 & 0
\end{pmatrix}
$$

used in probabilistic teleportation is unitary. You may assume that a and b are real.

3. In analogy to the Solved Example 4, design a dense coding based quantum game in which participants with access to quantum devices will always win, but the probability of winning for participants with classical devices is very small.

4. Show that the following set of 8 unitary operations can be used for dense coding using GHZ-like state ($|\lambda\rangle_{GHZ-like} = \frac{1}{\sqrt{2}}(|010\rangle+|100\rangle+|001\rangle+|111\rangle)$):

$$G_{GHZ-like} = \{I\otimes I,\ I\otimes X,\ I\otimes iY,\ I\otimes Z,\ Z\otimes I,\ Z\otimes X,\ Z\otimes iY,\ Z\otimes Z\}.$$

5. Show the unitary operators that are used for dense coding of GHZ-like state in the previous problem can also be used for dense coding using Q_4 state, given that $|Q_4\rangle = \frac{1}{2}(|0000\rangle+|0101\rangle+|1000\rangle+|1110\rangle)$.

Chapter 8

Quantum cryptography

The word cryptography originates from two Greek words: kryptos (secret) and graphein (writing). Thus cryptography is the art of communicating a message in secret manner or in other words, it is the art of rendering a message unintelligible to any unauthorized party. Its history is very old and exciting. From the beginning of human civilization there was a need to hide information from unauthorized people. This need gave birth to cryptography. Initial cryptographic protocols were very simple. With time they have become much more sophisticated and secure. However, all the classical cryptographic protocols developed so far are secure under some assumptions. In contrast, the protocols of quantum cryptography are unconditionally secure. This particular aspect of quantum cryptography makes it very interesting. This chapter aims to provide a flavor of the intrinsic beauty of various aspects of secure quantum communication. Before we start a journey into the fascinating world of secure quantum communication, it would be apt to become familiar with the jargon used in this field and also with some interesting protocols of classical cryptography. Let us first do that.

8.1 Jargon related to cryptography

To develop an elementary perception about cryptography, we may consider a practical situation in which the king of Atan wants to invade his neighboring state Acos. Now the king of Atan sends some information to his generals present in the battlefield through an emissary. The king must not write his instructions in normal language as he wants it to be understood by his generals only. So the king uses some kind of coding to hide his message. For example, we may think of the following simple ideas: The king may shift each letter of the English alphabet by 3 places, $A \mapsto D$, $B \mapsto E$ and so on, or the king may replace the letters randomly, say $A \mapsto W$, $B \mapsto N$, $\cdots$, $Z \mapsto Q$.

If the generals know the mapping (transformation) used by the king then they can easily decode the message. This kind of encoding is known as substitution cipher or Caesarian cipher as this technique was introduced by Julius Caesar. Substitution cipher is one of the simplest techniques of encoding. Today we use much more sophisticated techniques to encode our message. In general, the art of encoding the message is known as cryptography. If we assume that the emissary of Atan is caught and the king of Acos has obtained the message, then the king of Acos has to try to understand what is written in that secret message. To do so, he has to break the code used by the king of Atan. This is important for the security of his country. This practical need has developed the art of code breaking, which is known as cryptanalysis. Every success in cryptanalysis (cryptography) has motivated the cryptographers (crypto-analyzers) to improve the encryption (code-breaking) techniques. Consequently, both cryptography and cryptanalysis techniques have evolved with time and together they form a subject called cryptology, which is the combination of cryptography and cryptanalysis. We may say, cryptology=cryptography+cryptanalysis.

A cryptosystem or a cipher is an algorithm which combines the message to be encrypted with some additional information known as the key and produces a cryptogram. The most important thing in cryptology is the key. If the key is secure then the cryptogram is secure. This fundamental principle of cryptography is known as Kerckhoff's principle. This principle implies that the security of a cipher depends only on the secrecy of the key and not on the secrecy of the protocol used for encryption. In other words, a cryptosystem can be secure even if everything about the system, except the key, is public knowledge. Now we may elaborate on the concept of key through a very simple example. Consider that Alice has a message string 1101001 and a key 0101101. To encrypt the message Alice does bit-wise XOR operation between her key and the message and gets the encrypted message as 1000100. To obtain the original message Bob uses the same key and performs bit-wise XOR operation between his key and the encrypted message. This decrypts the message and thus Bob obtains the secret message encoded by Alice. The example is also shown in the box below.

Alice's message	1101001
Alice's key	0101101
Encrypted message after bit-wise $\oplus$ operation	1000100
Bob's key	0101101
Decrypted message after bit-wise $\oplus$ operation	1101001

In the above example, both Alice and Bob use the same key. If in a cryptographic protocol the same key is used for both encryption and decryption then such a protocol is called symmetric key cryptographic protocol. Similarly, if in a cryptographic protocol different keys are used for encoding and decoding then the protocol is called asymmetric key cryptographic protocol. Now we know that Alice and Bob need one or more key(s) for encryption and decryption processes. How do they get the key(s)? There are two simple possibilities:

1. **Private key cryptographic system:** Alice and Bob may have met previously and shared a key (or a trusted third party Charlie may have carried it from Alice to Bob). They may use the previously shared key at a later time when it is required. This type of cryptographic system is referred to as a private key cryptographic system. Usually private key cryptographic systems use a symmetric key. Until 1976 only this type of cryptographic system was in existence.

2. **Public key cryptographic system:** The key is generated through discussions over public channels. This type of cryptographic scheme is known as public key cryptography and is usually asymmetric. In 1976, Whitfield Diffie and Martin Hellman [100] introduced the concept of public key cryptography, which allows two people to exchange confidential information even if they have never met and the communication channel used by them is monitored by undesired person/persons. This was a path-breaking idea. The actual implementation of this idea was done by Ronald Rivest, Adi Shamir, and Leonard Adleman in 1978 [101]. The specific implementation is known as RSA and it is frequently used today.

Here we would like to note that a cipher is secure iff it is impossible to unlock the cryptogram without the key. In practice, this impossibility requirement is often weakened to the extent that the system is just extremely hard to crack. In the context of computational complexity in Chapter 2, we mentioned what we mean by a computationally hard problem. A hard problem cannot be solved in polynomial time. But it can be solved in some finite time. The amount of time required to solve a hard problem depends on the size of the problem. Here the computational task is decryption of the message and length of the problem is the size of the key. Exact computational task depends on the specific protocol. For example the task may be factorization of a large composite number. The appropriate size of the key depends on the amount of time for which the message is required to be protected. This point will be clear if we consider a specific example. Assume that the war has already started between Atan and Acos. The king of Atan has instructed his fighter plane pilots to bomb the capital of Acos at 7:30 am the next day. This specific information is required to be safe only until 7:30 am the next day. Thus an encrypted message is protected if the time required to decipher it exceeds its lifetime. The longer we wish to keep a message secret the larger key we have to use for encryption.

Cryptosystems can be classified in different ways. We have already classified cryptographic systems, first as symmetric key and asymmetric key cryptography and then as private key cryptography and public key cryptography. Actually the classification depends on the property chosen for classification. For example, on the basis of ways in which plain text is processed, cryptosystems may be classified as block cipher and stream cipher. As it appears from the names, block cipher encrypts and decrypts messages in blocks whereas the stream cipher encrypts and decrypts one bit or one byte of message at a time. On the basis of number of keys used we may classify cryptographic systems in three classes: (A) no key is used, (B) one key is used, and (C) two or more keys are used. Hash functions based classical cryptosystem belongs to class A. Protocols of quantum secure direct communications also belong to this class. All symmetric key cryptosystems are in class B. Most of the private key cryptosystems are in this class. They may be intentionally prepared to work in class C but that would not give any advantage. All asymmetrical and classical public key cryptographic systems are in class C. Thus the most important classical cryptographic system of present time, i.e., RSA is also in class C.

It is not our purpose to describe a large number of classical cryptographic protocols, but it would be apt to briefly describe a few of them.

8.2   Some interesting classical ciphers

We have described substitution cipher in the previous section and will describe Caesarian square in the solved examples (cf. Solved Example 1). Here are a few more interesting examples of classical ciphers and a glimpse of the history associated with them. We hope this will increase your interest in cryptography.

1. The famous ancient Greek historian Herodotus mentioned some interesting tricks used by Greeks around 400 BC. He mentioned that the Persians had occupied the kingdom of the king Histiaeus. Histiaeus wanted to instruct his people to revolt against the foreign occupation. To send this message, he shaved the head of a slave and tattooed a message on the slave's head. When the slave's hair grew back the king sent him to deliver the message. The receiver of the message shaved the head of the slave to reveal the king's message. This ancient trick is very simple and slow. Another trick mentioned by Herodotus was used by Demaratus who was the king of Sparta. He sent a message to Greece by writing it on a wooden panel and covering that by wax. As wax tablets were commonly used at that time as reusable writing surfaces, nobody suspected it. The receiver melted the wax to reveal the message.

2. Enigma is a class of electro-mechanical cipher machines invented during World War I and used until World War II. During World War II Germans used this type of machines to encrypt their secret messages.

But the German military ciphers were already broken by Polish scientist in 1932, and just before the commencement of World War II Polish scientists transferred their technology to British and French intelligence. Later on, a concentrated effort by Alan Turing and others at the British Government Code and Cypher School at Bletchley Park led to the design of an electro-mechanical decryption device called Bombe, which was successful in deciphering some of the German military messages. Cryptanalysis played an important role in the result of World War II and consequently in the future of humankind.

The classical ciphers discussed so far are historically relevant but are not used now. Today we mostly use the RSA encryption system. Consequently, it would be apt to describe the RSA encryption and decryption schemes with appropriate importance. In the following subsection we describe RSA and its limitations. The discussion will help us to clearly understand many important issues. For example, it will clarify: (i) How does public key cryptography work? (ii) What are the limitations of classical public key encryption schemes? (iii) Why is Shor's algorithm so important? (iv) Why should we use quantum cryptography?

8.2.1 RSA and its limitations

All the examples of classical cryptography that have been discussed until now are examples of private key cryptography. In contrast, RSA is a public key cryptographic system. We can specifically describe RSA protocol in the following 5 steps.

RSA 1: Bob starts with two large prime numbers p and q. He computes their product $n = pq$ and another product $\phi(n) = (p-1)(q-1)$, which is known as totient.

RSA 2: Now he chooses a nontrivial coprime[1] e of $\phi(n)$ and announces e and n publicly. So $\{n, e\}$ is the public key which will be used for encryption.

RSA 3: Bob computes $d = \frac{1 \bmod \phi(n)}{e}$ and uses that to form his private key. The key that he uses later for decryption is $\{d, n\}$, of which n is provided in public channel so essentially d is the private key.
Here it is apt to draw your attention to the point that Eve[2] knows e and n but to construct d she needs $\phi(n)$ and to obtain $\phi(n)$ she needs p and q. In other words, as n is publicly known, if Eve can factorize it and obtain p, q then she can follow the same procedure as used by Bob and obtain d. So if factorization of n can be done

---

[1] By nontrivial coprime we mean that $e \neq 1$ and $e \neq \phi(n)$.

[2] It is a convention to consider Alice, Bob and Eve as sender, receiver and unauthorized user (intruder), respectively.

efficiently then RSA is no more secure and here lies the importance of Shor's algorithm. The moment a scalable quantum computer is built, all RSA encrypted messages will be decrypted by eavesdroppers who will use the quantum computer to implement Shor's algorithm.

RSA 4: Alice encrypts a message m as $c = m^e \bmod n$ and sends it back to Bob over a public channel.

RSA 5: Bob decrypts the message as $c^d \bmod n = m$.

Note that in this protocol all communications between Alice and Bob are public and no pre-sharing of the key is needed. Of course an authentication is needed to ensure that to encode the message Alice is really using the public key generated by Bob. Otherwise, Eve may generate a fake public key and ask Alice to encode her message using that fake key. Now we have seen that Bob can decrypt the message as in addition to the publicly available information he has d, and to obtain d anybody other than Bob needs to factorize n. In Section 2.4 we mentioned that classically factorization is a problem of NPI class. So if p and q are sufficiently large then Eve cannot factorize n in reasonable time (say within the time when the encrypted message is important). Here is an intrinsic assumption that Eve uses number field sieve or a similar method. But this assumption may fail if Eve has an efficient algorithm, which is not known to Alice and Bob or if Eve randomly tries a prime number as factor of n and it works. Thus the security of information in RSA protocol relies on some intrinsic assumptions and consequently, the security is not unconditional. In contrast, quantum keys are unconditionally secure and they don't require any such assumption. This makes quantum cryptography very special. But before we proceed to describe protocols of quantum cryptography let us provide a simple example of RSA encryption and decryption process described above.
Example 8.1:

Step 1: Let $p = 19$ and $q = 17$. So Bob computes $n = pq = 323$ and $\phi(n) = 18 \times 16 = 288$.

Step 2: Bob chooses $e = 13$ as coprime to 288 and sends e and n to Alice using public channel.

Step 3: d is computed by Bob using $d = \frac{1 \bmod \phi(n)}{e} = \frac{1+x\phi(n)}{e} = \frac{1+288x}{13}$. For $x = 6$, we have $d = 133$.

Step 4: If Alice's message is 23, then she encodes her message as $c = 23^{13} \bmod 323 = 180$ and publicly announces the value of c.

Step 5: Bob decodes the message as $m = c^d \bmod n = 180^{133} \bmod 323 = 23$.

This is how RSA works[3]. It is an excellent protocol. However, the security is not unconditional and it is not secure if we can build a scalable quantum computer. Thus there are limitations of RSA (similar limitations exist for all existing classical cryptographic schemes) and these limitations motivate us to look for unconditionally secure cryptographic schemes. Interestingly, quantum mechanics provides us such schemes. In the remaining part of this chapter we will describe different quantum mechanical proposals to overcome the limitations of classical cryptography and to obtain unconditionally secure quantum cryptographic systems.

8.3 Different aspects of quantum cryptography

It is beyond the scope of the present book to describe the enormously rich and fascinating history of classical cryptography. However, we have tried to provide a flavor of it in the previous section. In this section we will try to introduce different aspects of quantum cryptography.

In Section 1.3 we discussed the interesting publication history of the 1970 work of S. Wiesner. Publication of that paper in 1983 [11] was the beginning of quantum cryptography. In 1984, C. H. Bennett and G. Brassard [18] provided a complete protocol of unconditionally secure quantum key distribution (QKD). The protocol of Bennett and Brassard, which is now known as BB84 protocol, drew considerable attention from the cryptographic community since the unconditional security of the key obtained in this protocol was unachievable in classical cryptography. Naturally, since 1984 several new protocols for different cryptographic tasks have been proposed.

In QKD protocols only a key is generated and distributed by quantum means. Subsequently, the key is used in some classical encryption techniques to encrypt the message. As the generated key is unconditionally secure, so Kerckhoff's principle ensures the unconditional security of the message. Most of the initial works on quantum cryptography [9, 18, 91, 102], were concentrated around QKD. Eventually, it came to be understood that quantum states can also be employed for other cryptographic tasks, for example, for quantum secret sharing (QSS) of classical secrets [90]. It was also realized that secure quantum communication is possible without prior generation of key. In other words, it was observed that it is possible to design protocols of direct secure quantum communication. To be precise, in 1999 a protocol for deterministic secure quantum communication (DSQC), using entangled photon pairs, was first proposed by Shimizu and Imoto

---

[3]To convince yourself you may practice with different choices of p and q. It would be convenient to compute mod by using Mathematica, MATLAB or other similar programs. For example, in Mathematica you can quickly compute $180^{133} \bmod 323$ using the command $\text{Mod}\left[180^{133}, 323\right]$.

[103]. Although it was found to be insecure, it indicated that the prior generation of key (i.e., QKD) can be circumvented and protocols for unconditionally secure direct quantum communication of the message can be designed. Subsequently, many such protocols were proposed. These protocols of secure direct quantum communication can be broadly divided into two classes: (a) those for quantum secure direct communication (QSDC) [104, 105, 106] and (b) those for DSQC [107, 108].

In DSQC, the receiver (Bob) can read out the secret message only after receipt of a *pre-key*: additional classical information of at least one bit for each qubit transmitted by the sender (Alice). In contrast, when no such additional classical information (pre-key) is required for decryption, then the secure quantum communication protocol is referred to as QSDC protocol [109]. We have already mentioned that a conventional QKD protocol generates the unconditionally secure key by quantum means but then uses classical cryptographic resources to encode the message. No such classical means are required in DSQC and QSDC protocols. This interesting feature of DSQC and QSDC protocols has motivated several groups to study different aspects of DSQC and QSDC protocols in detail [[109] and reference therein].

In all these QSDC and DSQC protocols, the meaningful information (secret message) travels only from Alice to Bob[4]. Thus the flow of information is unidirectional (one-way) only. In other words in these protocols, Alice and Bob cannot simultaneously transmit their different secret messages to each other (dialogue) and consequently the development of these protocols naturally leads to a question: Is it possible to extend these protocols for bidirectional quantum communication in which both Alice and Bob will be able to communicate (in a secure way) using the same quantum channel? Such bidirectional protocols are quantum dialogue protocols, where information can flow simultaneously along two directions (i.e., from Alice to Bob and from Bob to Alice). Such protocols are actually an essential requirement of our everyday communication problems. This can be visualized more clearly if we consider the analogy of a telephone. The possibility of extending the DSQC and QSDC protocols and the absolute need of bidirectional quantum communication motivated the quantum communication community to investigate the possibility of designing quantum dialogue protocols. The first protocol of quantum dialogue was proposed by Ba An [110] using Bell states in 2004. Subsequently, it was found that the protocol is not secure under intercept-resend attack [111]. However, the modification made in [111] failed to solve the problem since it lost the feature of dialogue (i.e., direct communication). In this connection, satisfactory improvements to the initial quantum dialogue protocol of Ba An [110] was obtained by Ba An himself in 2005 [112]. Here it is important to

---

[4]The protocol may be a two-way protocol like the ping-pong protocol [105] or LM05 protocol [106] but the meaningful information (message) is transmitted from Alice to Bob only.

note that quantum dialogue protocols are not as secure as QKD, DSQC and QSDC protocols are as quantum dialogue always involves some information leakage. We will return to this point later. We have already learned about different aspects of quantum cryptography, e.g., QKD, DSQC, QSDC and quantum dialogue (QD). Now we will briefly recall the principles of quantum mechanics that are intrinsically used to design the protocols of QKD, DSQC, QSDC and QD.

8.3.1 Quantum cryptography: The art of getting positive results from the negative rules of quantum mechanics

From the introductory days of quantum mechanics many people were unhappy with its probabilistic nature and with the limitations intrinsic to this probabilistic theory. But quantum mechanics took with one hand and has paid back with the other hand. The null results of quantum mechanics are found useful in providing unconditional security. Actually security of all secure quantum communication protocols essentially arises from (a) noncommutativity and (b) nonrealism. Before we elaborately discuss the security of QKD protocols, let us quickly recall the relevant null results of quantum mechanics.

1. **Nocloning theorem:** Nocloning theorem states that one cannot duplicate an unknown quantum state. This implies a restriction on Eve that she cannot make a copy of the qubit sent by Alice and keep it for future use. In general, in any secure quantum communication protocol the encrypted information or the key is sent in two or more pieces. Each piece is nonrevealing of the information and the protocol ensures that Eve does not get simultaneous access to all the pieces, and that makes the protocols unconditionally secure. We will elaborate on this point later, when we describe the origin of security in specific protocols.

   Another important point that we would like to mention is that often people argue that we don't need to use QKD until a scalable quantum computer is discovered, as RSA is more or less safe until then. This argument is wrong as one can make copies of the secret classical information communicated now and decipher them at a later time (say after 50 years) when a scalable quantum computer is built. Such a situation is not desirable, as every government wishes to keep some information secure for a much longer period. Consequently, even today it is reasonable to use QKD devices for the encryption of information having a longer lifetime.

2. **Noncommutativity:** One cannot simultaneously and accurately measure the values of two noncommuting observables. Its significance in QKD can be elaborated through the following example.

Example 8.2: Consider the single qubit measurement operators in computational basis $\{|M_0\rangle = |0\rangle\langle0|, |M_1\rangle = |1\rangle\langle1|\}$ and in diagonal basis $\{|M_0\rangle = |+\rangle\langle+|, |M_1\rangle = |-\rangle\langle-|\}$. Now it is easy to check that

$$[M_0, M_1] = M_0 M_1 - M_1 M_0 = |0\rangle\langle0|1\rangle\langle1| - |1\rangle\langle1|0\rangle\langle0| = 0,$$

$$[M_+, M_-] = M_+ M_- - M_- M_+ = |+\rangle\langle+|-\rangle\langle-| - |-\rangle\langle-|+\rangle\langle+| = 0,$$

but

$$
\begin{aligned}
[M_+, M_0] &= M_+ M_0 - M_0 M_+ \\
&= |+\rangle\langle+|0\rangle\langle0| - |0\rangle\langle0|+\rangle\langle+| \\
&= \frac{|+\rangle\langle0|-|0\rangle\langle+|}{\sqrt{2}} \neq 0.
\end{aligned}
$$

Similarly, $[M_+, M_1] \neq 0$, $[M_-, M_0] \neq 0$, and $[M_-, M_1] \neq 0$. Thus we cannot simultaneously and accurately measure a state in computational basis and diagonal basis. This implies that we cannot simultaneously measure the polarization of a photon in vertical-horizontal basis and also in diagonal basis.

Example 8.3: Consider the measurement operators corresponding to a basis set $\{|a_i\rangle\}$ and corresponding to another basis set $\{|b_j\rangle\}$. Now accurate and simultaneous measurement of a state in these bases is possible if

$$\left[M_{a_i}, M_{b_j}\right] = |a_i\rangle\langle a_i|b_j\rangle\langle b_j| - |b_j\rangle\langle b_j|a_i\rangle\langle a_i| = 0$$

for all values of i and j. This implies $\langle a_i|b_j\rangle = 0$ for all values of i and j. If this condition is not satisfied then $\{|a_i\rangle\}$ and $\{|b_j\rangle\}$ are called mutually unbiased bases (MUBs) or nonorthogonal bases.

The noncommutativity principle leads to the uncertainty principle and it also leads to a theorem elaborated on the next page. The theorem states that two nonorthogonal states cannot be discriminated with certainty, which is obvious from the last two examples. This provides the backbone of BB84, B92, LM05 and all other conjugate coding based protocols.

3. **Nonrealism:** This is the most fundamental source of quantum security. Wave function collapse or state vector reduction principle is a manifestation of the nonrealisitic (nondeterministic) nature of quantum mechanics. Because of this principle one cannot perform a measurement without perturbing the system. Since Eve cannot make a copy of the qubit sent by Alice, she has to measure the qubit to know what information Alice has sent. The moment she measures a qubit, the wave function (superposition state) collapses to one of the possible states and the system is perturbed. Later Alice and Bob can measure their states and compare the outcomes to find whether the state was perturbed due to the measurement of Eve or not. This

is how eavesdropping is checked. The ability to detect any eaves-dropping attempt isolates quantum cryptography from its classical counterpart and makes it unconditionally secure. The security of the Goldenberg Vaidman (GV) protocol of QKD and all other orthogonal state based protocols arises from nonrealism alone. In contrast, BB84 and similar conjugate coding based protocols use both nonrealism and noncommutativity. These ideas will be clarified when we describe the protocols.

Theorem: Two nonorthogonal states cannot be discriminated with certainty.

Logical proof: If $|\psi_1\rangle$ and $|\psi_2\rangle$ are not orthogonal then $|\psi_2\rangle$ can always be decomposed into a nonzero component parallel to $|\psi_1\rangle$ and components orthogonal to $|\psi_1\rangle$. Consequently, even if your projective measurement yields $|\psi_1\rangle$, you will not be sure whether it is $|\psi_1\rangle$ or $|\psi_2\rangle$. This is easy to visualize in two dimensions. This inability in discriminating two nonorthogonal states is an important resource for quantum cryptography. Let us try to visualize it. Assume that Alice prepares one copy of $|\psi_1\rangle$ or $|\psi_2\rangle$ at random and sends it to Bob. Bob projects it to $|\psi_1\rangle$ or $|\psi_2\rangle$. If he projects it to the correct state then the result should be perfectly correlated with the initial state. But if Eve exists in between and Eve also projects the state communicated by Alice to $|\psi_1\rangle$ or $|\psi_2\rangle$, then Eve's measurement will break the above-mentioned perfect correlation between Alice and Bob. Consequently, Alice and Bob will be able to detect the existence of Eve by comparing the measurement outcomes of Bob with the initial states sent by Alice. The probability of detecting Eve will be maximum if we choose the nonorthogonal states in such a way that $\langle\psi_1|\psi_2\rangle = \frac{1}{\sqrt{2}}$. This is exactly what we do in the BB84 protocol.

With this background of classical cryptography and quantum mechanical principles that yield quantum security we are now ready to learn quantum cryptographic protocols. To begin with let us try to develop a simple minded protocol of QKD in the next section. Limitations of that will help us to systematically introduce several unconditionally secure protocols of QKD in the subsequent sections.

8.4   Let us develop protocols of QKD

We have already learned some basics of quantum cryptography in the previous two sections and in Chapter 3 we discussed the idea of entanglement. Let us now combine these ideas and introduce a simple minded protocol of QKD.

Protocol 1:

Step 1: Alice prepares $|\psi^+\rangle^{\otimes n}$, i.e., Alice prepares n copies of Bell state[5] $|\psi^+\rangle = \frac{1}{\sqrt{2}} [|0_A 0_B\rangle + |1_A 1_B\rangle] = \frac{1}{\sqrt{2}} [|+_A +_B\rangle + |-_A -_B\rangle]$. She keeps the first qubit of all the entangled pairs with her (home qubit) and sends all the second qubits (travel qubits) to Bob.

Step 2: After Bob confirms that he has received all the qubits sent to him, Alice (Bob) measures all the qubits available with her (him) in the computational basis. The measurement will destroy the entanglement and create a random symmetric key.

Alice can send the qubits all at once or one after the other. Sending one after another is much more practical. The only restriction imposed here is that Alice and Bob must be able to establish a one-to-one correspondence between the transmitted and the received qubits. This restriction is applicable to all the protocols described in this chapter.

Since quantum measurement randomly collapses a superposition state into a possible basis state, so a particular measurement in computational basis made by Alice or Bob would collapse $|\psi^+\rangle$ randomly into $|00\rangle_{AB}$ or $|11\rangle_{AB}$. As a result of such measurements both Alice and Bob will obtain truly random strings of 0 and 1. Thus the generated key would be random[6]. Further, after a particular measurement in the computational basis the states of Alice and Bob are always the same. Consequently, the random strings obtained by Alice and Bob will be perfectly correlated. Thus the generated key is random and symmetric. If we start from $|\phi^\pm\rangle^{\otimes n} = \frac{1}{\sqrt{2^n}} [|0_A 1_B\rangle \pm |1_A 0_B\rangle]^{\otimes n}$ then the generated key of Bob will be conjugate to that of Alice and it will be a straightforward exercise for them to generate a symmetric key from that.

In simple words we can describe this protocol as: Alice prepares n copies of Bell states. She keeps one photon[7] of each Bell state with herself (home photon) and sends the other (travel photon) to Bob and after Bob confirms

---

[5]It is easy check that

$$
\begin{aligned}
|\psi^+\rangle &= \tfrac{1}{\sqrt{2}} [|00\rangle + |11\rangle] \\
&= \tfrac{1}{2\sqrt{2}} [(|+\rangle + |-\rangle) \otimes (|+\rangle + |-\rangle) + (|+\rangle - |-\rangle) \otimes (|+\rangle - |-\rangle)] \\
&= \tfrac{1}{\sqrt{2}} [|++\rangle + |--\rangle].
\end{aligned}
$$

[6]Also true for measurements on diagonal basis.

[7]In this chapter we have used photon and qubit as synonymous. This is reasonable as all the existing experimental implementations of QKD utilize photon as qubit.

that he has received all the photons, Alice (Bob) measures her (his) states and that creates a random symmetric key. This protocol appears very simple and consequently a question arises in our mind: Is there anything wrong with this simple minded approach? The answer is yes. What is wrong? It will be clear only if we try to attack this protocol. Let us try to do that.

8.4.1   Let us attack Protocol 1

Assume that Eve measures the travel qubit, notes the outcome and allows the collapsed state to go to Bob. Now all measurements of Alice and Bob will be perfectly correlated to Eve's result. Thus Eve has a copy of the key. At this point it seems a serious flaw. We have to modify protocol 1 and make it stronger, and to do so we must try to detect Eve. Now the question is: Is it possible to detect Eve? The answer is yes, but Alice and Bob have to change their strategy. They may adopt the following strategy.

Modified strategy: Alice and Bob choose two or more nonorthogonal bases, i.e., mutually unbiased bases (MUBs). Bob measure his qubits using these MUBs at random and announces which basis he has used to measure a particular qubit. Alice measures the partner particles (qubits) using the same basis. They use part of the string generated in this process as a verification string and compare the outcomes of the verification string to calculate the value of a correlation function. For a particular choice of entangled state and MUBs quantum mechanics will provide a particular value of correlation function. If it matches, then there is no eavesdropping. If it does not match then eavesdropping is occurring.

Now we may modify Protocol 1 using the above described strategy and obtain Protocol 2, which works as follows:

Protocol 2:

Step 1: Same as **Step 1** of Protocol 1.

Step 2: Bob randomly measures all the incoming qubits in $\{0, 1\}$ or $\{+, -\}$ basis and announces which basis he has used to measure a particular state.

Note that the measurements of Bob will destroy the entanglement and Bob will obtain a random string of $|0\rangle, |1\rangle, |+\rangle$ and $|-\rangle$.

Step 3: Alice measures the partner qubits using the same basis (as is used and announced by Bob).

Thus Alice also obtains a random string of $|0\rangle, |1\rangle, |+\rangle$ and $|-\rangle$. The strings of Alice and Bob are expected to be symmetric in an ideal case. Conventionally, one attributes the binary value 0 to states $|0\rangle$ and $|+\rangle$ and binary value 1 to the other two states. Thus in an ideal scenario (in absence of Eve and noise) a symmetric and random key is generated.

Step 4: Bob uses part of the generated string as verification string and publicly announces the results of measurement of those qubits along with their positions. Alice compares these results with her own results. If errors more than the tolerable limit are detected then the protocol is discarded. Otherwise, a perfectly symmetric, random and unconditionally secure quantum key is obtained.

Note that in an ideal scenario, the outcomes of Alice and Bob would be the same and they would be able to generate a secure key. The presence of Eve would lead to errors (mismatches). Since nocloning theorem prohibits Eve to make a copy of the travel qubit[8], Eve has to measure the travel qubits to know the key. After the measurement she has to note the result or make a copy of it and send it to Bob. This type of attack is known as measurement-resend attack. Now since the choice of basis is random, half of the time Eve and Bob will choose different bases and that would lead to a situation where 25% of Alice's outcomes would be different from that of Bob even when the same basis is used by Alice and Bob. Let us visualize it with a specific example.

Example 8.4: Suppose Alice has prepared $|\psi^+\rangle = \frac{1}{\sqrt{2}} [|0_A 0_B\rangle + |1_A 1_B\rangle] = \frac{1}{\sqrt{2}} [|+_A +_B\rangle + |-_A -_B\rangle]$. Eve has measured the state in $\{+, -\}$ basis and has obtained $|+\rangle$ as outcome. Eve has sent the collapsed state to Bob. The measurement of Eve has also reduced the state of Alice to $|+\rangle$. Now 50% of the time Bob will choose $\{0, 1\}$ basis (i.e., a basis that is different from the one used by Eve). Say after measuring his qubit in $\{0, 1\}$ basis Bob has obtained $|1\rangle$. As Bob announces the basis, Alice measures her state $|+\rangle$ using the same basis as is used by Bob (in this case $\{0, 1\}$ basis). Now only 50% of the time Alice's state $|+\rangle$ would reduce to $|1\rangle$ and the outcome of Alice and Bob would match. In the remaining 50% of the time Alice would get $|0\rangle$, and her outcome would not match with that of Bob. Thus if Eve measures all the travel qubits, then on half of the occasions the basis sets of Eve and Bob would not match and on half of those occasions (where Eve and Bob have chosen different bases), Alice's and Bob's results would not match, consequently 25% of the time Alice's and Bob's results would be different. This difference is a signature of Eve. This is how Eve can be detected. This protocol will work for other Bell states, too[9].

---

[8] A qubit that remains in the sender's port is called a home qubit and a qubit that is transmitted from a sender to the receiver (i.e., a qubit that travels through the channel) is called a travel qubit.

[9] Alice can start with any other Bell states as

$$|\psi^-\rangle = \frac{1}{\sqrt{2}} [|00\rangle - |11\rangle] = \frac{1}{\sqrt{2}} [|+-\rangle + |-+\rangle],$$

$$|\phi^+\rangle = \frac{1}{\sqrt{2}} [|01\rangle + |10\rangle] = \frac{1}{\sqrt{2}} [|++\rangle - |--\rangle],$$

The security of this protocol originates from the use of nonorthogonal bases as it is secure because of Eve's inability to deterministically discriminate two nonorthogonal states (say $|0\rangle$ and $|+\rangle$). The simple strategy adopted here intrinsically contains the backbone of a few more famous protocols of QKD, e.g., BB84 protocol, Ekert's protocol and B92 protocol. In Protocol 2, we have generated a random string of $|0\rangle$, $|1\rangle$, $|+\rangle$ and $|-\rangle$ by randomly measuring Bell states using $\{0,1\}$ or $\{+,-\}$ basis. As the security of the protocol arises from the inability of Eve to discriminate or clone the nonorthogonal states, the information available to Eve would remain the same if we can devise another mechanism in which we ensure that Eve has access to only a random string of $|0\rangle$, $|1\rangle$, $|+\rangle$ and $|-\rangle$. Just modify the above protocol in such a way that Alice prepares the same entangled states but instead of sending an entangled qubit to Bob, she measures her qubits first randomly in $\{0,1\}$ or $\{+,-\}$ basis and then sends all the second qubits to Bob. If we apply this modification then we obtain a modified protocol which works as follows:

Protocol 3:

Step 1: Alice prepares $|\psi^+\rangle^{\otimes n}$ and randomly measures all the first qubits in $\{0,1\}$ or $\{+,-\}$ basis and sends all the second qubits to Bob.
As the measurement has already destroyed the entanglement, Alice essentially sends Bob a random string of $|0\rangle, |1\rangle, |+\rangle$ and $|-\rangle$. A copy of the same string remains with Alice.

Step 2: Same as **Step 2** of Protocol 2. Bob's random number generator is independent from that of Alice.

Step 3: Alice informs Bob in which cases Bob's basis are the same as that of Alice. They keep those qubits that are measured using the same basis and discard the rest.
50% of the time Bob's basis will be same as that of Alice. The strings kept by Alice and Bob are expected to be perfectly symmetric in an ideal case.

Step 4: Same as **Step 4** of Protocol 2.

Note that here we are not sending entangled qubits, we are just using them to generate a random string of four states from two nonorthogonal bases. These bases are chosen in such a way that the elements of a basis set are maximally conjugate with the elements of the other set. To be precise, $\langle 0|+\rangle = \langle 0|-\rangle = \langle 1|-\rangle = \langle 1|+\rangle = \frac{1}{\sqrt{2}}$. This can even be done without using entanglement. This simplification reduces our simple minded protocol to

---

and

$$|\phi^-\rangle = \frac{1}{\sqrt{2}}\left[|01\rangle - |10\rangle\right] = \frac{1}{\sqrt{2}}\left[|+-\rangle - |-+\rangle\right].$$

In all these cases outcomes of Alice and Bob will be perfectly correlated/anticorrelated in the absence of Eve.

an exactly equivalent but much more famous protocol which is known as BB84 protocol.

8.4.2 Protocol 4: BB84 protocol

BB84 1: Alice prepares a random string of $|0\rangle, |1\rangle, |+\rangle$ and $|-\rangle$ and sends that to Bob.

BB84 2: Same as **Step 2** of Protocol 3.

BB84 3: Same as **Step 3** of Protocol 3.
    Here they keep those qubits which are measured by Bob using the same basis as was used by Alice to prepare them.

BB84 4: Same as **Step 4** of Protocol 3.

It is straightforward to see that Protocol 3 and Protocol 4 are the same[10]. The only difference lies in the process of generation of random string of $|0\rangle, |1\rangle, |+\rangle$ and $|-\rangle$. Thus starting from a simple minded protocol, we have obtained BB84 protocol. The process of reduction seems interesting. But the story does not end here: we can also use three sets of nonorthogonal states (instead of two sets), then Protocol 2 described above transforms to Ekert's protocol. In Ekert's protocol we assume that there exists a source of entangled photon between Alice and Bob. It sends one photon to Alice and the other to Bob. It is equivalent to our simple minded approach. Now if Alice and Bob use three MUBs to measure the qubits and find the correlation function's value to detect eavesdropping, then the protocol is called Ekert's protocol. When it uses two MUBs it is equivalent to BB84.

We have already discussed that if Eve attacks all the travel qubits then it introduces a 25% error which is quite high and Eve will definitely be detected. To avoid this possible detection, Eve may change her strategy and attack a lesser number of travel qubits. For example, she may measure 10% of the travel qubits. Then following the same logic as above, we can obtain that Eve's attack will introduce a 2.5% error. It would be difficult for Alice and Bob to identify Eve in this case as there are some errors due to noise, too. So Alice and Bob also modify their strategy and take some steps for error correction and privacy amplification. The simplest way to realize these two ideas is to use XOR operations between two bits. For example, they may start by an error correction scheme in which Alice randomly chooses two bits and announces their coordinates and their XOR

---

[10]A usual convention is to denote computational basis and diagonal basis as $\oplus$ basis and $\otimes$ basis, respectively. We may better visualize the BB84 protocol if we assume that in $\oplus$ basis binary 1 and 0 corresponds to photon with polarization angle 0° and 90°, respectively. Similarly, in $\otimes$ basis binary 1 and 0 corresponds to photon with polarization angle 45° and 135°, respectively. Thus **BB84 1** may be visualized as a process in which Alice prepares and sends a set of photons, each of which is randomly polarized in one of the four possible polarization angles, i.e. 0°, 45°, 90°, and 135°.

value (i.e., their sum modulo 2). Bob performs the same operation using the same bits and if he obtains the same result then he replies accept, otherwise he replies discard. When Bob's response is accept then Alice and Bob keep the first bit of the pair and discard the second; in the other case (i.e., when Bob's response is discard) they discard both the bits. Still Eve can have some information. To reduce the information available to Eve, privacy amplification protocol which is analogous to the error correction protocol described above, is applied by Alice and Bob. In the simplest privacy amplification protocol Alice randomly chooses two bits and computes their XOR value but announces their coordinates only. She does not announce the XOR value. Now Bob also computes the XOR value and they replace the two bits by their XOR value. This reduces the key size but increases the privacy of the key. This is why it is called privacy amplification protocol. There exist many sophisticated privacy amplification protocols but we have described the simplest possible one as it clearly describes the idea of privacy amplification.

If the error introduced by Eve is below the tolerance limit, then Eve has measured only a small fraction of the qubits. But still Eve has some information. Now for privacy amplification when Alice chooses two arbitrary bits then it's highly unlikely that Eve has both the bits, and unless she has both the bits she has no information about their XOR value. For example, assume that Alice has chosen 13th and 37th bits and computed the XOR value. Eve knows that the 13th bit is 0 but Eve does not know the 37th bit. In this case Eve cannot guess anything about the XOR value as $0 \oplus 0 = 0$ and $0 \oplus 1 = 1$. Thus privacy amplification reduces Eve's information. If required, one may apply repeated cycles of privacy amplification protocol to reduce Eve's information to an arbitrarily small value.

We could have directly described BB84 protocol but we have followed a different path to provide you a perception of how cryptanalysis can motivate us to improve our cryptographic protocol. To be precise, here we have seen that an excellent protocol (BB84) can be developed from a simple minded idea and the strategy of attack and counter strategy to detect the attack. This is how things work in cryptography. The path followed here is imaginary, but through this simple path we have also seen that cryptanalysis is very important for the development of cryptology. Here we would also like to note that eavesdroppers are not really bad people, they have often saved their countries from the enemies.

8.4.2.1 Elementary idea of decoy state

BB84 protocol has a few implementational issues. It works in ideal situation. It requires a single photon source but no such source exists at the moment. Consequently, photon number splitting (PNS) attack is possible. In PNS attack Eve measures the number of photons of each pulse. When it is one, she just blocks it. When it is more than one then she splits the photons and keeps at least one photon with herself and allows the rest of the photons to reach Bob. At a later time when Bob announces the basis used by him to measure a particular qubit, Eve uses the same basis to measure the corresponding qubit that she has obtained by PNS attack. Eve will remain undetected as this attack does not perturb the state obtained by Bob.

Here we would like to draw your attention to the fact that Eve is restricted by natural laws only but Alice and Bob are restricted by existing technology. No law of nature stops Eve from detecting the number of photons without destroying the state of the photons (as the corresponding operators commute). However, in practice all the photon number counters destroy the state of the photon. If it was not so then Alice could have used the opposite strategy and counted the photon number in every pulse and stopped the multiphoton pulses and sent the single photon pulses only. This strategy could have solved the problem of single photon source. Alice adopts the following strategy to circumvent PNS attack and to distinguish Eve and noise.

Alice intentionally and randomly mixes multiphoton pulses (decoy-pulses) with her signal pulses, which are generated from an approximate single photon source. As Eve cannot detect which one is a signal pulse and which one is a decoy pulse, she applies PNS attack to both. Eavesdropping will cause a considerable loss in signal pulse as it contains a single photon most of the time. But it will not cause similar loss to decoy-pulse. On the other hand, the effect of noise will be similar to both kinds of pulses. Consequently, if the loss of the decoy pulses is found to be considerably less than that of signal pulses, then we conclude that Eve is present and the whole protocol is aborted. Otherwise we continue.

The notion of decoy qubits is broadened in the context of DSQC and QSDC where additional qubits prepared in $|0\rangle, |1\rangle, |+\rangle$ and $|-\rangle$ or in Bell states are randomly inserted in the message string. The additional qubits inserted for detection of eavesdropping are called decoy qubits but they are not essentially multiphoton pulses.

Now we have an unconditionally secure protocol of QKD. But still there are some simple questions that have not been addressed. The fundamental question is: What minimum resources are required to achieve unconditional security? It is difficult to answer this question at this point but we may try to answer a few related questions. To begin with, we may note that in Protocol 2-Protocol 4, we have used four states, ($|0\rangle, |1\rangle, |+\rangle$ and $|-\rangle$) and we have mentioned that Ekert's protocol uses six states. This observation encourages us to ask: Do we need at least four states to obtain unconditional security by using conjugate coding? The answer is no. This is easy to visualize as Eve cannot deterministically discriminate between $|0\rangle$ and $|+\rangle$, $|0\rangle$ and $|-\rangle$, $|1\rangle$ and $|+\rangle$, $|1\rangle$ and $|-\rangle$. Alice can prepare her random string by using one state from one basis set and the other from a nonorthogonal basis set. The scheme would remain equally secure. For example, assume that Alice has a random number generator, which generates 0 and 1 randomly. If Alice's random number generator yields 0 then she sends $|0\rangle$ to Bob and if the random number generator yields 1 then she sends $|+\rangle$ to Bob. Thus Alice sends Bob a random string of $|0\rangle$ and $|+\rangle$. The encoding is pre-decided but not unique. We mean that Alice and Bob know that $|0\rangle$ and $|+\rangle$ are associated with bit values 0 and 1, but this mapping is not unique; they could have agreed to use $|0\rangle$ and $|-\rangle$ or $|1\rangle$ and $|-\rangle$, etc. to encode the bit values. Such a protocol where only two nonorthogonal states are used was first introduced by Bennett in 1992 [91] and is known as B92 protocol. Here we would like to note that straightforward use of BB84 protocol with two states will not work. However, simple modifications of the BB84 protocol can covert it to B92 protocol. Let us see how:

8.5 Protocol 5: B92 protocol

B92 1: Alice sends Bob a random string of $|0\rangle$ and $|+\rangle$. We may assume that $|0\rangle$ corresponds to bit value 0 and $|+\rangle$ corresponds to bit value 1.

This step is the same as **BB84 1** with the only difference that instead of a random sequence of 4 states ($|0\rangle, |1\rangle, |+\rangle$ and $|-\rangle$) Alice sends a random sequence of four states ($|0\rangle$ and $|+\rangle$).

B92 2: Bob measures all the incoming qubits randomly in one of the bases $\{0, 1\}$ or $\{+, -\}$.

This step is the same as **BB84 2** with the only difference that Bob does not announce the basis used to make a measurement.

B92 3: Bob keeps all those qubits where his measurement outcome is $|1\rangle$ or $|-\rangle$ and discards all other qubits. He announces which qubits are to be kept. Following Bob's announcement, Alice also keeps corresponding qubits (partner qubits) and discards the rest.

If Bob's measurement yields $|0\rangle$ he will not be able to conclude whether Alice has sent $|0\rangle$ or $|+\rangle$ as 50% of the time the $|+\rangle$ state measured in computational basis will collapse to $|0\rangle$. Consequently, if Bob's measurement yields $|0\rangle$ then he cannot conclude anything about the encoding of Alice. Further, if Alice sends $|0\rangle$ then Bob can never get it as $|1\rangle$. This is so because if Bob chooses computational basis he will always get it as $|0\rangle$ and if he chooses diagonal basis then with equal probability he will obtain $|+\rangle$ or $|-\rangle$ state. Thus Bob's measurement can yield $|1\rangle$ if and only if Alice has sent $|+\rangle$. Therefore, whenever Bob gets $|1\rangle$ he can conclude that Alice has sent $|+\rangle$. Similarly, whenever Bob's measurement yields $|-\rangle$ he can conclude that Alice has sent him $|0\rangle$. He cannot conclude anything whenever his measurement yields $|+\rangle$. As Bob knows the encoding, the remaining string can be used to generate a random symmetric key.

B92 4: Same as **BB84 4** with the only difference that Alice's outcomes and Bob's outcomes are expected to be related as $|0\rangle_A \to |-\rangle_B$ and $|+\rangle_A \to |1\rangle_B$. Any deviation from this mapping would be considered as eavesdropping.

We have mentioned the similarity of B92 with B84 to make it visible that some simple modification of BB84 can lead to B92. Further, to visualize security of the protocol, we may provide a specific example.

Example 8.5: Consider that Alice has sent $|0\rangle$. 50% of the time Eve will measure it in wrong basis. Say in a particular case she has measured it as $|+\rangle$ and has sent it to Bob. If Bob measures it in diagonal basis then he will get it as $|+\rangle$ and he will discard the state. If Bob measures it in computational basis and obtains $|1\rangle$ then he keeps it and communicates the result to Alice who will immediately recognize that it does not satisfy the expected relation between Alice's input and Bob's outcome (i.e., it does not satisfy $|0\rangle_A \to |-\rangle_B$). This indicates eavesdropping. Thus the eavesdropping effort will be detected by Alice and Bob. If sufficiently few errors are detected then only the key is used for encryption. Consequently, ability to detect any attempt of eavesdropping makes B92 protocols unconditionally secure.

All the protocols described until now (i.e., Protocol 2-Protocol 5) use nonorthogonal states (conjugate coding). This leads to a question: Is conjugate coding essential for QKD? The answer is no. To be precise, GV protocol does not require nonorthogonal states. This orthogonal state based protocol is described in the next section.

8.6 GV protocol: QKD with orthogonal states

The GV protocol was introduced by Goldenberg and Vaidman [102] in 1995. This is a unique protocol and is fundamentally different from all the protocols described until now. The GV protocol is our 6th protocol, which works as follows.

Protocol 6:

Let $|a\rangle$ and $|b\rangle$ be two localized wave packets and linear combinations of these two wave packets yield two orthogonal states

$$|\psi_0\rangle = \tfrac{1}{\sqrt{2}} (|a\rangle + |b\rangle) \tag{8.1}$$

and

$$|\psi_1\rangle = \tfrac{1}{\sqrt{2}} (|a\rangle - |b\rangle), \tag{8.2}$$

which represent bit values 0 and 1 respectively. Wave packets $|a\rangle$ and $|b\rangle$ are sent by Alice to Bob along two different channels. Alice sends Bob either $|\psi_0\rangle$ or $|\psi_1\rangle$, but in this communication procedure $|a\rangle$ and $|b\rangle$ are not sent together. Say, $|a\rangle$ is sent first and $|b\rangle$ is delayed by time τ which is greater than the traveling time (θ) of particles (wave packets) from Alice to Bob.

Thus $|b\rangle$ enters the channel only after $|a\rangle$ reaches Bob. Consequently, both the wave packets (i.e., the entire superposition) are never found together in the transmission channel. Now Bob delays $|a\rangle$ by τ and recreates the superposition state after he receives $|b\rangle$.

Alice and Bob perform the following two tests to detect Eve (using a classical channel).

1. They compare the sending time t_s with the receiving time t_r for each particle. Since the traveling time is θ and the delay time is τ, we must have $t_r = t_s + \theta + \tau$. This ensures that Eve cannot delay $|a\rangle$ and wait for $|b\rangle$ to reach her so that she can appropriately superpose $|a\rangle$ and $|b\rangle$. Although Eve cannot delay a wave packet, she may replace a wave packet. The second test detects such an attack.

2. Alice and Bob look for changes in the data by comparing a portion of the transmitted bits with the same portion of the received bits.

Here we would like to note that the sending time in GV protocol has to be random. Otherwise, at the known arrival time of $|a\rangle$, Eve can prepare a fake state in $|\psi_0\rangle$ and send the fake $|a\rangle$ to Bob without causing any delay. Eve keeps the original $|a\rangle$ and the fake $|b\rangle$ wave packets with her. Later, when original $|b\rangle$ is received by Eve then she measures the original state. If she finds the original state in $|\psi_0\rangle$ then she sends the fake wave packet $|b\rangle$ stored with her to Bob. Otherwise, she corrects the phase of the fake wave packet and sends $-|b\rangle$ to Bob. If we assume that the time required for Eve's measurement is negligible, then this procedure neither introduces any additional delay nor introduces any detectable error in the state received by Bob. Consequently, the sending time in GV protocol is required to be random.

GV protocol is essentially a QKD protocol. In its original form it cannot be directly used for secure direct quantum communication. This point can be understood easily. Let us assume that instead of a random sequence of bits, Alice transmits a meaningful message to Bob by sending a sequence of $|\psi_0\rangle$ and $|\psi_1\rangle$ by following GV protocol. Now when Alice sends $|a\rangle$ then Eve can keep it with her and send a fake $|a\rangle$ to Bob without causing any delay. Later, when $|b\rangle$ is sent by Alice then Eve will keep that also with her and send a fake $|b\rangle$ to Bob. Now Eve can appropriately superpose $|a\rangle$ and $|b\rangle$ and obtain the information encoded by Alice. Of course the security check will reveal the existence of Eve but by then Eve has already obtained the encoded information. This leakage is not a problem with QKD because if existence of Eve is found then Alice and Bob will not use that key for future encryption. This orthogonal state based protocol is fundamentally different from the conjugate coding based protocols as it uses only single basis for encryption and decryption. There is another difference of this protocol with the conjugate coding based QKD schemes. Here the random number is generated at Alice's end and then shared, but in BB84 and similar protocols the randomness is generated over the channel by appropriate measurements.

So far we have seen that quantum resources may be used to generate a secure quantum key. This secure key can be used for secure communication using an existing classical protocol. At this point quantum cryptographic protocols look like hybrid protocols, as they need classical encryption technique and quantum key. Now it's natural to look for a quantum protocol that can be used for secure direct communication. We will introduce such protocols in the following sections.

8.7   Ping-pong and modified ping-pong protocols

So far we have only described the QKD protocols. Now we will show that quantum resources can be used for cryptographic tasks which are more general than just creation and distribution of keys (QKD). To begin with we will describe a protocol of QSDC. This protocol was introduced by Bostrom and Felbinger in 2002 [105]. It is popularly known as the ping-pong (PP) protocol. The original protocol works as follows:

8.7.1   Protocol 7: Ping-pong protocol

PP1 Bob prepares the state $|\psi^+\rangle^{\otimes n}$, where $|\psi^+\rangle \equiv \frac{1}{\sqrt{2}}(|00\rangle + |11\rangle)_{AB}$, and transmits all the first qubits (say A) of the Bell pairs to Alice, keeping the other half (B) with himself.

PP2 Alice randomly chooses a set of $\frac{n}{2}$ qubits from the string received by her to form a verification string, on which the BB84 subroutine[11] to detect eavesdropping is implemented by measuring in the MUBs $\{0, 1\}$ or $\{+, -\}$. If sufficiently few errors are found, they proceed to the next step; else, they return to the previous step.

PP3 Alice randomly chooses half of the unmeasured qubits as verification string for the return path and encodes her message in the remaining $\frac{n}{4}$ qubits. To encode a message, Alice does nothing if she wants to encode 0 on a message qubit, and applies a X gate if she wants to encode 1. After the encoding operation, Alice sends all the qubits in her possession to Bob.

---

[11]BB84 subroutine means eavesdropping is checked using conjugate coding in a manner similar to that followed in BB84 protocol. Explicitly, BB84 subroutine implies that Alice (Bob) randomly chooses a set of qubits from the string received by her (him) and forms a verification string. She (He) measures either all or a set of qubits of verification string randomly in $\{0, 1\}$ or $\{+, -\}$ basis and announces which basis she (he) has used to measure a particular qubit, position of that qubit in the string and outcome. Bob (Alice) also measures the corresponding qubit using the same basis (if needed) and compares his (her) results with Alice (Bob) to detect eavesdropping.

PP4 Alice discloses the coordinates of the verification qubits after receiving authenticated acknowledgment of receipt of all the qubits from Bob. Bob applies the BB84 subroutine on the verification string and computes the error rate.

PP5 If the error rate is tolerably low, then Bob performs Bell-state measurements on the remaining Bell pairs, and decodes the message.

If in **PP3** Alice has encoded 0 then Bob will obtain $|\psi^+\rangle$ (the same as he had sent) in **PP5**, otherwise he will receive $|\phi^+\rangle$. Since $|\psi^+\rangle$ and $|\phi^+\rangle$ are orthogonal a Bell measurement will deterministically distinguish $|\psi^+\rangle$ and $|\phi^+\rangle$ and consequently decode the message encrypted by Alice. In this two-way protocol the travel qubit moves from Bob to Alice and comes back. This movement of the travel qubit is similar to a table tennis (ping-pong) ball which moves back and forth between two sides of the table. Keeping this analogy in mind, the above protocol is called ping-pong protocol. This protocol can be modified in many different ways. Here we show a few of them. We start with Cai Li (CL) protocol [113] which can be obtained by a very simple modification of PP protocol.

8.7.2 The modified PP protocols

It is easy to observe that in the original PP protocol full power of dense coding is not used. Alice could have used I, X, iY and Z to encode $00, 01, 10$ and 11 respectively and that would have increased the efficiency of ping-pong protocol. This is so because the same amount of communication would have successfully carried two bits of classical information. This fact was first formally included in a modified PP protocol proposed by Cai and Li in 2004 [113]. Here it would be apt to note that the purpose of Bostrom and Felbinger's work was to show that the capacity of quantum communication is not limited to QKD. It may also be used for direct communication. The inclusion of dense coding in CL protocol just makes it more efficient and opens up the possibility of using other entangled states (say GHZ, cluster, etc.) for QSDC. In principle any entangled state where dense coding is possible may be used to design a ping-pong type protocol for QSDC.

8.7.2.1 Protocol 8: PP protocol with full power of dense coding: CL protocol

The CL protocol is just like PP, except that step **PP3** is replaced by:

CL3 Alice applies unitary operations I, X, iY and Z to encode 2 bits, of value 00, 01, 10 and 11 respectively.

In PP and CL protocols the encoding is done by Alice by applying unitary operators on a state to convert it into an orthogonal state of it so that the initial state and the final state can be deterministically discriminated by Bob. Now this kind of encoding does not require entanglement. For example, X can convert $|0\rangle$ and $|1\rangle$ into their orthogonal states $|1\rangle$ and $|0\rangle$ respectively. Further, eavesdropping is checked by using BB84 subroutine which does not require entanglement. Consequently, it seems that PP type QSDC protocol can also be designed without using entanglement. Such an effort was made in 2005 by Lucamarini and Mancini [106]. The protocol proposed by them is known as LM05 protocol and it can be viewed as a modified PP protocol. The protocol is described in the following subsection.

8.7.2.2 Protocol 9: LM05 protocol

The LM05 protocol, which is a PP-type protocol without entanglement, works as follows:

LM1 Bob prepares and sends Alice a random string of $|0\rangle, |1\rangle, |+\rangle$ and $|-\rangle$.

LM2 Same as **PP2**.

LM3 Alice encodes a key/message as follows. Alice does nothing (i.e., she applies identity operator I) if she wants to encode 0 and she applies $iY = ZX$ on her qubit if she wants to encode 1. After encoding of her message she returns the qubit to Bob.

The encoding will transform the initial states into the orthogonal states as follows:

$$
\begin{aligned}
iY|0\rangle &= ZX|0\rangle = Z|1\rangle = -|1\rangle \\
iY|1\rangle &= ZX|1\rangle = Z|0\rangle = |0\rangle \\
iY|\pm\rangle &= \tfrac{iY}{\sqrt{2}}(|0\rangle \pm |1\rangle) = \tfrac{1}{\sqrt{2}}(-|1\rangle \pm |0\rangle) = \pm|\mp\rangle.
\end{aligned}
\tag{8.3}
$$

LM4 Same as **PP4**.

LM5 If the error rate is tolerably low then Bob deterministically decodes Alice's message by measuring the qubit in the same basis that he had used to prepare the initial state in **LM1**.

On the basis of the discussions so far we may note a few interesting observations on PP and LM05 protocols. Before we proceed further let us note them here.

1. In PP, CL and LM05 protocols Bob does not require classical communication from Alice for decoding of the message. Thus these are QSDC protocols.

2. In these protocols Alice can do the encoding without knowing the incoming state.

3. If we use conjugate coding then this task (QSDC) cannot be achieved with two states. We need at least four states for two-way direct communication. If we start with a random string of two states prepared

using one element from each of two MUBs (as we do in B92) then that would be sufficient to verify the presence of Eve in the channel. However, in the encoding process Alice would need to map each of the initial states into two orthogonal states (otherwise Bob will not be able to decode the message) and that would transform the protocol into a four state protocol. It would be interesting to see what happens when we don't use conjugate coding.

In a completely orthogonal-state-based protocol like GV, both error checking and encoding are done by using orthogonal states. PP and CL employ orthogonal encoding, but error checking uses the BB84 subroutine with two (or more) noncommuting bases (MUBs). To turn PP or CL into a GV-like protocol, which we call PP^{GV} and CL^{GV}, respectively, the error checking in them must be modified to employ only measurement in a single basis, given by that of the orthogonal encoding states, which is the Bell basis. In the following subsection we briefly describe the recent proposal of Yadav, Srikanth and Pathak [114] and explain the working of PP^{GV} and CL^{GV} protocols.

8.7.3 Protocol 10: PP^{GV} protocol

To implement PP^{GV} and CL^{GV} protocols, both particles (qubits) belonging to a pair must be sent across the channel. Although they are sent in such a way that both are not to be available at the same time in the channel, per the GV requirement, there is a new danger here not present in the original GV protocol. Eve can measure the position of the incoming qubits without disturbing the encoded information. This is essentially equivalent to having timing information in the GV protocol. Knowing the position of the qubits, Eve replaces the first qubit with a dummy, measures it jointly with the second qubit, and based on the state determined, sends a second dummy accordingly. Clearly, an error check cannot reveal this attack.

To counter this, Bob randomly re-orders his qubits before transmission, revealing the reordering information only after Alice's receipt. Knowing the particle coordinates does not reveal to Eve which are the partner particles, and thus does not help Eve to determine the encoded information, which, had it been successful, could have been used to decode the key/message and prepare noiseless dummies. We observe that reordering also means that the GV time control, which was required to physically separate legitimate particle pairs from Eve's perspective, is no longer needed. Thus reordering serves both the purpose of temporal separation and transmission timing randomization in the original GV protocol. Let us now specifically describe how PP can be converted to PP^{GV}.

To obtain a GV-like protocol, we follow the approach of Yadav, Srikanth and Pathak [114] and modify PP by replacing steps **PP1**, **PP2** and **PP4** as follows:

PP^{GV}1 Bob prepares the state $|\psi^+\rangle^{\otimes n}$. He keeps half of the second qubits of the Bell pairs with himself. On the remaining $\frac{3n}{2}$ qubits he applies a random permutation operation $\Pi_{\frac{3n}{2}}$ and transmits them to Alice. n of the transmitted qubits are Bell pairs and the remaining $\frac{n}{2}$ are the partner particles of the particles which remained with Bob.

PP^{GV}2 After receiving Alice's authenticated acknowledgment, Bob announces $\Pi_n \in \Pi_{\frac{3n}{2}}$, the coordinates of the transmitted Bell pairs. Alice measures them in the Bell basis to determine if they are each in the state $|\psi^+\rangle$. If the error detected by Alice is within the tolerable limit, they continue to the next step. Otherwise, they discard the protocol and restart from **PP^{GV}1**.

PP^{GV}4 Alice discloses the coordinates of the verification qubits after receiving Bob's authenticated acknowledgment of receipt of all the qubits. Bob combines the qubits of verification string with their partner particles already in his possession and measures them in the Bell basis to compute the (return trip) error rate.

The other steps in PP remain the same. Similarly, in the CL protocol, replacing step **CL1**, **CL2** and **CL4** by **PP^{GV}1**, **PP^{GV}2** and **PP^{GV}4** gives the GV version of CL, denoted **CL^{GV}** which is our **Protocol 11.**

Briefly, security in PP^{GV} and CL^{GV} arise as follows. The reordering has the same effect as time control and time randomization in GV. Eve is unable to apply a 2-qubit operation on legitimate partner particles to determine the encoding in spite of their orthogonality. Any correlation she generates by interacting with individual particles will diminish the observed correlations between Alice and Bob because of restrictions on shareability of quantum correlations [114].

At best, Eve can either entangle a sufficiently large system with the transmitted $\frac{3n}{2}$ qubits, or replace them altogether by $\frac{3n}{2}$ dummies. The problem Eve faces is that through only local operations she should be able to collapse any n of the transmitted particles into an ordered state of $\frac{n}{2}$ pairs in the state $|\psi^+\rangle$, which has exponentially low probability of success.

PP^{GV} (CL^{GV}) is a two-way four-step QSDC protocol. Note that PP^{GV}, is a QSDC protocol that requires only two encoding states. We have already shown that two nonorthogonal states are sufficient for QKD, but at least four states are required for implementation of DSQC or QSDC using conjugate coding. The above protocol of Yadav, Srikanth and Pathak is the first ever two-state protocol of QSDC and also the first ever orthogonal state based protocol of QSDC. These observations were motivating enough to explore the possibility of converting other existing conjugate-coding-based protocols of DSQC and QSDC to GV type protocol.

All the protocols of direct secure quantum communication described so far are two way protocols. However, it is possible to modify them into one-way protocols. A very interesting one-way protocol can be obtained by modifying CL protocol. This specific protocol was introduced by Deng, Long and Liu in 2003 [115]. The protocol is known as DLL protocol and it can be viewed as modified CL protocol in which the two-way, two-step quantum communication is transformed to a one-way two-step communication that uses dense coding. In the next section we describe DLL protocol and have subsequently modified that to a GV type protocol which we refer to as DLL^{GV}.

8.8   DLL and modified DLL protocols

Before we describe the protocol we first note that after the first step of PP protocol Alice and Bob share entanglement. To share an entanglement it is not required to be created by Bob, even Alice can create that and send a qubit to Bob as was done in the first step of Protocol 2. Let us modify the first step of PP protocol and see what happens:

8.8.1   Protocol 12: DLL protocol

DLL1 Alice prepares the state $|\psi^+\rangle^{\otimes n}$, where $|\psi^+\rangle \equiv \frac{1}{\sqrt{2}}(|00\rangle + |11\rangle)_{AB}$, and transmits all the second qubits (say B) of the Bell pairs to Bob, keeping the other half (A) with herself.

DLL2 Bob randomly chooses a set of $\frac{n}{2}$ qubits from the string received by him to form a verification string, on which the BB84 subroutine to detect eavesdropping is implemented by measuring in the noncommuting bases $\{0, 1\}$ or $\{+, -\}$. If sufficiently few errors are found, they then proceed to the next step; else, they return to **DLL1**.

DLL3 Alice randomly chooses half of the qubits in her possession to form the verification string for the next round of communication, and encodes her message in the remaining $\frac{n}{4}$ qubits. To encode a 2-bit key message, Alice applies one of the four Pauli operations I, X, iY, Z on her qubit. After the encoding operation, Alice sends all the qubits in her possession to Bob.

DLL4 Alice discloses the coordinates of the verification qubits after receiving authenticated acknowledgment of receipt of all the qubits from Bob. Bob applies a BB84 subroutine to the verification string and computes the error rate.

DLL5 If the error rate is tolerably low, then Bob decodes the encoded states via a Bell-state measurement on the remaining Bell pairs.

Presenting the DLL protocol in this way helps us to visualize the inherent symmetry among PP, CL and DLL protocols. This is a one-way two-step QSDC protocol. The way it is converted from PP essentially indicates that any two-way QSDC protocol can be converted to a one-way two-step QSDC protocol. Thus all states for which dense coding is reported may yield equivalent protocols of QSDC. It is easy to observe that this protocol looks similar to PP protocol with dense coding (i.e., CL protocol). However there is a fundamental difference between a two-way protocol and a two-step one-way protocol which uses the same resources and encoding operations. The difference lies in the fact that in a two-way protocol home qubit always remains at sender's port but in a one-way two-step protocol both the qubits travel through the channel. At this specific point we observe a symmetry between DLL protocol and GV protocol. Here the superposition is broken into two pieces in such a way that the entire superposed (entangled) state is never available in the transmission channel but only the entire superposition (i.e., the superposed state or entangled state) contains meaningful information. Visualization of this intrinsic symmetry helps us to generalize DLL protocol to obtain an orthogonal version of the same.

8.8.1.1   Protocol 13: The modified DLL protocol (DLL^{GV})

Based on the reasoning analogous to the one used for turning PP to PP^{GV}, we may propose the following GV-like version of DLL, which may be called DLL^{GV} in accordance with the recent work of Yadav, Srikanth and Pathak [114]. As before, we retain the steps of DLL, replacing only steps **DLL1**, **DLL2** and **DLL4** as follows:

DLL^{GV}1 Alice prepares the state $|\psi^+\rangle^{\otimes n}$. She keeps half of the first qubits of the Bell pairs with herself. On the remaining $\frac{3n}{2}$ qubits she applies a random permutation operation $\Pi_{\frac{3n}{2}}$ and transmits them to Bob; n of the transmitted qubits are Bell pairs while the remaining $\frac{n}{2}$ are the entangled partners of the particles remaining with Alice.

DLL^{GV}2 After receiving Bob's authenticated acknowledgment, Alice announces $\Pi_n \in \Pi_{\frac{3n}{2}}$, the coordinates of the transmitted Bell pairs. Bob measures them in the Bell basis to determine if they are each in the state $|\psi^+\rangle$. If the error detected by Bob is within a tolerable limit, they continue to the next step. Otherwise, they discard the protocol and restart from **DLL^{GV}1**.

DLL^{GV}4 Same as **PP^{GV}4**, except that the 'return trip' is replaced by Alice's second onward communication.

So two-way protocols of QSDC are now converted to one-way protocols. But still we need two steps. This motivates us to ask: Do we always need at least two steps for secure direct quantum communications? Apparently it looks so because if we send both the qubits of an entangled pair together then Eve may do Bell measurement and find out the message. Even if Eve is detected afterward it would not be of any use because she has already obtained the message. There exists an excellent trick called rearrangement of particle order, which can be used to circumvent this problem. We have already used this trick in implementing PP^{GV}, CL^{GV} and DLL^{GV}. Further, we have not yet described any protocol of DSQC. Let us now describe a one-step one-way protocol of DSQC recently introduced by Shukla, Pathak and Srikanth [116]. Apart from being DSQC and one-step protocol this protocol implements secure quantum communication using arbitrary quantum states.

8.9   DSQC protocol and its modifications

This protocol is designed to achieve DSQC using any arbitrary quantum state $|a_i\rangle$. Given a specific n-qubit state $|a_i\rangle$ we can always construct a basis set $\{|a_j\rangle\} : |a_i\rangle \in \{|a_j\rangle\}$. In the beginning, Alice publicly announces the basis set $\{|a_j\rangle\}$ to be used by her, the initial state $|a_i\rangle$ she is going to prepare and the set of unitary operators $\{U_j : U_j|a_i\rangle = |a_j\rangle\}$ to be used by her for encoding. There are various alternative ways to construct $\{U_j\}$. The following particularly symmetric (in fact Hermitian) set $\{U_j\}$ may be used as a specific example

$$
\begin{aligned}
U_i &= I, \\
U_{j\neq i} &= |a_i\rangle\langle a_j| + |a_j\rangle\langle a_i| + \sum_{k=1;k\neq i,j}^{M} |a_k\rangle\langle a_k|.
\end{aligned}
\tag{8.4}
$$

For the practical implementation purpose, our assumption is that Alice has devices to prepare states in the basis $\{|a_j\rangle\}$, to implement the set of unitary operators $\{U_j\}$, and that Bob has devices to make measurements in $\{|a_j\rangle\}$ basis. The main part of the protocol of DSQC works as follows.

8.9.1   Protocol 14: The DSQC protocol with arbitrary state

DSQC1: Alice prepares $|a_i\rangle^{\otimes N}$, which is an Nn-qubit state (as $|a_i\rangle$ is an n-qubit state). Qubits of $|a_i\rangle^{\otimes N}$ are indexed as $p_1, p_2, \cdots, p_{Nn}$. Thus p_s is the sth qubit of $|a_i\rangle^{\otimes N}$ and $\{p_{nl-n+1}, p_{nl-n+2}, \cdots, p_{nl} : l \leq N\}$ are the n qubits of the lth copy of $|a_i\rangle$.

DSQC2: Alice encodes her n-bit classical secret message by applying one of the n-qubit unitaries $\{U_j\} = \{U_1, U_2, \cdots, U_{M=2^n}\}$. The encoding scheme, which is predefined and known to Bob, is such that $U_1, U_2, U_3, \cdots, U_M$ are used to encode $0_10_2\cdots0_n$, $0_10_2\cdots1_n$, $0_10_2\cdots1_{n-1}0_n, \cdots, 1_11_2\cdots1_n$ respectively. Thus the coded states $U_j|a_i\rangle = |a_j\rangle$ are mutually orthogonal.

DSQC3: Using all the qubits of her possession, Alice creates an ordered sequence $P_B = [p_1, p_2, p_3, p_4, \cdots, p_{Nn-1}, p_{Nn}]$. She prepares Nn decoy qubits d_i with $i = 1, 2, \cdots, Nn$ such that $d_i \in \{|0\rangle, |1\rangle, |+\rangle, |-\rangle\}$

and concatenates them with P_B to yield a larger sequence $P_{B'} = [p_1, p_2, p_3, p_4, ..., p_{Nn-1}, p_{Nn}, d_1, d_2, d_3, d_4, ..., d_{Nn-1}, d_{Nn}]$. Thereafter Alice applies a permutation operator Π_{2Nn} on $P_{B'}$ to create a random sequence $P_{B''} = \Pi_{2Nn} P_{B'}$ and sends that to Bob. The actual order is known to Alice only.

DSQC4: After receiving Bob's authenticated acknowledgment of receipt of all the qubits, Alice announces $\Pi_{Nn} \in \Pi_{2Nn}$, the coordinates of the decoy qubits. The BB84 subroutine to detect eavesdropping is then implemented on the decoy qubits by measuring them in the nonorthogonal bases $\{|0\rangle, |1\rangle\}$ or $\{|+\rangle, |-\rangle\}$. If sufficiently few errors are found, then they go to the next step; else, they return to **DSQC1**. All intercept resend attacks will be detected in this step and even if eavesdropping has happened Eve will not obtain any meaningful information about the encoding operation executed by Alice as the encoded sequence is rearranged.

DSQC5: Alice discloses the coordinates of the remaining qubits.

DSQC6: Bob measures his qubits in $\{|a_j\rangle\}$ basis and deterministically decodes the information encoded by Alice.

The above protocol is clearly a protocol of DSQC as Alice needs to announce the actual order of the sequence. Rearrangement of particle ordering present in the DSQC protocol may be avoided by sending the information encoded states in n steps and by checking eavesdropping after each step. To be precise, consider that Alice sends a sequence of all the first qubits first with N decoy photons. If sufficiently few errors are found, only then she sends the sequence of all the second qubits, and so on. In such a situation the DSQC protocol presented above will be reduced to a QSDC protocol as no classical information will be required for decoding.

8.9.1.1   Protocol 15: A QSDC protocol from Protocol 14

The previous protocol can be easily generalized to a QSDC protocol. To do so, we need to modify **DSQC3-DQSC5** in the above protocol. Therefore, the modified protocol may be described as follows:

QSDC1: Same as **DSQC1**.

QSDC2: Same as **DSQC2**.

QSDC3: Alice prepares n sequences: $P_{Bs} = [p_s, p_{s+n}, \cdots, p_{s+(N-1)n} : 1 \leq s \leq n]$. She also prepares Nn decoy qubits as in **DSQC3** and inserts N of the decoy qubits randomly into each of the n sequences prepared by her. This creates n extended sequences (P_{B1+N}, P_{B2+N}, $\cdots$, P_{Bn+N}) each of which contains $2N$ qubits. Then she sends the first sequence P_{B1+N} to Bob. Receiving Bob's authenticated acknowledgment of receipt of $2N$ qubits, she announces the positions of the decoy qubits in P_{B1+N}. BB84 subroutine is then implemented on the decoy qubits to check eavesdropping and if sufficiently few errors are found then Alice sends P_{B2+N} to Bob and they check for eavesdropping and the process continues till the error free (i.e., within the tolerance limit) transmission of P_{Bn+N}. If at any stage of this step errors more than the tolerable rate are detected then they truncate the protocol and return to **QSDC1**; else, they go to the next step.

QDDC4: Same as **DSQC6**.

Since Eve cannot obtain more than 1 qubit of an n-partite state (as we are sending the qubits one by one and checking for eavesdropping after each step) she has no information about the encoded state and consequently this direct quantum communication protocol is secure. Thus the rearrangement of particle order is not required if we do the communication in multiple steps. Further, this protocol does not require any classical communication for the decoding operation. Consequently, it is a QSDC protocol. Its qubit efficiency will be naturally higher than the previous protocol. This is so because here Alice does not need to disclose the actual sequence and consequently the amount of classical communication required for decoding of the message is reduced. But this increase in qubit efficiency[12] is associated with a cost. This QSDC protocol will be slower compared to its DSQC counterpart as Alice has to communicate in multiple steps and has to check eavesdropping in each of the steps. Protocol 14 and 15 can also be generalized to corresponding orthogonal state based protocols. The same was recently reported in [116]. We will not describe them here.

Starting from a two-way QSDC protocol we have obtained a one-way DSQC protocol, but still the information flows from Alice to Bob only. This leads to a question: Can we construct a protocol of two way quantum communication where information flows simultaneously from both Alice to Bob and Bob to Alice? In the beginning of this chapter we mentioned about such protocols and also mentioned that such a protocol is called quantum dialogue protocol. In the next section we describe two protocols of quantum dialogue.

8.10   Protocols of quantum dialogue

The first protocol of quantum dialogue was proposed by Ba An [110, 112] using Bell states in 2004. Let us first describe Ba An's original scheme of quantum dialogue.

---

[12]Qubit efficiency will be defined in Solved Example 6 of this chapter.

8.10.1   Protocol 16: Ba An protocol

This simple protocol works in the following steps:

BA1 Bob prepares a large number of copies of a Bell state $|\phi^+\rangle = \frac{|01\rangle+|10\rangle}{\sqrt{2}}$. He keeps the first qubit of each Bell state with himself as home qubit and encodes his secret message $00, 01, 10$ and 11 by applying unitary operations U_1, U_2, U_3 and U_4 respectively on the second qubit. Without loss of generality we may assume that $U_1 = I$, $U_2 = \sigma_x = X$, $U_3 = i\sigma_y = iY$ and $U_4 = \sigma_z = Z$.

BA2 Bob sends the second qubit (travel qubit) to Alice and confirms that Alice has received a qubit.

BA3 Alice encodes her secret message by using the same set of encoding operations as was used by Bob and sends back the travel qubit to Bob. After receiving the encoded travel qubit Bob measures it in Bell basis.

BA4 Alice announces whether it was run in message mode (MM) or in control mode (CM). In MM, Bob decodes Alice's bits and announces his Bell basis measurement result. Alice uses that result to decode Bob's bits. In CM, Alice reveals her encoding value to Bob to check the security of their dialogue.

It is easy to recognize that this is a modification of PP protocol [105] and the operations used for encoding are the operators usually used for dense coding and the protocol starts with an initial state $|\psi\rangle_{initial} = |\phi^+\rangle$. Recently this protocol was generalized by Shukla, Banerjee, Kothari and Pathak [117]. The generalized protocol of quantum dialogue is described below.

8.10.2   Protocol 17: Generalized protocol of quantum dialogue

The generalized protocol works using n-qubit basis set $\{|\phi_i\rangle\}$ and m-qubit $(m \leq n)$ unitary operators $\{U_1, U_1, \cdots, U_{2^n}\}$ such that $U_i|\phi_1\rangle = |\phi_i\rangle$ and the set of operators $\{U_1, U_2, \cdots, U_{2^n}\}$ forms a group under multiplication (without global phase). The protocol works as follows:

QD1 Bob prepares $|\phi_1\rangle^{\otimes N}$, and encodes his classical secret message by applying m-qubit unitary operators $\{U_1, U_2, \cdots, U_{2^n}\}$. For example, to encode $0_1 0_2 \cdots 0_n, 0_1 0_2 \cdots 1_n, 0_1 0_2 \cdots 1_{n-1} 0_n, \cdots, 1_1 1_2 \cdots 1_n$ he applies $U_1, U_2, U_3, \cdots, U_{2^n}$, respectively. The information encoded states are mutually orthogonal to each other as discussed above.

QD2 There are two possibilities: (i) $m < n$, i.e., dense coding is possible and (ii) $m = n$, i.e., dense coding is not possible for the set of quantum states and the set of unitary operators used for encoding. If $m < n$, then Bob uses the m qubits on which encoding is done as travel qubits and the remaining $n-m$ qubits as home qubits and keeps them with himself in an ordered sequence $P_B = [p_1(h_1, h_2, ..., h_{n-m}), p_2(h_1, h_2, ..., h_{n-m}), \cdots, p_N(h_1, h_2, ..., h_{n-m})]$, where the subscript $1, 2, \cdots, N$ denotes the order of a n-partite state $p_i = \{h_1, h_2, ..., h_{n-m}, t_1, t_2, \cdots, t_m\}$, which is in one of the n-partite state $|\phi_j\rangle$ (value of j depends on the encoding). Symbol h and t are used to indicate home qubit (h) and travel qubit (t) respectively. If dense coding is not possible then Bob has to use all qubits as travel qubits. In general, he uses all the travel qubits to prepare an ordered sequence $P_{A0} = [p_1(t_1, t_2, \cdots, t_m), p_2(t_1, t_2, \cdots, t_m), ..., p_N(t_1, t_2, \cdots, t_m)]$. He prepares Nm decoy qubits d_i with $i = 1, 2, \cdots, Nm$ such that $d_i \in \{|0\rangle, |1\rangle, |+\rangle, |-\rangle\}$ and concatenates them with P_{A0} to yield a larger sequence $P_{A1} = [p_1(t_1, t_2, \cdots, t_m), p_2(t_1, t_2, \cdots, t_m), \cdots, p_N(t_1, t_2, \cdots, t_m), d_1, d_2, d_3, d_4, \cdots, d_{Nm-1}, d_{Nm}]$. Thereafter, he applies a permutation operator Π_{2Nm} on P_{A1} to create a random sequence $P_{A2} = \Pi_{2Nm} P_{A1}$ and sends that to Alice.

QD3 After receiving Alice's authenticated acknowledgment of receipt of all the qubits, Bob announces $\Pi_{Nm} \in \Pi_{2Nm}$, the coordinates of the decoy qubits. The BB84 subroutine to detect eavesdropping is then implemented on the decoy qubits by measuring them in the nonorthogonal bases $\{|0\rangle, |1\rangle\}$ or $\{|+\rangle, |-\rangle\}$. If sufficiently few errors are found, then they go to the next step; else, they return to **QD1.**

QD4 Bob announces the order of the remaining qubits.

QD5 After knowing the actual order, Alice transforms the sequence into actual order and encodes her information using the same encoding scheme as was used by Bob. That creates a new sequence P_{A2}. Alice prepares Nm decoy qubits in a random sequence of $\{|0\rangle, |1\rangle, |+\rangle, |-\rangle\}$, and concatenates them with P_{A2} to yield a larger sequence P_{A3}. Thereafter, she applies a permutation operator Π_{2Nm} on P_{A3} to create a random sequence $P_{A4} = \Pi_{2Nm} P_{A3}$ and sends the sequence P_{A4} to Bob.

QD6 After receiving Bob's authenticated acknowledgment of receipt of all the qubits, Alice announces $\Pi_{Nm} \in \Pi_{2Nm}$, the coordinates of the decoy qubits. The BB84 subroutine to detect eavesdropping is then implemented on the decoy qubits by measuring them in the nonorthogonal bases $\{|0\rangle, |1\rangle\}$ or $\{|+\rangle, |-\rangle\}$. If sufficiently few errors are found, then they go to the next step; else, they return to **QD1.** This makes the protocol safe from all kinds of eavesdropping strategies in the return path.

QD7 Alice announces the order of the remaining qubits.

QD8 Bob reorders the sequence to obtain P_{A2}, recombines it with P_B and measures each n-partite state in $\{|\phi_i\rangle\}$ basis. As he already knows the unitary operators applied by him or the state $|\phi_i\rangle$ sent by him, he can now easily decode the message encoded by Alice. After the measurement Bob publicly announces the final states that he has obtained.

QD9 As Alice knows her encoding into a particular state, she uses that information to decode the secret message of Bob.

Now we would like to note that if $m = n$, then in **QD5** after knowing the actual order Alice could have decoded the message of Bob and in that case the public announcement of Bob in **QD8** and the entire **QD9** would be redundant. In such case, the protocol essentially gets decomposed into two protocols of DSQC: one from Alice to Bob and the other from Bob to Alice. That is not really in accordance with the true spirit of the quantum dialogue protocols.

Assume that Bob applies U_B to encode his bit and Alice applies U_A to encode her bit. In this case the initial state $|\phi_1\rangle$ will be transformed to $U_A U_B |\phi_1\rangle = |\phi_k\rangle$, where $|\phi_k\rangle \in \{|\phi_i\rangle\}$. This is so because we have chosen the unitary operators in such a way that $\{U_i\}$ forms a group under multiplication which implies that $U_A U_B \in \{U_i\}$. Now Eve knows initial state and final state from the announcement of Bob. Consequently she knows $U_A U_B$ but this information is not enough to decrypt the message encoded by Alice or Bob as she does not know U_A and U_B. On the other hand, Alice (Bob) knows U_A (U_B) so she (he) can decrypt the message encoded by Bob (Alice). Because of this leakage of correlation information, the proposed protocol of quantum dialogue is not as secure as DSQC and QSDC protocols are. However, this protocol can be used to provide a solution to the socialist millionaire problem. This interesting problem is discussed in the following subsection.

8.10.2.1   Applications of quantum dialogue protocols in the socialist millionaire problem

Protocol 17 can be easily extended to secure multi-party computation (SMC) tasks. As an example, here we will briefly describe how our results can be used to solve a specific SMC task known as the socialist millionaire problem [118]. In this problem two millionaires wish to compare their wealth but they do not want to disclose the amount of their wealth to each other. This problem is also referred to as the problem of private comparison of equal information. Now if we consider that Alice and Bob are the millionaires who want to compare their wealth and Charlie is a semi-honest third party, then our protocol works as follows: Charlie prepares an n-qubit entangled state in one of the possible mutually orthogonal states $\{|\phi_1\rangle, |\phi_2\rangle, \cdots, |\phi_i\rangle, \cdots, |\phi_{2^n}\rangle\}$ (say he prepares it in $|\phi_i\rangle$) and keeps the home qubits with himself and sends the travel qubits to Alice, who encodes her information (the value of her wealth) by applying unitary operations $\{U_1, U_2, U_3, \cdots, U_{2^n}\}$ as per the encoding rule. Then Alice sends the encoded qubits to Bob. As Bob has access to only travel qubits and since he does not know the initial state he cannot obtain the information encoded by Alice. Now Bob also encodes his information (the value of his wealth) by using the same set of encoding operations $\{U_1, U_2, U_3, \cdots, U_{2^n}\}$ and sends the qubits to Charlie. Now if $\{U_1, U_2, U_3, \cdots, U_{2^n}\}$ forms a group under multiplication then Charlie will obtain one of the mutually orthogonal states. Charlie can measure the final state using $\{|\phi_1\rangle, |\phi_2\rangle, \cdots, |\phi_i\rangle, \cdots, |\phi_{2^n-1}\rangle\}$ as basis and deterministically obtain the final state $|\phi_f\rangle$. Until this point this protocol is similar to the quantum dialogue protocol. The difference between the two protocols is that instead of Alice, now Charlie prepares the initial state and Charlie knows nothing about Alice and Bob's encoding. He knows only the final state and initial state, so his knowledge is the same as that of Eve in the previous protocol. If Charlie finds that the initial state and the final state are the same (i.e., $|\phi_i\rangle = |\phi_f\rangle$) then the classical information encoded by Alice and Bob are the same. In all other cases the classical information encoded by Alice and Bob are different. If we consider the encoded information as the amount of their assets then it solves the socialist millionaire problem. Neither Alice nor Bob (not even Charlie) knows how much assets are there in possession of the other, unless the amount is the same. Since classical broadcasting is not required here, the intrinsic problem of information leakage in quantum dialogue protocol is not present here.

8.11 Protocol 18: Quantum secret sharing

The first protocol of QSS was introduced by Hillery, Buzek and Bertaiume in 1999 [90]. A variant of it was experimentally demonstrated by Tittel, Zbinden, and Gisin in 2001 [119]. We will briefly introduce QSS to show the relevance of QIS in the context of secure quantum communication. In this chapter we have repeatedly seen that splitting of information into two or more non-revealing pieces is at the heart of security of all quantum cryptographic protocols. Let us now consider a simpler situation which requires QIS. Consider that Alice is boss of a company and she lives in Amsterdam. Bob and Charlie are her agents in Berlin. Alice wants to send them a secret message to perform a job. However, one of them may be dishonest and Alice does not know who is dishonest. In this situation, Alice may use QIS scheme described in Section 7.4 and send the information in two pieces so that neither Bob nor Charlie can read the message of Alice without the help of the other. However, there exist possibilities of eavesdropping. For example, consider that Bob is dishonest and he captures the qubit sent to Charlie, too. If Bob measures Charlie's qubit in the diagonal basis and sends the qubit to Charlie then using the final expression of (7.4), Bob will be able to get all of the information without any help of Charlie. So Alice needs to add some error checking schemes to the existing QIS scheme. One way to achieve this is as follows:

QSS1: Alice prepares $|\psi\rangle_{GHZ}^{\otimes n}$, where $|\psi\rangle_{GHZ} = \frac{1}{\sqrt{2}}(|000\rangle + |111\rangle)_{ABC}$. As $|\psi\rangle_{GHZ}^{\otimes n}$ is a $3n$-qubit state, qubits of $|\psi\rangle_{GHZ}^{\otimes n}$ may be indexed as $p_1, p_2, \cdots, p_{3n}$. Thus p_s is the sth qubit of $|\psi\rangle_{GHZ}^{\otimes n}$ and $\{p_{3l-2}, p_{3l-1}, p_{3l} : l \leq n\}$ are the 3 qubits of the lth copy of $|\psi\rangle_{GHZ}^{\otimes n}$.
This step is similar to **DSQC1**.

QSS2: Using all the first qubits of her possession, Alice creates an ordered sequence $P_A = [p_1, p_4, p_7, \cdots, p_{3n-2}]$. Similarly, she prepares an ordered sequence with all the second qubits as $P_B = [p_2, p_5, p_8, \cdots, p_{3n-1}]$ and another ordered sequence $P_C = [p_3, p_6, p_9, \cdots, p_{3n}]$. She prepares $2n$ decoy qubits d_i with $i = 1, 2, \cdots, 2n$ such that $d_i \in \{|0\rangle, |1\rangle, |+\rangle, |-\rangle\}$ and concatenates first n of them with P_B to yield a larger sequence $P_{B'} = [p_2, p_5, p_8, \cdots, p_{3n-1}, d_1, d_2, \cdots, d_n]$. Similarly using P_C and the remaining decoy qubits she creates $P_{C'} = [p_3, p_6, p_9, \cdots, p_{3n}, d_{n+1}, d_{n+2}, \cdots, d_{2n}]$. Thereafter, Alice applies a permutation operator Π_{2n} on $P_{B'}$ and $P_{C'}$ to create random sequences $P_{B''} = \Pi_{2n}P_{B'}$ and $P_{C''} = \Pi_{2n}P_{C'}$ and sends $P_{B''}$ and $P_{C''}$ to Bob and Charlie respectively. The actual order is known to Alice only.

QSS3: After receiving Bob and Charlie's authenticated acknowledgments of receipt of all the qubits, Alice announces $\Pi_n \in \Pi_{2n}$, the coordinates of the decoy qubits in each sequence. The BB84 subroutine to detect eavesdropping is then implemented on the decoy qubits by measuring them in the nonorthogonal bases $\{|0\rangle, |1\rangle\}$ or $\{|+\rangle, |-\rangle\}$. If sufficiently few errors are found in both the sequences, then they go to the next step; else, they return to **QSS1**.
This will ensure that the initial GHZ state is appropriately distributed among Alice, Bob and Charlie without any eavesdropping.

QSS4: Alice discloses the coordinates of the remaining qubits and Bob and Charlie rearrange their sequences accordingly.
The remaining part of the protocol is the same as QIS described in Section 7.4. Now Alice, Bob and Charlie share n 4-qubit states of the form (7.4). For example, if we consider a particular 4-qubit state then Alice's quantum secret which is to be shared is $\alpha|0\rangle + \beta|1\rangle$.

QSS5: Alice measures her qubits in Bell basis and announces the results. Without loss of generalization we may assume that Alice has asked Bob to prepare the secret state transmitted by her.

QSS6: Charlie measures his qubits in diagonal basis and communicates the result to Bob.

QSS7: Bob applies appropriate unitary operators (as described in Table 7.3) in accordance with the measurement outcomes of Alice and Charlie, and reconstructs the secret quantum state transmitted by Alice.

8.12 Discussion

From Protocols 2-18, we have described various aspects of quantum cryptography. The presented protocols are experimentally realizable and it's interesting to note that a few commercial products have already been launched by id Quantique (see http://www.idquantique.com) and MagiQ (see http://www.

magiqtech.com). Due to the problem of decoherence, it does not seem that a large quantum computer will be built in the near future. However, good quantum cryptographic solutions are already in the market. For example, id Quantique's CLAVIS^2 and MagiQ's QPN-8505 and Q-Box are very interesting products. Another nice and simple example of quantum device is a quantum random number generator. Assume that a single photon encounters a beam splitter (as shown in Fig. 1.3b) and two detectors are placed along two paths. When a detector clicks it implies that the state is measured and the qubit is collapsed to one of the two possible states. As the collapse of a state is a completely random process, so the detectors will click completely randomly. We will note 0 (1) if the detector along reflected (transmitted) path clicks. This simple idea will give us a true random number generator. id Quantique's product QUANTIS which is a true random number generator, works on this principle. Interestingly, classically there does not exist any true random number generator, but all classical cryptographic devices require random number generator. At the moment it is straightforward to use the quantum random number generators in the classical cryptographic devices and transform them to hybrid devices. But we are not interested in this type of hybrid device; we need an unconditionally secure key. Although some commercial products implementing QKD are available, still there are certain technological issues. For example, there does not exist any single photon source. The efficiency of the single photon detectors is low. Further, the available single photon detectors are most efficient around 800 nm but the attenuation of existing optical fibers is minimum around the teleportation range 1300-1550 nm. We need new materials to design efficient detectors in the teleportation range. As we mentioned, the quantum efficiencies of the existing single photon detectors are low which implies that the loss is high at present. Consequently, at present it is difficult to faithfully implement DSQC, QSDC and quantum dialogue protocols. However, QKD can be implemented faithfully. This is so because if we lose some qubits in QKD then only the key size reduces, but in the case of DSQC and QSDC if we lose some qubits then we lose part of the meaningful message.

Finally we finish this text with an optimistic view that in our lifetime we will be able to beat decoherence and will be able to build a scalable quantum computer. We may justify our optimism using a quote from Bennett *et al.*'s paper of 1991 [120], where they described the initial situation of quantum cryptography as follows: "Initially quantum cryptography was thought of by everyone (including ourselves) mostly as a work of science fiction because the technology required to implement it was out of reach ...." However, by now we already have commercial quantum cryptographic products. We hope something similar will happen with the technology required to circumvent the challenge posed by decoherence. Even if we don't succeed in performing quantum computing using a large enough quantum computer, secure quantum communication will definitely succeed. Further, the whole process of studying quantum computation and quantum communication is continuously improving our understanding of nature.

This book started with the story of a curious man. Eventually, he became interested in knowing the answers to the questions that appeared in his mind. In the process he followed the logical flow of the present textbook and ended up reading this book. He found it interesting up to a point but he slept at some point. When he woke up he realized that he had had a nice and exciting dream. In his dream he won a jackpot of 10 million US$ in a TV show by using GHZ states (see Solved Example 4 of Chapter 7 and [121]). He deposited the money in his classical bank account at Zurich, Switzerland but suddenly he heard an unsupported news story that somebody had discovered a huge quantum computer. He calculated that the security of his RSA based classical bank account may be broken in two hours. But he was in Calgary, Canada, which is far from Zurich. He was very anxious for a while but soon he realized that there is an option to reach Zurich within an hour. He went to the Calgary teleportation station and get himself teleported to Zurich. Immediately he closed the account. He withdrew the money and deposited it in a quantum bank. He was relieved and went to rest in a hotel. But he could not sleep as many people were shouting and protesting in front of the classical bank as they had lost their money. At that noise he woke up and realized that none of this had happened; he had been reading this textbook and fell asleep.

Cartoon 8.1: When do you expect it to happen?

8.13   Solved examples

1. A primitive classical encryption system known as Caesarian square works for plain text messages having square number of characters without punctuations and space. In this encryption first space and punctuations are dropped from the horizontal message and then the message is written vertically in the square pattern. After that a horizontal message is created by copying row by row of the square pattern. Now use this technique to encrypt: How are you?
   Solution: Removing the space and punctuation the plain text is "howareyou" which has 9 characters so we can write it vertically in the square form as

$$
\begin{array}{ccc}
h & a & y \\
o & r & o \\
w & e & u
\end{array}
$$

   and then rewrite row by row to obtain the encrypted message as hayoroweu.

2. Decrypt the messages (a) inatainhmrpaabak, (b) iioanvmle, given that the message is encrypted using Caesarian square.
   Solution: (a) Since there are 16 characters we can write the message in square form as

$$
\begin{array}{cccc}
i & n & a & t \\
a & i & n & h \\
m & r & p & a \\
a & b & a & k
\end{array}
$$

   Now to decrypt first we rewrite it column by column (inverse operation of the encryption) and obtain iamanirbanpathak; inserting appropriate spaces it reads I am Anirban Pathak. This is the decrypted message.
   (b) There are 9 characters in the message so we can write it in square form as

$$
\begin{array}{ccc}
i & i & o \\
a & n & v \\
m & l & e
\end{array}
$$

   and then joining column by column we obtain iaminlove, which can be easily decrypted as "I am in love." You may find it easy to decrypt but remember that when a method of this type is used then the procedure adopted for encryption is kept secret.

3. In BB84 protocol Alice prepares an ensemble $\{p_1 = \frac{1}{4}, |\psi_1\rangle = |0\rangle, p_2 = \frac{1}{4}, |\psi_2\rangle = |1\rangle, p_3 = \frac{1}{4}, |\psi_3\rangle = |+\rangle, p_4 = \frac{1}{4}, |\psi_4\rangle = |-\rangle\}$. Write down the density operator and compute the amount of ignorance for Eve

(Von Neumann entropy).
Solution: Here

$$\rho = \tfrac{1}{4}|0\rangle\langle 0| + \tfrac{1}{4}|1\rangle\langle 1| + \tfrac{1}{4}|+\rangle\langle +| + \tfrac{1}{4}|-\rangle\langle -|$$

$$= \tfrac{1}{4}\left\{ \begin{pmatrix} 1 \\ 0 \end{pmatrix}(1 \;\; 0) + \begin{pmatrix} 0 \\ 1 \end{pmatrix}(0 \;\; 1)\right.$$

$$+ \tfrac{1}{2}\begin{pmatrix} 1 \\ 1 \end{pmatrix}(1 \;\; 1) + \tfrac{1}{2}\begin{pmatrix} 1 \\ -1 \end{pmatrix}(1 \;\; -1)\left.\right\}$$

$$= \tfrac{1}{4}\left\{ \begin{pmatrix} 1 & 0 \\ 0 & 0 \end{pmatrix} + \begin{pmatrix} 0 & 0 \\ 0 & 1 \end{pmatrix} + \tfrac{1}{2}\begin{pmatrix} 1 & 1 \\ 1 & 1 \end{pmatrix} + \tfrac{1}{2}\begin{pmatrix} 1 & -1 \\ -1 & 1 \end{pmatrix}\right\}$$

$$= \tfrac{1}{2}\begin{pmatrix} 1 & 0 \\ 0 & 1 \end{pmatrix}.$$

The density matrix is already diagonal and we can easily conclude that the eigenvalues of ρ are $\lambda_1 = \lambda_2 = \tfrac{1}{2}$ and consequently amount of ignorance for Eve is $S(\rho) = \sum_{i=1}^{2} -\lambda_i \log_2 \lambda_i = -\tfrac{1}{2}\log_2 \tfrac{1}{2} - \tfrac{1}{2}\log_2 \tfrac{1}{2} = 1$ bit.

4. In a brute force attack, a code breaker tries to guess a key at random. If the attacker can check 1 billion keys per second then in the worst case how long will it take to break a 128 bit key by brute force attack?
   Solution: If the key is of size N bit, then there are 2^N different possible keys, in other words we say that the key-space is of size 2^N. In the worst case the code breaker has to check all 2^N possibilities and if he can check M possibilities per second then it will take $t = \frac{2^N}{M}$ seconds. Here $N = 128$ and $M = 10^9$, so it will take $t = \frac{2^{128}}{10^9}$ sec $= 3.4 \times 10^{29}$ sec $= 1.1 \times 10^{22}$ years. This is much more than the age of the universe.

5. In an alphabetic substitution code the letters $A, B, \cdots, Z$ of the alphabet are randomly replaced by 26 numbers $01, 02, \cdots, 26$. How big is the key space? If Eve can check 1 billion keys per second then in the worst case scenario, how long will it take to break the key by brute force attack?
   Solution: The key can be generated in $26! = 4 \times 10^{26}$ ways so the size of the key space is 4×10^{26}. As Eve can check 10^9 keys per second it will take 4×10^{17} seconds $= 12.8 \times 10^9$ years.

6. In the existing literature, two analogous but different parameters are used for analysis of efficiency of quantum communication protocols. The first one is simply defined as

$$\eta_1 = \frac{c}{q}, \tag{8.5}$$

where c denotes the total number of transmitted classical bits (message bits) and q denotes the total number of qubits used. This simple measure does not include the classical communication that is required for decoding of information in a DSQC protocol. Consequently it is a weak measure. Another measure [122] that is frequently used and which includes the classical communication is given as

$$\eta_2 = \frac{c}{q+b},\qquad(8.6)$$

where b is the number of classical bits exchanged for decoding of the message (classical communications used for checking of eavesdropping is not counted). Now use these definitions to compute qubit efficiency of Protocols 14 and 15.

Solution: In Protocol 14, n bits of classical information are sent by n-qubits and an equal number (i.e., n) of decoy qubits so we have $c = n$ and $q = 2n$. Further to disclose the actual order we need n bits of classical information. Thus $b = n$. Therefore, for this DSQC protocol (Protocol 14) we have $\eta_1 = \frac{1}{2}$ and $\eta_2 = \frac{1}{3}$, and similarly for Protocol 15 we have $\eta_1 = \eta_2 = \frac{1}{2}$ as in this case $b = 0$, $c = n$ and $q = 2n$.

7. In B92 protocol if Alice sends the following string of qubits:

$$|0\rangle, |0\rangle, |+\rangle, |0\rangle, |+\rangle, |+\rangle, |0\rangle, |+\rangle, |0\rangle, |0\rangle, |+\rangle, |0\rangle$$

and Bob's measurements collapse them to following string

$$|-\rangle, |+\rangle, |0\rangle, |0\rangle, |1\rangle, |+\rangle, |-\rangle, |+\rangle, |0\rangle, |-\rangle, |1\rangle, |0\rangle,$$

which of the above qubits are to be kept for error checking and generation of key?

Solution: As in B92 protocol Bob keeps only those qubits where his measurement outcome is either $|-\rangle$ or $|1\rangle$, therefore, in the above case they will keep the qubit number 1, 5, 7, 10 and 11.

8. In a specific run of B92 protocol after error checking Bob's states are $|-\rangle, |1\rangle, |1\rangle, |-\rangle, |-\rangle, |1\rangle, |-\rangle, |1\rangle, |-\rangle, |-\rangle$. Can you find the key that is distributed in this specific run of the B92 protocol?

Solution: We know that Bob's outcome $|-\rangle(|1\rangle)$ corresponds to Alice's input state $|0\rangle$ ($|+\rangle$) which implies the encoded bit value is 0 (1). Thus the key that is distributed in this specific run of the B92 protocol is 0110010100.

8.14   Further reading

1. An excellent history of development of quantum cryptography and nice introduction to the essential ideas can be found at C. H. Bennett *et al.*, Experimental quantum cryptography, J. Cryptology, 5 (1992)

3-28. The article is also available at http://cs.uccs.edu/~cs691/crypto /BBBSS92.pdf. Another article along the same lines is G. Brassard, Brief history of quantum cryptography: a personal perspective, Proceedings of IEEE Information Theory Workshop on Theory and Practice in Information Theoretic Security, Awaji Island, Japan (2005) 19-23, quant-ph/0604072v1.

2. For a very lucid and exciting description of the history of cryptography and its impact on socio-economic situations, interested readers may see Simon Singh, The code book: The science of secrecy from ancient Egypt to quantum cryptography, Anchor Books, New York, 1999.

3. N. Gisin, G. Ribordy, W. Tittel and H. Zbinden, Quantum cryptography, Rev. Mod. Phys. **74** (2002) 145-195, quant-ph/0101098. This provides an excellent review of quantum cryptography until 2002.

4. We have not elaborated on Ekert's protocol: Artur K. Ekert, Quantum cryptography based on Bell's theorem, Phys. Rev. Lett. **67** (1991) 661–663. This is an important and well studied protocol of QKD.

5. We have not described N09 protocol introduced by T.-G. Noh in 2009. This interesting protocol, which is often referred to as counterfactual cryptographic protocol, is also orthogonal-state-based protocol. See T.-G. Noh, Counterfactual quantum cryptography, Phys. Rev. Lett. **103** (2009) 230501, quant-ph/0809.3979.

6. A. Avella, G. Brida, D. Carpentras, A. Cavanna, I. P. Degiovanni, M. Genovese, M. Gramegna and P. Traina, Review on recent groundbreaking experiments on quantum communication with orthogonal states, quant-ph/1206.1503v1. This is an excellent recent review focused on GV protocol and N09 protocol.

7. K. Svozil, Staging quantum cryptography with chocolate balls, Am. J. Phys. **74** (2006) 800.

8. E. Gerjuoy, Shor's factoring algorithm and modern cryptography, An illustration of the capabilities inherent in quantum computers, Am. J. Phys. **73** (2005) 521.

9. D. Gottesman and H-K. Lo, From quantum cheating to quantum security, Physics Today, Nov. 2000, 22.

10. S. Loepp and W. K. Wootters, Protecting information: From classical error correction to quantum cryptography, Cambridge University Press, Cambridge, UK (2006). In the QKD protocols Alice distributes a key to Bob. In contrast, in quantum key agreement (QKA), Alice

and Bob jointly create a shared key by discussion over an insecure public channel. Thus both Alice and Bob contribute his/her part to the shared key and neither Alice nor Bob can determine the entire key alone. Several protocols of QKA have recently been proposed in recent past. Just as an example, interested readers may see S.-K. Chong, C.-W. Tsai and T. Hwang, Int. J. Theor. Phys. **50** (2011) 1793-1802.

8.15 Exercises

1. Using the concept of entanglement swapping and a GHZ like state design a protocol of DSQC where the actual information-encoded-state is never transmitted through the channel.

2. Decrypt the messages (a)

   wnkiradleqeboriyluyutetsladtouiekniicnocntsoocnuoutnloarwmrpsnle

   and (b) paadukgbsiootsoy, given that the messages are encrypted using Caesarian square.

3. Eve uses the POVM described in (3.40) to measure the qubits sent by Alice to Bob in B92 protocol. Show that B92 protocol is secure under this attack.

4. Output of a laser is a quantum state $|\alpha\rangle = \sum_n \dfrac{\alpha^n \exp\left(-\frac{|\alpha|^2}{2}\right)}{\sqrt{n!}} |n\rangle$ which is known as coherent state. It is known to satisfy $a|\alpha\rangle = \alpha|\alpha\rangle$, where a is the usual annihilation operator. Average photon number is $\langle\alpha|a^\dagger a|\alpha\rangle = |\alpha|^2 = \bar{n}$. Find an expression for the probability of getting n photon in a pulse having $\bar{n}$ photon on the average. If the attenuated laser pulse contains 0.1 photon on average, then what % of them can be attacked by Eve using PNS attack strategy?

5. Explain how the use of decoy qubits can help us to detect PNS attacks of Eve.

6. Explicitly describe the working of RSA protocol assuming that Bob starts with prime numbers 31 and 19. Also assume that Alice's secret is 18.

7. Show that DLL protocol can be viewed as a special case of Protocol 14 with $|a_i\rangle = |\psi^+\rangle$.

8. Compute the qubit efficiency η_1 and η_2 of Protocol 17. Also find the bounds on the possible values of the qubit efficiency that can be achieved in this protocol. For definition of η_2 see Solved Example 6.

9. What advantages will you have if you can construct the following?
   (a) An almost perfect single photon source.
   (b) An efficient single photon detector which works at teleportation range.
   (c) An optical fiber which has minimum attenuation at 800 nm.

10. Assume that in LM05 protocol Bob has received $|0\rangle, |+\rangle, |-\rangle, |1\rangle,$ $|-\rangle, |1\rangle, |0\rangle, |0\rangle$ in message mode. Corresponding qubits were initially prepared by him as $|0\rangle, |-\rangle, |-\rangle, |0\rangle, |+\rangle, |0\rangle, |1\rangle, |0\rangle$. Find the sequence of bit values that is encoded by Alice.

11. Dense coding of GHZ state is shown in Tables 7.8 and 7.9. Use them to construct CL type protocol of QSDC. Also compute the qubit efficiency of the protocol designed by you.

12. Computational basis and diagonal basis can be represented as $\oplus$ and $\otimes$ respectively. Now consider that in BB84 protocol Alice wanted to transmit 001011010, which is a raw key, to Bob. To do so, she has prepared a sequence of qubits using the following sequence of bases: $\otimes, \otimes, \oplus, \otimes, \oplus, \otimes, \oplus, \otimes, \oplus, \oplus$. Now assume that Bob has measured the qubits using the following sequence of bases: $\otimes, \oplus, \otimes, \otimes, \oplus, \oplus, \otimes, \oplus, \oplus,$ $\otimes$. Find the shifted key.

Bibliography

[1] N. Gershenfeld, The physics of information technology, Cambridge University Press, Cambridge, UK (2002).

[2] C. E. Shannon, A mathematical theory of communication, The Bell System Technical Journal, **27** (1948) 379–423 and 623–656.

[3] R. Landauer, Information is physical, Proc. Workshop on Physics and Computation PhysComp 92 (IEEE Comp. Sci. Press, Los Alamitos, CA, 1993) 1-4.

[4] A. Steane, Quantum computing, Rep. Prog. Phys. **61** (1998) 117-173, quant-ph/9708022.

[5] R. Landauer, Irreversibility and heat generation in the computing process, IBM J. Res. Dev. **5** (1961) 183-191.

[6] C. H. Bennett, Logical reversibility of computation, IBM J. Res. Dev. **17** (1973) 525-532.

[7] S. Lloyd, Rolf Landauer (1927-99): Head and heart of the physics of information, Nature **400** (1999) 720-720.

[8] V. Vedral, Introduction to quantum information science, Oxford University Press, New York (2006).

[9] A. K. Ekert, Quantum cryptography based on Bell's theorem, Phys. Rev. Lett. **67** (1991) 661-663.

[10] R. Feynman, Simulating physics with computers, Int. J. Theo. Phys. **21** (1982) 467-488.

[11] S. Wiesner, Conjugate coding, ACM SIGACT News **15** (1983) 78-88.

[12] P. Benioff, The computer as a physical system: A microscopic quantum mechanical Hamiltonian model of computers as represented by Turing machines, J. Stat. Phys. **22** (1980) 563-591.

[13] P. Benioff, Quantum mechanical models of Turing machines that dissipate no energy, Phys. Rev. Lett. **48** (1982) 1581–1585.

[14] P. Benioff, Quantum mechanical Hamiltonian models of Turing machines, J. Stat. Phys. **29** (1982) 515-546.

[15] P. Benioff, Quantum mechanical Hamiltonian models of discrete processes that erase their own histories: Application to Turing machines, Int. J. Theo. Phys. **21** (1982) 177-201.

[16] W. K. Wootters and W. H. Zurek, A single quantum cannot be cloned, Nature **299** (1982) 802-803.

[17] D. Dieks, Communication by EPR devices, Phys. Lett. A **92** (1982) 271-272.

[18] C. H. Bennett and G. Brassed, Quantum cryptography: Public key distribution and coin tossing, Proceedings of the IEEE International Conference on Computers, Systems, and Signal Processing, Bangalore, India (1984) 175-179.

[19] D. Deutsch, Quantum theory, the Church-Turing principle and the universal quantum computer, Proceedings of the Royal Society of London; Series A, Mathematical and Physical Sciences, **400** (1985) 97-117.

[20] C. H. Bennett and S. J. Wiesner, Communication via one- and two-particle operations on Einstein-Podolsky Rosen states, Phys. Rev. Lett. **69** (1992) 2881-2884.

[21] C. H. Bennett *et al.*, Teleporting an unknown quantum state via dual classical and Einstein-Podolsky-Rosen Channels, Phys. Rev. Lett. **70** (1993) 1895-1899.

[22] P.W. Shor, Polynomial-time algorithms for prime factorization and discrete logarithms on a quantum computer, in Proc. 35th Annual Symp. on Foundations of Computer Science, (1994) Santa Fe, IEEE Computer Society Press; quant-ph/9508027.

[23] B. Schumacher, Quantum coding, Phys. Rev. A **51** (1995) 2738–2747.

[24] P. W. Shor, Scheme for reducing decoherence in quantum computer memory, Phys. Rev. A **52** (1995) R2493-R2496.

[25] R. Laflamme, C. Miquel, J. P. Paz and W. H. Zurek, Perfect quantum error correcting code, Phys. Rev. Lett. **77** (1996) 198–201, quant-ph/9602019v1.

[26] P. W. Shor, Fault-tolerant quantum computation, in Proc. 37th Annual Symposium on Foundations of Computer Science (1996) 55-65, quant-ph/9605011.

[27] L. K. Grover, Quantum mechanics helps in searching for a needle in a haystack, Phys. Rev. Lett. **79** (1997) 325-328, quant-ph/9706033.

[28] D. Bouwmeester, J.-W. Pan, K. Mattle, M. Eibl, H. Weinfurter and A. Zeilinger, Experimental quantum teleportation, Nature **390** (1997) 575-579.

[29] I. L. Chuang, N. Gershenfeld and M. Kubinec, Experimental implementation of fast quantum searching, Phys. Rev. Lett. **80** (1998) 3408–3411.

[30] L. M. K. Vandersypen, M. Steffen, M. H. Sherwood, C. S. Yannoni, G. Breyta, and I. L. Chuang, Implementation of a three-quantum-bit search algorithm, Appl. Phys. Lett. **76** (2000) 646-648; quant-ph/9910075v2.

[31] D. P. DiVincenzo, The physical implementation of quantum computation, Fortschritte der Physik, **48** (2000) 771-783, quant-ph/0002077v3.

[32] L. M. K. Vandersypen, M. Steffen, G. Breyta, C. S. Yannoni, R. Cleve and I. L. Chuang, Experimental realization of an order-finding algorithm with an NMR quantum computer, Phys. Rev. Lett. **85** (2000) 5452–5455, quant-ph/0007017v2.

[33] L. M. K. Vandersypen, M. Steffen, G. Breyta, C. S. Yannoni, M. H. Sherwood and I. L. Chuang, Experimental realization of Shor's quantum factoring algorithm using nuclear magnetic resonance, Nature **414** (2001) 883-887, quant-ph/0112176v1.

[34] Z. Yuan, C. Gobby, and A. J. Shields, Quantum key distribution over distances as long as 101 km, (2003) DOI: 10.1109/QELS.2003.1276483.

[35] H. Häffner *et al.*, Scalable multiparticle entanglement of trapped ions, Nature **438** (2005) 643-646, quant-ph/0603217.

[36] A. Mirza and F. Petruccione, Realizing long-term quantum cryptography, J. Opt. Soc. Am. B, **27** (2010) A185-A188.

[37] X. M. Jin *et al.*, Experimental free-space quantum teleportation, Nature Photonics, **4** (2010) 376-381.

[38] Juan Yin *et al.*, Quantum teleportation and entanglement distribution over 100-kilometre free-space channels, Nature **488** (2012) 185–188.

[39] A. Barenco, Quantum physics and computers, Contemporary Phys. **37** (1996) 375-389, quant-ph/9612014v2.

[40] C. H. Bennett, Quantum information and computation, Physics Today, October 1995, 24-30.

[41] A. Ekert, P. Hayden and H. Inamori, Basic concepts in quantum computation, quant-ph/0011013.

[42] J. Preskill, Quantum computing: pro and con, Proc. R. Soc. Lond. A, **454** (1998) 469-486, quant-ph/9705032v3.

[43] V. Vedral and M. B. Plenio, Basics of quantum computation, Prog. Quant. Electronics, **22** (1998) 1-39, quant-ph/9802065v1.

[44] J. Preskil, Lecture notes for information for physics 219, http://theory.caltech.edu/people/preskill/ph229/#lecture. This web page also contains links to some interesting introductory articles.

[45] D. Aharonov, (2001), Lecture notes for quantum computing, http://www.cs.huji.ac.il/~doria/. Some of the materials are in Hebrew.

[46] U. Vazirani, Quantum computation course, http://www.cs.berkeley.edu/~vazirani/quantum.html.

[47] I. Chunag, MIT MAS.961: Quantum information science course homepage is available at http://www.media.mit.edu/quanta/mas961/index.php.

[48] S. Lloyd's lecture notes are available at http://web.mit.edu/2.111/www/.

[49] A. Cabello, Bibliographic guide to the foundations of quantum mechanics and quantum information, quant-ph/0012089v12. This contains more than 10,000 references related to quantum mechanics and quantum information.

[50] A. Turing, On computable numbers, with an application to the Entscheidungsproblem, Proceedings of the London Mathematical Society, ser. 2, **42** (1936-7) 230-265; corrections, Ibid, **43** (1937) 544-546.

[51] A. Church, An unsolvable problem of elementary number theory, Am. J. Math. **58** (1936) 345-363.

[52] R. Solovay and V. Strassen, A fast Monte-Carlo test for primality, SIAM J. Comput. **6** (1977) 84-85.

[53] D. Aharonov, A simple proof that Toffoli and Hadamard are quantum universal, quant-ph/0301040.

[54] P. Kaye, R. Laflamme and M. Mosca, An introduction to quantum computing, Oxford University Press, New York, (2007).

[55] L. Gurvits, Classical complexity and quantum entanglement, J. Comput. Syst. Sci. **69** (2004) 448-484.

[56] J. S. Bell, On the Einstein Podolsky Rosen paradox, Physics **1** (1964) 195-200.

[57] A. Aspect, J. Dalibard, G. Roger, Experimental test of Bell's inequalities using time-varying analyzers, Phys. Rev. Lett. **49** (1982) 1804-1807.

[58] V. Scarani, Feats, features and failures of the PR-box, Quantum Mechanics, AIP Conference Proceedings **844** (2006) 309-320, quant-ph/0603017v2.

[59] A. J. Leggett, Nonlocal hidden-variable theories and quantum mechanics: an incompatibility theorem, Foundations of Phys. **33** (2003) 1469-1493.

[60] G. Brassard, C. Crépeau, R. Jozsa and D. Langlois, A quantum bit commitment scheme provably unbreakable by both parties, Proceedings of 34th Annual IEEE Symposium on the Foundations of Computer Science, (1993) 362-371.

[61] A. Peres, How the no-cloning theorem got its name, Fortschr. Phys. **51** (2003) 458-461, quant-ph/0205076v1.

[62] M. A. Nielsen and I. L. Chuang, Quantum computation and quantum information, Cambridge University Press, New Delhi, India (2008).

[63] C. P. Williams and S. H. Clearwater, Explorations in quantum computing, Springer Verlag, New York (1998).

[64] M. Mohammadi and M. Eshghi, On figures of merit in reversible and quantum logic designs, Quant. Info. Process, **8** (2009) 297-318.

[65] M. Haghparast *et al.*, Optimized reversible multiplier circuit, J. Circuits Syst. Comp. **18** (2009), 1-13.

[66] A. Banerjee and A. Pathak, Reversible multiplier circuit, Proccedings of 3rd International Conference on Emerging Trends in Engineering and Technology (ICETET), (2010) 781-786, DOI: 10.1109/ICETET.2010.70.

[67] J. L. O'Brien, G. J. Pryde, A. G. White, T. C. Ralph and D. Branning, Demonstration of an all-optical quantum controlled-NOT gate, Nature **426** (2003) 264-267, quant-ph/0403062.

[68] L. Isenhower, Demonstration of a neutral atom controlled-NOT quantum gate, Phys. Rev. Lett. **104** (2010) 010503, quant-ph/0907.5552.

[69] T. Monz *et al.*, Realization of the quantum Toffoli gate with trapped ions, Phys. Rev. Lett. **102** (2009) 040501, quant-ph/0804.0082.

[70] A. Feorov *et al.*, Implementation of a Toffoli gate with superconducting circuits, Nature **481** (2012) 170-172, quant-ph/1108.3966.

[71] A. J. Poustie and K. J. Blow, Demonstration of an all-optical Fredkin gate, Opt. Commun. **174** (2000) 317–320.

[72] D. Maslov, G. W. Dueck, N. Scott, Reversible logic synthesis benchmarks page, http://webhome.cs.uvic.ca/~dmaslov/ (2009).

[73] J. Stoke and D. Suter, Quantum computing: A short course from theory to experiment, Wiley Vch, Weinheim, Germany (2004).

[74] S. Hallgern, Polynomial-time quantum algorithms for Pell's equation and the principal ideal problem, Journal of the ACM (JACM) **54** (2007) 4.

[75] P. W. Shor, Why haven't more quantum algorithms been found?, Journal of the ACM (JACM), **50** (2003) 87-90.

[76] I. L. Chuang et al., Experimental realization of a quantum algorithm, Nature, **393** (1998) 143-146, quant-ph/9801037v2.

[77] D. Wei *et al.*, NMR experimental realization of seven-qubit D-J algorithm and controlled phase-shift gates with improved precision, Chinese Science Bulletin **48** (2003) 239-243.

[78] Z. Wu *et al.*, Experimental demonstration of the Deutsch-Jozsa algorithm in homonuclear multispin systems, Phys. Rev. A **84** (2011) 042312.

[79] E. M. López *et al.*, Experimental realization of Shor's quantum factoring algorithm using qubit recycling, Nature Photonics **6** (2012) 773–776.

[80] A. Politi, J. C. F. Matthews and J. L. O'Brien, Shor's quantum factoring algorithm on a photonic chip, Science **325** (2009) 1221-1222, quant-ph/0911.1242.

[81] N. Xu *et al.*, Quantum factorization of 143 on a dipolar-coupling nuclear magnetic resonance system, Phys. Rev. Lett. **108** (2012) 130501.

[82] E. Lucero *et al.*, Computing prime factors with a Josephson phase qubit quantum processor, Nature Phys. **8** (2012) 719–723, quant-ph/1202.5707.

[83] R. Landauer, Is quantum-mechanically coherent computation useful?, Proc. Drexel-4 Symposium on Quantum Nonintegrability-Quantum-Classical Correspondence, Philadelphia, PA, **8** (1994) 1052-1055 (eds. D. H. Feng and B.-L. Hu, International Press, Boston).

[84] D. P. DiVincenzo and D P Loss, Quantum information is physical, Superlattices and Microstructures **23** (1998) 419-432, cond-mat/9710259.

[85] P. L. Knight, A. Beige and W. J. Munro, Hiding from environment: Decoherence-free subspaces in quantum information processing, ftp://ftp.cordis.europa.eu/pub/ist/docs/fet/qip2-eu-22.pdf.

[86] T. v. d. Sar *et al.*, Decoherence-protected quantum gates for a hybrid solid-state spin register, Nature **484**, (2012) 82–86, cond-mat/1202.4379.

[87] J. M. Chow et al., Universal quantum gate set approaching fault-tolerant thresholds with superconducting qubits, Phys. Rev. Lett. **109** (2012) 060501.

[88] J. Zhang, R. Laflamme and D. Suter, Experimental implementation of encoded logical qubit operations in a perfect quantum error correcting code, Phys. Rev. Lett. **109** (2012) 100503, quant-ph/1208.479.

[89] P. Schindler, Experimental repetitive quantum error correction, Science **332** (2011) 1059-1061.

[90] M. Hillery, V. Buzek and A. Bertaiume, Quantum secret sharing, Phys. Rev. A **59** (1999) 1829-1833.

[91] C. H. Bennett, Quantum cryptography using any two nonorthogonal states, Phys. Rev. Lett. **68** (1992) 3121-3124.

[92] X.-W. Wang, D.-Y. Zhang, S.-Q. Tang, X.-G. Zhan, K.-M.You, Hierarchical quantum information splitting with six-photon cluster states, Int. J. Theor. Phys. **49** (2010) 2691–2697.

[93] C. Shukla and A. Pathak, Hierarchical quantum communication, Phys. Lett. A (2013) DOI:10.1016/j.physleta.2013.04.010, quant-ph/1301.0498.

[94] A. K. Pati, Minimum classical bit for remote preparation and measurement of a qubit, Phys. Rev. A **63** (2000) 014302, quant-ph/9907022.

[95] N. B. An and J. Kim, Joint remote state preparation, J. Phys. B. **41** (2008) 095501.

[96] Q. Zhang *et al.*, Experimental quantum teleportation of a two-qubit composite system, Nature Phys. **2** (2006) 678-682, quant-ph/0609129.

[97] Y.-F. Huang *et al.*, Experimental teleportation of a quantum controlled-NOT gate, Phys. Rev. Lett. **93** (2004) 240501, quant-ph/0408007v1.

[98] K. Mattle, H. Weinfurter, P. G. Kwiat, and A. Zeilinger, Dense coding in experimental quantum communication, Phys. Rev. Lett. **76** (1996) 4656-4659.

[99] X. Fang. *et al.*, Experimental implementaton of dense coding using nuclear magnetic resonance, Phys. Rev. A **61**, (2000) 022307, quant-ph/9906041v2.

[100] W. Diffie and M. Hellman, New directions in cryptography, IEEE Trans. Info. Theor. **22** (1976) 644-654.

[101] R. L. Rivest, A. Shamir and L. Adleman, A method for obtaining digital signatures and public-key cryptosystems, Comm. ACM **21** (1978) 120-126.

[102] L. Goldenberg and L. Vaidman, Quantum cryptography based on orthogonal states, Phys. Rev. Lett. **75** (1995) 1239-1243, quant-ph/9502021v1.

[103] K. Shimizu and N. Imoto, Communication channels secured from eavesdropping via transmission of photonic Bell states, Phys. Rev. A **60** (1999) 157-166.

[104] G. L. Long and X. S. Liu, Theoretically efficient high-capacity quantum-key-distribution scheme, Phys. Rev. A **65** (2002) 032302, quant-ph/0012056v3.

[105] K. Bostrom and T. Felbinger, Deterministic secure direct communication using entanglement, Phys. Rev. Lett. **89** (2002) 187902, quant-ph/0209040.

[106] M. Lucamarini and S. Mancini, Secure deterministic communication without entanglement, Phys. Rev. Lett. **94** (2005) 140501, quant-ph/0405083.

[107] J. Liu *et al.*, Revisiting quantum secure direct communication with W state, Chin. Phys. Lett. **23** (2006) 2652.

[108] X. H. Li *et al.*, Deterministic secure quantum communication without maximally entangled states, J. Korean Phys. Soc. **49** (2006) 1354-1359, quant-ph/0606007.

[109] G. L. Long *et al.*, Quantum secure direct communication and deterministic secure quantum communication, Front. Phys. China, **2** (2007) 251-272.

[110] N. B. An, Quantum dialogue, Phys. Lett. A **328** (2004) 6-10, quant-ph/0406130v1.

[111] Z. X. Man, Z. J. Zhang, and Y. Li, Quantum dialogue revisited, Chin. Phys. Lett. **22** (2005) 22-24.

[112] N. B. An, Secure dialogue without prior key distribution, J. Kor. Phys. Soc. **47** (2005) 562-567.

[113] Q.-y. Cai and B.-w. Li, Phys. Rev. A, Improving the capacity of the Boström-Felbinger protocol, **69** (2004) 054301.

[114] P. Yadav, R. Srikanth and A. Pathak, Generalization of the Goldenberg-Vaidman QKD protocol, quant-ph/1209.4304v1.

[115] F.-G. Deng, G. L. Long, and X.-S. Liu, Two-step quantum direct communication protocol using the Einstein-Podolsky-Rosen pair block, Phys. Rev. A **68** (2003) 042317, quant-ph/0308173v1.

[116] C. Shukla, A. Pathak and R. Srikanth, Beyond the Goldenberg-Vaidman protocol: Secure and efficient quantum communication using arbitrary, orthogonal, multi-particle quantum states, Int. J. Quant. Infor. **10** (2012), 1241009, quant-ph/1210.2583v1.

[117] C. Shukla, V. Kothari, A. Banerjee and A. Pathak, On the group-theoretic structure of a class of quantum dialogue protocols, Phys. Lett. A **377** (2013) 518-527, quant-ph/1203.5931.

[118] W. Liu, Y. B. Wang and Z. T. Jiang, An efficient protocol for the quantum private comparison of equality with W state, Opt. Commun. **284** (2011) 3160-3163.

[119] W. Tittel, H. Zbinden and N. Gisin, Experimental demonstration of quantum secret sharing, Phys. Rev. A **63** (2001) 042301.

[120] C. H. Bennett *et al.*, Experimental quantum cryptography, Journal of Cryptology, **5** (1992) 3-28.

[121] A. M. Steane and W. v. Dam, Physicists triumph at guess my number, Physics Today, February 2000, 35-39.

[122] A. Cabello, Quantum key distribution in the Holevo Limit, Phys. Rev. Lett. **85** (2000) 5635-5638, quant-ph/0007064.

Index

T - #0111 - 111024 - C340 - 234/156/16 - PB - 9780367379872 - Gloss Lamination